Hole's *essentials of* Human Anatomy & Physiology

tenth edition

David Shier   **Jackie Butler**   **Ricki Lewis**

Terry R. Martin

Kishwaukee College

McGraw-Hill Higher Education

Boston   Burr Ridge, IL   Dubuque, IA   New York   San Francisco   St. Louis
Bangkok   Bogotá   Caracas   Kuala Lumpur   Lisbon   London   Madrid   Mexico City
Milan   Montreal   New Delhi   Santiago   Seoul   Singapore   Sydney   Taipei   Toronto

Mc Graw Hill

McGraw-Hill
Higher Education

LABORATORY MANUAL TO ACCOMPANY HOLE'S ESSENTIALS OF HUMAN ANATOMY & PHYSIOLOGY, TENTH EDITION

1 2 3 4 5 6 7 8 9 0 QPD/QPD 0 9 8

ISBN 978-0-07-296567-4
MHID 0-07-296567-3

Publisher: *Michelle Watnick*
Senior Sponsoring Editor: *James F. Connely*
Director of Development: *Kristine Tibbetts*
Developmental Editor: *Fran Schreiber*
Marketing Manager: *Lynn M. Breithaupt*
Senior Project Manager: *Jayne L. Klein*
Lead Production Supervisor: *Sandy Ludovissy*
Designer: *Laurie B. Janssen*
Cover/Interior Designer: *Christopher Reese*
(USE) Cover Image: *Greg Epperson, Gettyimages*
Senior Photo Research Coordinator: *John C. Leland*
Photo Research: *Toni Michaels/PhotoFind, LLC*
Supplement Producer: *Mary Jane Lampe*
Compositor: *Precision Graphics*
Typeface: *10/12 Garamond Book*
Printer: *Quebecor World Dubuque, IA*

The credits section for this book begins on page 372 and is considered an extension of the copyright page.

Some of the laboratory experiments included in this text may be hazardous if materials are handled improperly or if procedures are conducted incorrectly. Safety precautions are necessary when you are working with chemicals, glass test tubes, hot water baths, sharp instruments, and the like, or for any procedures that generally require caution. Your school may have set regulations regarding safety procedures that your instructor will explain to you. Should you have any problems with materials or procedures, please ask your instructor for help.

www.mhhe.com

Contents

Preface v
To the Student viii
Correlation of Textbook Chapters and Laboratory Exercises x

Fundamentals of Human Anatomy and Physiology

Laboratory Exercise 1
Scientific Method and Measurements 1
Laboratory Report 1 5

Laboratory Exercise 2
Body Organization and Terminology 9
Laboratory Report 2 17

Laboratory Exercise 3
Chemistry of Life 21
Laboratory Report 3 25

Laboratory Exercise 4
Care and Use of the Microscope 29
Laboratory Report 4 31

Cells

Laboratory Exercise 5
Cell Structure and Function 39
Laboratory Report 5 43

Laboratory Exercise 6
Movements Through Cell Membranes 45
Laboratory Report 6 49

Laboratory Exercise 7
Cell Cycle 53
Laboratory Report 7 57

Tissues

Laboratory Exercise 8
Epithelial Tissues 61
Laboratory Report 8 63

Laboratory Exercise 9
Connective Tissues 65
Laboratory Report 9 67

Laboratory Exercise 10
Muscle and Nervous Tissues 69
Laboratory Report 10 71

Integumentary System

Laboratory Exercise 11
Integumentary System 73
Laboratory Report 11 77

Skeletal System

Laboratory Exercise 12
Bone Structure 79
Laboratory Report 12 83

Laboratory Exercise 13
Organization of the Skeleton 85
Laboratory Report 13 89

Laboratory Exercise 14
Skull 91
Laboratory Report 14 97

Laboratory Exercise 15
Vertebral Column and Thoracic Cage 101
Laboratory Report 15 107

Laboratory Exercise 16
Pectoral Girdle and Upper Limb 109
Laboratory Report 16 115

Laboratory Exercise 17
Pelvic Girdle and Lower Limb 119
Laboratory Report 17 125

Laboratory Exercise 18
Joint Structure and Movements 129
Laboratory Report 18 131

Muscular System

Laboratory Exercise 19
Skeletal Muscle Structure 135
Laboratory Report 19 139

Laboratory Exercise 20
Muscles of the Face, Head, and Neck 141
Laboratory Report 20 143

Laboratory Exercise 21
Muscles of the Chest, Shoulder, and Upper Limb 145
Laboratory Report 21 149

Laboratory Exercise 22
Muscles of the Abdominal Wall and Pelvic Outlet 153
Laboratory Report 22 157

Laboratory Exercise 23
Muscles of The Hip and Lower Limb 159
Laboratory Report 23 167

Surface Anatomy

Laboratory Exercise 24
Surface Anatomy   169
Laboratory Report 24   177

Nervous System

Laboratory Exercise 25
Nervous Tissue and Nerves   181
Laboratory Report 25   187

Laboratory Exercise 26
Spinal Cord and Meninges   189
Laboratory Report 26   191

Laboratory Exercise 27
Reflex Arc and Reflexes   193
Laboratory Report 27   197

Laboratory Exercise 28
Brain and Cranial Nerves   199
Laboratory Report 28   203

Laboratory Exercise 29
Dissection of the Sheep Brain   205
Laboratory Report 29   211

Special Senses

Laboratory Exercise 30
Ear and Hearing   213
Laboratory Report 30   219

Laboratory Exercise 31
Eye Structure   221
Laboratory Report 31   227

Laboratory Exercise 32
Visual Tests and Demonstrations   229
Laboratory Report 32   235

Endocrine System

Laboratory Exercise 33
**Endocrine Histology and Diabetic
   Physiology   237**
Laboratory Report 33   245

Cardiovascular System*

Laboratory Exercise 34
Blood Cells and Blood Typing   249
Laboratory Report 34   255

Laboratory Exercise 35
Heart Structure   259
Laboratory Report 35   265

Laboratory Exercise 36
Cardiac Cycle   267
Laboratory Report 36   271

Laboratory Exercise 37
Blood Vessel Structure, Arteries, and Veins   275
Laboratory Report 37   287

Laboratory Exercise 38
Pulse Rate and Blood Pressure   291
Laboratory Report 38   293

Lymphatic System

Laboratory Exercise 39
Lymphatic System   295
Laboratory Report 39   299

Digestive System

Laboratory Exercise 40
Digestive Organs   301
Laboratory Report 40   311

Laboratory Exercise 41
Action of a Digestive Enzyme   313
Laboratory Report 41   315

Respiratory System

Laboratory Exercise 42
Respiratory Organs   317
Laboratory Report 42   321

Laboratory Exercise 43
**Breathing and Respiratory Volumes
   and Capacities   323**
Laboratory Report 43   329

Urinary System

Laboratory Exercise 44
Kidney Structure   331
Laboratory Report 44   337

Laboratory Exercise 45
Urinalysis   339
Laboratory Report 45   343

Reproductive System

Laboratory Exercise 46
Male Reproductive System   345
Laboratory Report 46   349

Laboratory Exercise 47
Female Reproductive System   351
Laboratory Report 47   357

Laboratory Exercise 48
Genetics   359
Laboratory Report 48   365

*Supplemental Laboratory Exercise

Laboratory Exercise 49
Blood Testing
Laboratory Report 49
(This lab is available at aris.mhhe.com. Click on the title
for *Hole's Essentials of Human Anatomy & Physiology* 10e)

Appendix 1 Preparation of Solutions   369
Appendix 2 Assessments of Laboratory Reports   370
Credits   372
Index   373

Preface

This laboratory manual was prepared to supplement the textbook *Hole's Essentials of Human Anatomy and Physiology,* Tenth Edition, by David Shier, Jackie Butler, and Ricki Lewis. As in the textbook, the laboratory manual is designed for students with minimal backgrounds in the physical and biological sciences pursuing careers in professional health fields.

The laboratory manual contains forty-nine laboratory exercises and reports closely integrated with the chapters of the textbook. The exercises are planned to illustrate and review the anatomical and physiological facts and principles presented in the textbook and to help students investigate some of these ideas in greater detail.

Often the laboratory exercises are short or are divided into several separate procedures. This allows an instructor to select those exercises or parts of exercises that will best meet the needs of a particular program. Also, exercises requiring a minimal amount of laboratory equipment have been included.

The laboratory exercises include a variety of special features designed to stimulate interest in the subject matter, to involve students in the learning process, and to guide them through the planned activities. These special features include the following:

Materials Needed

This section lists the laboratory materials required to complete the exercise and to perform the demonstrations and learning extensions.

Safety

A list of safety guidelines is included inside the front cover. Each lab session that requires special safety guidelines has a safety section following "Materials Needed." Your instructor might require some modifications of these guidelines.

Introduction

The introduction briefly describes the subject of the exercise or the ideas that will be investigated.

Purpose of the Exercise

The purpose provides a statement about the intent of the exercise—that is, what will be accomplished.

Learning Outcomes 1

The learning outcomes list in general terms what a student should be able to do after completing the exercise. Each learning outcome will have a matching assessment indicated by a corresponding icon 1 in the laboratory exercise or the laboratory report.

Procedure

The procedure provides a set of detailed instructions for accomplishing the planned laboratory activities. Usually these instructions are presented in outline form so that a student can proceed through the exercise in stepwise fashion. Often, the student is referred to particular sections of the textbook for necessary background information or for review of subject matter previously presented.

The procedures include a wide variety of laboratory activities and, from time to time, direct the student to complete various tasks in the laboratory reports.

Laboratory Reports

A laboratory report to be completed by the student immediately follows each exercise. These reports include various types of review activities, spaces for sketches of microscopic objects, tables for recording observations and experimental results, and questions dealing with the analysis of such data.

It is hoped that as a result of these activities, students will develop a better understanding of the structural and functional characteristics of their bodies and will increase their skills in gathering information by observation and experimentation. Some of the exercises also include demonstrations, learning extensions, and useful illustrations.

Demonstrations

Demonstrations appear in separate boxes. They describe specimens, specialized laboratory equipment, or other materials of interest that an instructor may want to display to enrich the student's laboratory experience.

Learning Extensions

Learning extensions also appear in separate boxes. They encourage students to extend their laboratory experiences. Some of these activities are open-ended in that they suggest the student plan an investigation or experiment and carry it out after receiving approval from the laboratory instructor. Some of the figures are illustrated as line art or in grayscale. This will allow colored pencils to be used as a visual learning activity to distinguish various structures.

The Use of Animals in Biology Education*

The National Association of Biology Teachers (NABT) believes that the study of organisms, including nonhuman animals, is essential to the understanding of life on Earth.

---

* Adopted by the Board of Directors in October 1995. This policy supersedes and replaces all previous NABT statements regarding animals in biology education.

NABT recommends the prudent and responsible use of animals in the life science classroom. NABT believes that biology teachers should foster a respect for life. Biology teachers also should teach about the interrelationship and interdependency of all things.

Classroom experiences that involve nonhuman animals range from observation to dissection. NABT supports these experiences so long as they are conducted within the long-established guidelines of proper care and use of animals, as developed by the scientific and educational community.

As with any instructional activity, the use of nonhuman animals in the biology classroom must have sound educational objectives. Any use of animals, whether for observation or dissection, must convey substantive knowledge of biology. NABT believes that biology teachers are in the best position to make this determination for their students.

NABT acknowledges that no alternative can substitute for the actual experience of dissection or other use of animals and urges teachers to be aware of the limitations of alternatives. When the teacher determines that the most effective means to meet the objectives of the class do not require dissection, NABT accepts the use of alternatives to dissection, including models and the various forms of multimedia. The Association encourages teachers to be sensitive to substantive student objections to dissection and to consider providing appropriate lessons for those students where necessary.

To implement this policy, NABT endorses and adopts the "Principles and Guidelines for the Use of Animals in Precollege Education" of the Institute of Laboratory Animals Resources (National Research Council). Copies of the "Principles and Guidelines" may be obtained from the ILAR (2101 Constitution Avenue, NW, Washington, DC 20418; 202–334–2590).

Illustrations

Diagrams from the textbook and diagrams similar to those in the textbook often are used as aids for reviewing subject matter. Other illustrations provide visual instructions for performing steps in procedures or are used to identify parts of instruments or specimens. Micrographs are included to help students identify microscopic structures or to evaluate student understanding of tissues.

In some exercises, the figures include line drawings suitable for students to color with colored pencils. This activity may motivate students to observe the illustrations more carefully and help them to locate the special features represented in the figures. Students can check their work by referring to the corresponding full-color illustrations in the textbook.

Frequent variations exist in anatomical structures among humans. The illustrations in the textbook and the laboratory manual represent normal (normal means the most common variation) anatomy.

Features of This Edition

Many of the changes in this new edition of the laboratory manual are a result of evaluations and suggestions from anatomy and physiology students. Many suggestions from users and reviewers of the ninth edition have been incorporated. Some features include the following:

1. Learning outcomes with icons ![1] have matching assessments with icons ![1] so students can be sure they have accomplished all of the content of the laboratory exercise.

2. For the first time a surface anatomy lab has been added to the lab manual. Lurana Bain LMT, massage therapy program coordinator and instructor at Elgin Community College, was the contributing author of this laboratory exercise. Surface anatomy is important for students selecting careers in massage therapy, physical therapy, nursing, and emergency medical services.

3. A new genetics lab has been added to this lab manual for the first time. It includes sections on human phenotypes, probability, and genetics problems.

4. The endocrine system lab has been expanded to include more histology and a new section on diabetic physiology.

5. The joints lab has been expanded to include additional material on movements.

6. The blood cells lab and the blood typing lab have been combined into a single laboratory exercise.

7. The blood vessels lab and the arteries and veins lab have been combined into a single laboratory exercise.

8. A new more colorful design has been incorporated into this lab manual. This facilitates locating labs, laboratory reports, and studying tables.

9. "Blood Testing" is a supplemental lab available at aris .mhhe.com (click on *Hole's Essentials of Human Anatomy & Physiology 10e*) instead of within the printed laboratory manual.

10. To meet the need for clearer and more definite safety guidelines, a revised and updated safety list is located inside the front cover, and safety sections with safety icons are found in appropriate labs.

11. "Critical Thinking Applications" are incorporated within most of the laboratory exercises to enhance valuable critical thinking skills that students will need throughout their lives.

12. The *Instructor's Manual for Laboratory Manual to Accompany Hole's Essentials of Human Anatomy and Physiology* has been revised and updated. It contains a "Student Safety Contract" and a "Student Informed Consent Form" for possible inclusion in the class. The forms could be adapted for a specific institution based upon school policy. The instructor's manual is located online at aris.mhhe.com. Click on the title for *Hole's Essentials of Human Anatomy & Physiology 10e*.

13. Two assessment tools (rubrics) for laboratory reports are included in Appendix 2.

Reviewers

I would like to express my sincere gratitude to all users of the laboratory manual who provided suggestions for its improvement. I am especially grateful for the contributions of the reviewers who kindly reviewed the manual and examined the manuscript of the new edition and gave their thoughtful comments and valuable suggestions. They include the following:

Tammy Atchison, *Pitt Community College*

Michael O. Casey, *Gaston College*

C.M. Churchill Jr., *University of Northern Colorado*

Betsy L. Diegel, *Davenport University*

Tom Dudley, *Angelina College*

Sharon J. Fugate, *Madisonville Community College*

Gary W. Hunt, *Tulsa Community College*

Babu P. Patlolla, *Alcorn State University*

Leba Sarkis, *Aims Community College*

Carl J. Shuster, *Madison Area Technical College*

Anthony Udeogalanya, *Medgar Evers College of the City University of New York*

Stephanie A. Shumate Vance, *Jefferson Community College*

Acknowledgments

I value all the support and encouragement by the staff at McGraw-Hill Higher Education, including Michelle Watnick, Jim Connely, Fran Schreiber, Jayne Klein, and Cathy Schmitt. Special recognition is granted to Colin Wheatley for his insight, confidence, wisdom, warmth, and friendship.

I am grateful for the professional talent of David Shier, Jackie Butler, and Ricki Lewis for all of their coordination efforts and guidance with the laboratory manual. A special thank-you is given to Lurana Bain for her contribution of the surface anatomy lab. The deepest gratitude is extended to John W. Hole, Jr., for his years of dedicated effort in early editions of this classic work and for the opportunity to revise his established laboratory manual.

I am particularly thankful to Dr. Norman Jenkins and Dr. David Louis, retired presidents of Kishwaukee College, and Dr. Thomas Choice, president of Kishwaukee College, for their support and confidence in my endeavors. I am appreciative for the expertise of Womack Photography. The professional reviews of the nursing procedures were provided by Kathy Schnier. I am also grateful to Michele Dukes, Troy Hanke, Jenifer Holtzclaw, Stephen House, Marcie Kessler, Angele Myska, Sparkle Neal, Robert Stockley, Shatina Thompson, Nancy Valdivia, Jana Voorhis, and DeKalb Clinic Chartered for their special contributions.

There have been valuable contributions from my students, who have supplied thoughtful suggestions and assisted in the clarification of details.

To my son Ross, I owe gratitude for his keen eye and creative suggestions. Foremost, I am appreciative to Sherrie Martin, my spouse and best friend, for advice, understanding, and devotion throughout the writing and revising.

Terry R. Martin
Kishwaukee College
21193 Malta Road
Malta, IL 60150

About the Author

This tenth edition is the fifth revision by Terry R. Martin of Kishwaukee College. Terry's teaching experience of over thirty years, his interest in students and love for college instruction, and his innovative attitude and use of technology-based learning enhance the solid tradition of John Hole's laboratory manual. Among Terry's awards are the 1972 Kishwaukee College Outstanding Educator, 1977 Phi Theta Kappa Outstanding Instructor Award, 1989 Kishwaukee College ICCTA Outstanding Educator Award, 1996 and 2004 Who's Who Among America's Teachers, 1996 Kishwaukee College Faculty Board of Trustees Award of Excellence, and 1998 Continued Excellence Award for Phi Theta Kappa Advisors. Terry's professional memberships include the National Association of Biology Teachers, Illinois Association of Community College Biologists, Human Anatomy and Physiology Society, Chicago Area Anatomy and Physiology Society (founding member), Phi Theta Kappa (honorary member), and Nature Conservancy. In addition to writing many publications, he co-produced with Hassan Rastegar a videotape entitled *Introduction to the Human Cadaver and Prosection,* published by Wm. C. Brown Publishers in 1989. Terry revised the *Laboratory Manual to Accompany Hole's Human Anatomy and Physiology,* Eleventh Edition, and authored *Human Anatomy and Physiology Laboratory Manual, Fetal Pig Dissection,* Third Edition. Terry serves as a Faculty Consultant for Advanced Placement Biology examination readings. During 1994, Terry was a faculty exchange member in Ireland. The author locally supports historical preservation, natural areas, scouting, and scholarship. We are pleased to have Terry continue the tradition of John Hole's laboratory manual.

To the Student

The Editor

The exercises in this laboratory manual will provide you with opportunities to observe various anatomical parts and to investigate certain physiological phenomena. Such experiences should help you relate specimens, models, microscope slides, and your body to what you have learned in the lecture and read about in the textbook.

The following list of suggestions may help to make your laboratory activities more effective and profitable.

1. Prepare yourself before attending the laboratory session by reading the assigned exercise and reviewing the related sections of the textbook. It is important to have some understanding of what will be done in the laboratory before you come to class.

2. Bring your laboratory manual and textbook to each laboratory session. These books are closely integrated and will help you complete most of the exercises.

3. Be on time. During the first few minutes of the laboratory meeting, the instructor often will provide verbal instructions. Make special note of any changes in materials to be used or procedures to be followed. Also listen carefully for information about special techniques to be used and precautions to be taken.

4. Keep your work area clean and your materials neatly arranged so that you can quickly locate needed items. This will enable you to efficiently proceed and will reduce the chances of making mistakes.

5. Pay particular attention to the purpose of the exercise, which states what you are to accomplish in general terms, and to the learning outcomes, which list what you should be able to do as a result of the laboratory experience. Then, before you leave the class, review the outcomes and make sure that you can meet all of the assessments.

6. Precisely follow the directions in the procedure and proceed only when you understand them clearly. Do not improvise procedures unless you have the approval of the laboratory instructor. Ask questions if you do not understand exactly what you are supposed to do and why you are doing it.

7. Handle all laboratory materials with care. These materials often are fragile and expensive to replace. Whenever you have questions about the proper treatment of equipment, ask the instructor.

8. Treat all living specimens humanely and try to minimize any discomfort they might experience.

9. Although at times you might work with a laboratory partner or a small group, try to remain independent when you are making observations, drawing conclusions, and completing the activities in the laboratory reports.

10. Record your observations immediately after making them. In most cases, such data can be entered in spaces provided in the laboratory reports.

11. Read the instructions for each section of the laboratory report before you begin to complete it. Think about the questions before you answer them. Your responses should be based on logical reasoning and phrased in clear and concise language.

12. At the end of each laboratory period, clean your work area and the instruments you have used. Return all materials to their proper places and dispose of wastes, including glassware or microscope slides that have become contaminated with human blood or body fluids, as directed by the laboratory instructor. Wash your hands thoroughly before leaving the laboratory.

Study Skills for Anatomy and Physiology

My students have found that certain study skills worked well for them while they were enrolled in Human Anatomy and Physiology. Although everyone has his or her learning style, there are techniques that work well for most students. Using some of the skills listed here could make your course more enjoyable and rewarding.

1. **Time management:** Prepare monthly, weekly, and daily schedules. Include dates of quizzes, exams, and projects on the calendar. On your daily schedule, budget several short study periods. Daily repetition alleviates cramming for exams. Prioritize your time so that you still have time for work and leisure activities. Find an appropriate study atmosphere with minimum distractions.

2. **Note taking:** Look for the main ideas and briefly express them in your own words. Organize, edit, and review your notes soon after the lecture. Add textbook information to your notes as you reorganize them. Underline or highlight with different colors the important points, major headings, and key terms. Study your notes daily, as they provide sequential building blocks of the course content.

3. **Chunking:** Organize information into logical groups or categories. Study and master one chunk of information at a time. For example, study the bones of the upper limb, lower limb, trunk, and head as separate study tasks.

4. **Mnemonic devices:** An *acrostic* is a combination of association and imagery to aid your memory. It is often in the form of a poem, rhyme, or jingle in which the first letter of each word corresponds to the first letters of the words you need to remember. **So Long Top Part, Here Comes The Thumb** is an example of

such a mnemonic device for remembering the eight carpals in the correct sequence. *Acronyms* are words formed by the first letters of the items to remember. *IPMAT* is an example of this type of mnemonic device to help you remember the phases of the cell cycle in the correct sequence. Try to create some of your own.

5. **Note cards/flash cards:** Make your own. Add labels and colors to enhance the material. Keep them with you in your pocket or purse. Study them often and for short periods. Concentrate on a small number of cards at one time. Shuffle your cards and have someone quiz you on their content. As you become familiar with the material, you can set aside cards that don't require additional mastery.

6. **Recording and recitation:** An auditory learner can benefit by recording lectures and review sessions with a cassette recorder. Many students listen to the taped sessions as they drive or just before going to bed. Reading your notes aloud can help also. Explain the material to anyone (even if there are no listeners). Talk about anatomy and physiology in everyday conversations.

7. **Study groups:** Small study groups that meet periodically to review course material and compare notes have helped and encouraged many students. However, keep the group on the task at hand. Work as a team and alternate leaders. This group often becomes a support group.

Best wishes on your anatomy and physiology endeavor.

Correlation of Textbook Chapters and Laboratory Exercises

Textbook Chapters	Related Laboratory Exercises
Chapter 1 Introduction to Human Anatomy and Physiology	Exercise 1 Scientific Method and Measurements
	Exercise 2 Body Organization and Terminology
Chapter 2 Chemical Basis of Life	Exercise 3 Chemistry of Life
Chapter 3 Cells	Exercise 4 Care and Use of the Microscope
	Exercise 5 Cell Structure and Function
	Exercise 6 Movements Through Cell Membranes
	Exercise 7 Cell Cycle
Chapter 4 Cellular Metabolism	
Chapter 5 Tissues	Exercise 8 Epithelial Tissues
	Exercise 9 Connective Tissues
	Exercise 10 Muscle and Nervous Tissues
Chapter 6 Integumentary System	Exercise 11 Integumentary System
Chapter 7 Skeletal System	Exercise 12 Bone Structure
	Exercise 13 Organization of the Skeleton
	Exercise 14 Skull
	Exercise 15 Vertebral Column and Thoracic Cage
	Exercise 16 Pectoral Girdle and Upper Limb
	Exercise 17 Pelvic Girdle and Lower Limb
	Exercise 18 Joint Structure and Movements
Chapter 8 Muscular System	Exercise 19 Skeletal Muscle Structure
	Exercise 20 Muscles of the Face, Head, and Neck
	Exercise 21 Muscles of the Chest, Shoulder, and Upper Limb
	Exercise 22 Muscles of the Abdominal Wall and Pelvic Outlet
	Exercise 23 Muscles of the Hip and Lower Limb
	Exercise 24 Surface Anatomy
Chapter 9 Nervous System	Exercise 25 Nervous Tissue and Nerves
	Exercise 26 Spinal Cord and Meninges
	Exercise 27 Reflex Arc and Reflexes
	Exercise 28 Brain and Cranial Nerves
	Exercise 29 Dissection of the Sheep Brain
Chapter 10 The Senses	Exercise 30 Ear and Hearing
	Exercise 31 Eye Structure
	Exercise 32 Visual Tests and Demonstrations
Chapter 11 Endocrine System	Exercise 33 Endocrine Histology and Diabetic Physiology
Chapter 12 Blood	Exercise 34 Blood Cells and Blood Typing
	Exercise 49 Blood Testing (available online)
Chapter 13 Cardiovascular System	Exercise 35 Heart Structure
	Exercise 36 Cardiac Cycle
	Exercise 37 Blood Vessel, Structure, Arteries, and Veins
	Exercise 38 Pulse Rate and Blood Pressure
Chapter 14 Lymphatic System and Immunity	Exercise 39 Lymphatic System
Chapter 15 Digestive System and Nutrition	Exercise 40 Digestive Organs
	Exercise 41 Action of a Digestive Enzyme
Chapter 16 Respiratory System	Exercise 42 Respiratory Organs
	Exercise 43 Breathing and Respiratory Volumes and Capacities
Chapter 17 Urinary System	Exercise 44 Kidney Structure
	Exercise 45 Urinalysis
Chapter 18 Water, Electrolyte, and Acid-Base Balance	
Chapter 19 Reproductive Systems	Exercise 46 Male Reproductive System
	Exercise 47 Female Reproductive System
Chapter 20 Pregnancy, Growth, Development, and Genetics	Exercise 48 Genetics

Laboratory Exercise 1

Scientific Method and Measurements

Scientific investigation involves a series of logical steps to arrive at explanations for various biological phenomena. This technique, called the *scientific method,* is used in all disciplines of science. It allows scientists to draw logical and reliable conclusions about phenomena.

The scientific method begins with *observations* related to the topic under investigation. This step commonly involves the accumulation of previously acquired information and/or your observations of the phenomenon. These observations are used to formulate a tentative explanation known as the *hypothesis.* An important attribute of a hypothesis is that it must be testable. The testing of the hypothesis involves performing a carefully controlled *experiment* to obtain data that can be used to support, reject, or modify the hypothesis. An *analysis of data* is conducted using sufficient information collected during the experiment. Data analysis may include organization and presentation of data as tables, graphs, and drawings. From the interpretation of the data analysis, *conclusions* are drawn. (If the data do not support the hypothesis, you must reexamine the experimental design and the data, and if needed develop a new hypothesis.) The final presentation of the information is made from the conclusions. Results and conclusions are presented to the scientific community for evaluation through peer reviews, presentations at professional meetings, and published articles. If many investigators working independently can validate the hypothesis by arriving at the same conclusions, the explanation becomes a **theory.** A theory verified continuously over time and accepted by the scientific community becomes known as a **scientific law** or **principle.** A scientific law serves as the standard explanation for an observation unless it is disproved by new information. The five components of the scientific method are summarized as

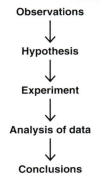

Observations
↓
Hypothesis
↓
Experiment
↓
Analysis of data
↓
Conclusions

Metric measurements are characteristic tools of scientific investigations. The English system of measurements is often used in the United States, so the investigator must make conversions from the English system to the metric system. A reference table for the conversion of English units of measure to metric units for length, mass, volume, time, and temperature is located inside the back cover of the laboratory manual.

Purpose of the Exercise

To become familiar with the scientific method of investigation, to learn how to formulate sound conclusions, and to provide opportunities to use the metric system of measurements.

Learning Outcomes

After completing this exercise, you should be able to

1. Convert English measurements to the metric system, and vice versa.
2. Measure and record upper limb lengths and heights of ten subjects.
3. Apply the scientific method to test the validity of a hypothesis concerning the direct, linear relationship between human upper limb length and height.
4. Design an experiment, formulate a hypothesis, and test it using the scientific method.

Procedure A—Using the Steps of the Scientific Method

1. Many people have observed a correlation between the length of the upper and lower limbs and the height (stature) of an individual. For example, a person who has long upper limbs (the arm, forearm, and hand combined) tends to be tall. Make some visual observations of other people in your class to observe a possible correlation.

2. From such observations, the following hypothesis is formulated: The length of a person's upper limb is equal to 0.4 (40%) of the height of the person. Test this hypothesis by performing the following experiment.

3. In this experiment, use a meterstick (fig. 1.1) to measure an upper limb length of ten subjects. For each measurement, place the meterstick in the axilla (armpit) and record the length in centimeters to the end of the longest finger (fig. 1.2). Obtain the height of each person in centimeters by measuring them without shoes against a wall (fig. 1.3). The height of each person can be calculated by multiplying each individual's height in inches by 2.54 to obtain his/her height in centimeters. Record all your measurements in Part A of Laboratory Report 1.

4. The data collected from all of the measurements can now be analyzed. The expected (predicted) correlation between upper limb length and height is determined using the following equation:

Height × 0.4 = expected upper limb length

The observed (actual) correlation to be used to test the hypothesis is determined by

Length of upper limb/height
= actual % of height

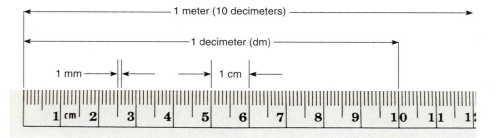

Metric ruler

Figure 1.1 Metric ruler with metric lengths indicated. A meterstick length would be 100 centimeters. (The image size is approximately to scale.)

Figure 1.2 Measurement of upper limb length.

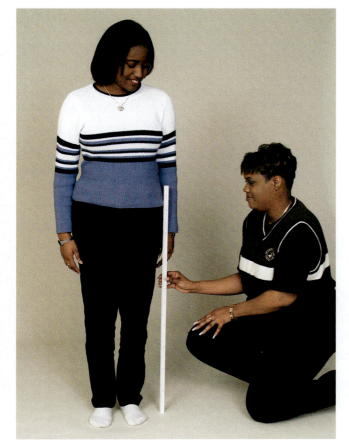

Figure 1.3 Measurement of height.

5. A graph is an excellent way to display a visual representation of the data. Plot the subjects' data in Part A of the laboratory report. Plot the upper limb length of each subject on the x-axis and the height of each person on the y-axis. A line is already located on the graph that represents a hypothetical relationship of 0.4 (40%) upper limb length compared to height. This is a graphic representation of the original hypothesis.

6. Compare the distribution of all of the points (actual height and upper limb length) that you placed on the graph with the distribution of the expected correlation represented by the hypothesis.

7. Complete Part A of the laboratory report.

Procedure B—Design an Experiment

Critical Thinking Application

You have probably concluded that there is some correlation of the length of body parts to height. Often when a skeleton is found, it is not complete, especially when paleontologists discover a skeleton. It is occasionally feasible to use the length of a single bone to estimate the height of an individual. Observe human skeletons and locate the radius bone in the forearm. Use your observations to identify a mathematical relationship between the length of the radius and height. Formulate a hypothesis that can be tested. Make measurements, analyze data, and develop a conclusion from your experiment. Complete Part B of the laboratory report.

Name _____

Date _____

Section _____

The ⬅ corresponds to the indicated outcome(s) found at the beginning of the laboratory exercise.

Scientific Method and Measurements

Part A Assessments

1. Record measurements for the upper limb length and height of ten subjects. Use a calculator to determine the expected upper limb length and the actual percentage (as a decimal or a percentage) of the height for the ten subjects. Record your results in the following table. ◄2

Subject	Measured Upper Limb Length (cm)	Height* (cm)	Height × 0.4 = Expected Upper Limb Length (cm)	Actual % of Height = Upper Limb Length (cm)/ Height (cm)
1.				
2.				
3.				
4.				
5.				
6.				
7.				
8.				
9.				
10.				

*The height of each person can be calculated by multiplying each individual's height in inches by 2.54 to obtain his/her height in centimeters. ◄1

2. Plot the distribution of data (upper limb length and height) collected for the ten subjects on the following graph. The line located on the graph represents the **expected** 0.4 (40%) ratio of upper limb length to measured height (the original hypothesis). (The x-axis represents upper limb length and the y-axis represents height.) Draw a line of *best fit* through the distribution of points of the plotted data of the ten subjects.
Compare the two distributions (expected line and the distribution line drawn for the ten subjects). ◀**3**

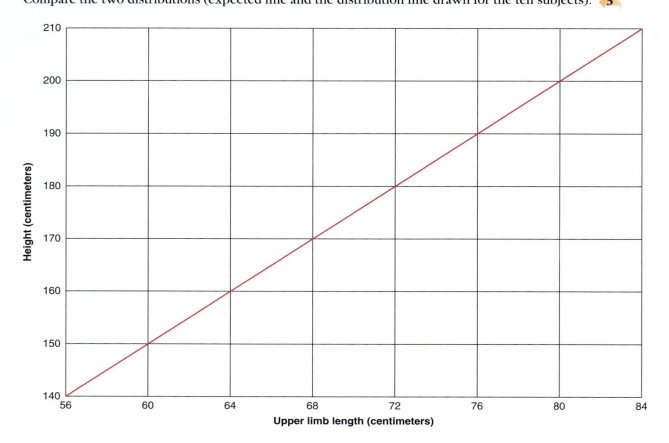

3. Does the distribution of the ten subjects' measured upper limb lengths support or reject the original hypothesis? _____ Explain your answer. ◀**3**

Part B Assessments

1. Describe your observations of a possible correlation between the radius length and height.

2. Write a hypothesis based on your observations. ◀**4**

3. Describe the design of the experiment that you devised to test your hypothesis. ◀4

4. Place your analysis of the data in this space in the form of a table and a graph. ◀4

5. Based on an analysis of your data, what conclusions can you make? Did these conclusions confirm or refute your original hypothesis? ◄4

6. Discuss your results and conclusions with classmates. What common conclusion can the class formulate about the correlation between radius length and height?

Laboratory Exercise

Body Organization and Terminology

Materials Needed

Textbook
Dissectible human torso model (manikin)
Variety of specimens or models sectioned along various planes

For Learning Extension:
Colored pencils

The major features of the human body include certain cavities, a set of membranes associated with these cavities, and a group of organ systems composed of related organs. To communicate effectively with each other about the body, scientists have devised names to describe these body features. They also have developed terms to represent the relative positions of body parts, imaginary planes passing through these parts, and body regions.

Purpose of the Exercise

To review the organizational pattern of the human body; to review its organ systems and the organs included in each system; and to become acquainted with the terms used to describe the relative position of body parts, body sections, and body regions.

Learning Outcomes

After completing this exercise, you should be able to

1. Locate and name the body cavities and identify the organs and membranes associated with each cavity.
2. Differentiate the general functions of the organ systems of the human body.
3. Associate the organs included within each system and locate the organs in a dissectible human torso model.
4. Select the terms used to describe the relative positions of body parts.
5. Match the terms used to identify body sections and identify the plane along which a particular specimen is cut.
6. Label body regions and associate the terms used to identify body regions.

Procedure A—Body Cavities and Membranes

1. Review the sections entitled "Body Cavities" and "Thoracic and Abdominopelvic Membranes" in chapter 1 of the textbook.
2. As a review activity, label figures 2.1 and 2.2.
3. Locate the following features on the reference plates near the end of chapter 1 of the textbook and on the dissectible human torso model (fig. 2.3):

body cavities
    cranial cavity
    vertebral canal (spinal cavity)
    thoracic cavity
        mediastinum (region between the lungs; includes pericardial cavity)
        pleural cavities
    abdominopelvic cavity
        abdominal cavity
        pelvic cavity

diaphragm

smaller cavities within the head
    oral cavity
    nasal cavity with connected sinuses
    orbital cavity
    middle ear cavity

membranes and cavities
    pleural cavity
        parietal pleura
        visceral pleura
    pericardial cavity
        parietal pericardium (covered by fibrous pericardium)
        visceral pericardium (epicardium)
    peritoneal cavity
        parietal peritoneum
        visceral peritoneum

4. Complete Part A of Laboratory Report 2.

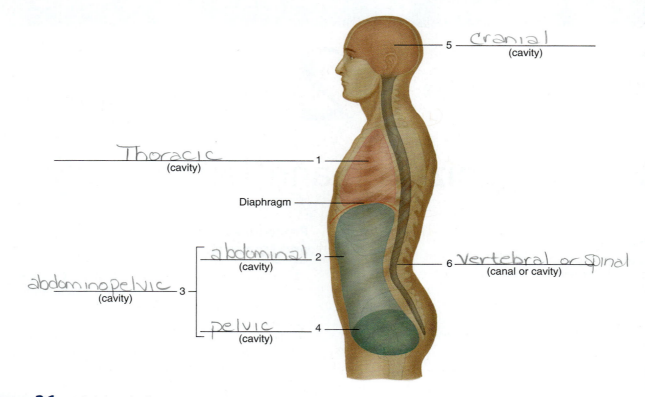

Figure 2.1 Label these body cavities. 1

Labels on figure:
- 5 — Cranial (cavity)
- 1 — Thoracic (cavity)
- Diaphragm
- 2 — abdominal (cavity)
- 3 — abdominopelvic (cavity)
- 4 — pelvic (cavity)
- 6 — Vertebral or spinal (canal or cavity)

Procedure B—Organ Systems

1. Review the section entitled "Organ Systems" in chapter 1 of the textbook.
2. Use the reference plates near the end of chapter 1 of the textbook and the dissectible human torso model (fig. 2.3) to locate the following organs:

integumentary system
  skin
  accessory organs such as hair and nails

skeletal system
  bones
  ligaments
  cartilages

muscular system
  skeletal muscles
  tendons

nervous system
  brain
  spinal cord
  nerves

endocrine system
  pituitary gland
  thyroid gland
  adrenal glands
  pancreas

  ovaries
  testes
  thymus

cardiovascular system
  heart
  arteries
  veins

lymphatic system
  lymphatic vessels
  lymph nodes
  thymus
  spleen

digestive system
  mouth
  tongue
  teeth
  salivary glands
  pharynx
  esophagus
  stomach
  liver
  gallbladder
  pancreas
  small intestine
  large intestine

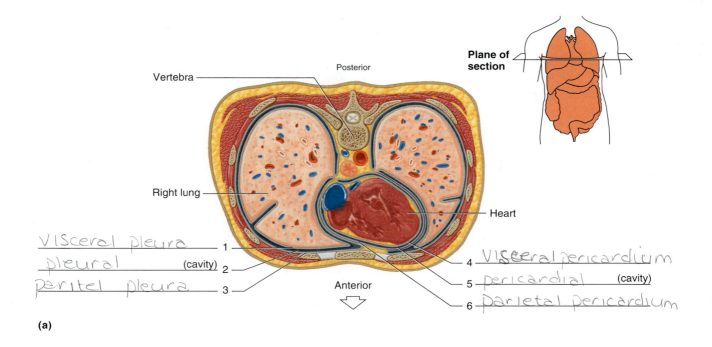

Posterior

Vertebra

Right lung

Heart

Plane of section

Visceral pleura — 1
pleural (cavity) — 2
Peritel pleura — 3

Visceral pericardium — 4
pericardial (cavity) — 5
Parietal pericardium — 6

Anterior

(a)

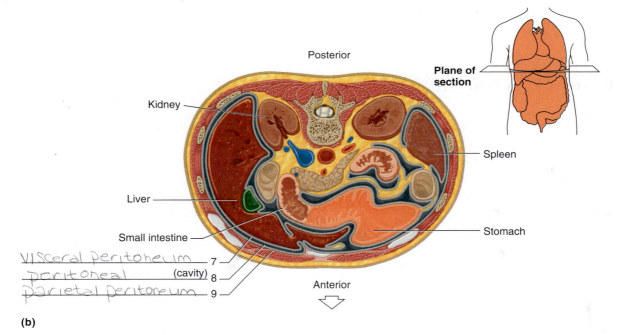

Posterior

Kidney

Spleen

Liver

Stomach

Small intestine

Plane of section

Visceral peritoneum — 7
peritoneal (cavity) — 8
Parietal peritoneum — 9

Anterior

(b)

Figure 2.2 Label the thoracic membranes and cavities in (*a*) and the abdominopelvic membranes and cavity in (*b*) as shown in these superior views of transverse sections. ◄ **1**

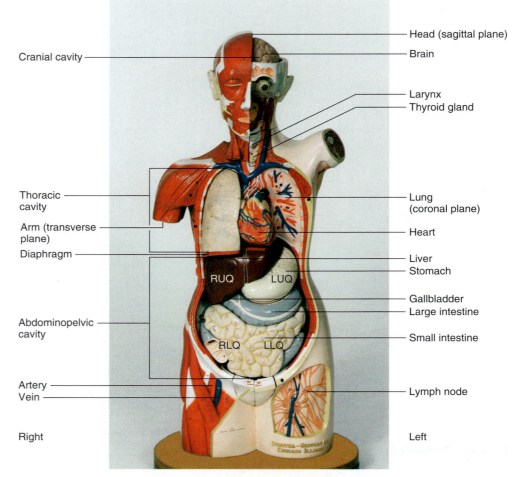

Figure 2.3 Dissectible human torso model with body cavities, abdominopelvic quadrants, body planes, and major organs indicated.

respiratory system
  nasal cavity
  pharynx
  larynx
  trachea
  bronchi
  lungs
urinary system
  kidneys
  ureters
  urinary bladder
  urethra
male reproductive system
  scrotum
  testes
  penis
  urethra

female reproductive system
  ovaries
  uterine tubes (oviducts; fallopian tubes)
  uterus
  vagina

3. Complete Part B of the laboratory report.

Procedure C—Relative Positions, Planes, Sections, and Regions

1. Observe the person standing in anatomical position (fig. 2.4). Anatomical terminology assumes the body is in anatomical position even though a person is often observed differently.
2. Review the section entitled "Anatomical Terminology" in chapter 1 of the textbook.
3. As a review activity, label figures 2.5, 2.6, and 2.7.
4. Examine the sectioned specimens on the demonstration table and identify the plane along which

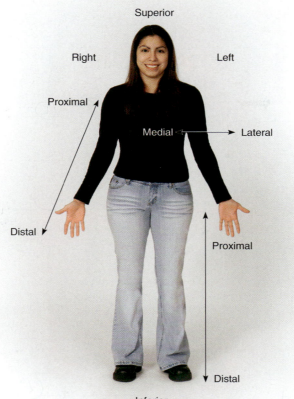

Figure 2.4   Anatomical position with directional terms indicated. The body is standing erect, face forward, with upper limbs at the sides and palms forward. This results in an anterior view of the body.

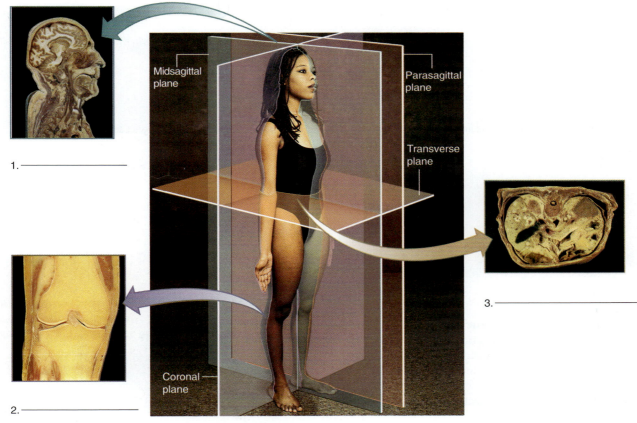

1. _____

2. _____

3. _____

Figure 2.5   Label the planes represented in this illustration. ↩**5**

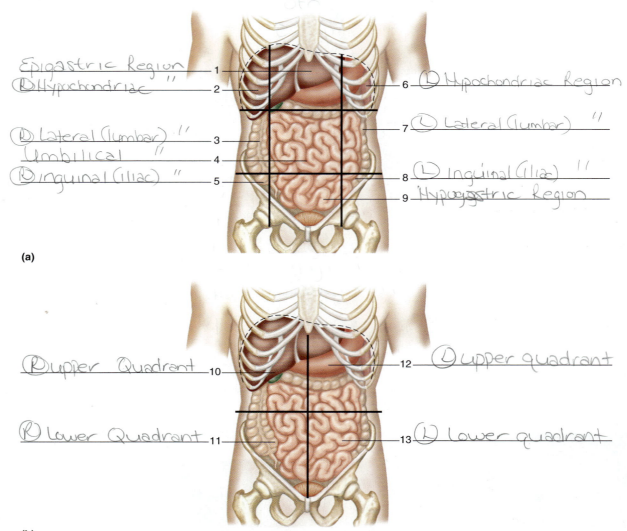

Epigastric Region ——1

ⒷHypochondriac '' ——2

ⒹLateral (lumbar) '' ——3

Umbilical '' ——4

Ⓑinguinal (Iliac) '' ——5

6—— ⓁHypochondriac Region

7—— Ⓛ Lateral (lumbar) ''

8—— Ⓛ inguinal (Iliac) ''

9—— Hypogastric Region

(a)

ⒷUpper Quadrant ——10

Ⓡ Lower Quadrant ——11

12—— Ⓛupper quadrant

13—— Ⓛ lower quadrant

(b)

Figure 2.6   Label (*a*) the regions and (*b*) the quadrants of the abdominopelvic cavity.

each is cut. Cylindrical structures, such as a long bone or a blood vessel, may be cut in cross section, oblique section, or longitudinal section. The same three sections can be demonstrated by three cuts of a banana (fig. 2.8).

5. Complete Parts C, D, E, and F of the laboratory report.

Learning Extension

Use different colored pencils to distinguish the body regions in figure 2.7.

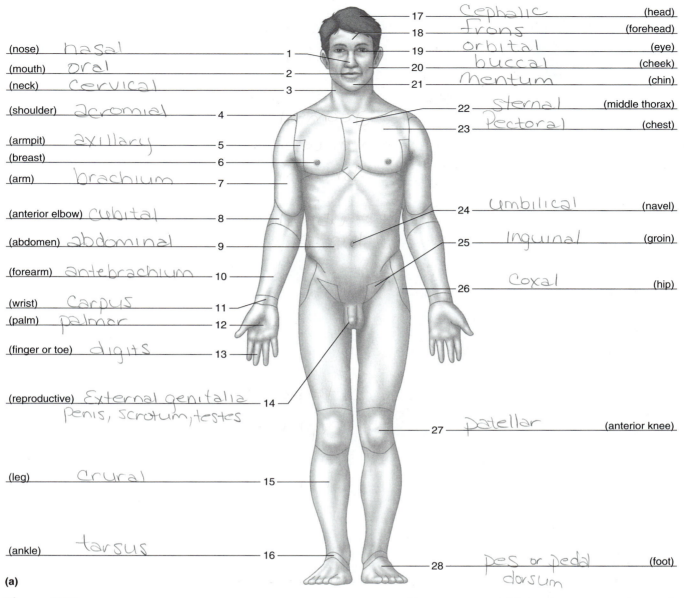

(nose) nasal — 1

(mouth) oral — 2

(neck) Cervical — 3

(shoulder) acromial — 4

(armpit) axillary — 5

(breast) — 6

(arm) brachium — 7

(anterior elbow) Cubital — 8

(abdomen) abdominal — 9

(forearm) antebrachium — 10

(wrist) Carpus — 11

(palm) palmar — 12

(finger or toe) digits — 13

(reproductive) External genitalia
Penis, Scrotum, testes — 14

(leg) Crural — 15

(ankle) tarsus — 16

17 — Cephalic (head)

18 — frons (forehead)

19 — orbital (eye)

20 — buccal (cheek)

21 — mentum (chin)

22 — sternal (middle thorax)

23 — Pectoral (chest)

24 — umbilical (navel)

25 — Inguinal (groin)

26 — Coxal (hip)

27 — patellar (anterior knee)

28 — pes or pedal (foot)
dorsum

(a)

Figure 2.7 Label these diagrams with terms used to describe body regions: (*a*) anterior regions; (*b*) posterior regions. **6**

#	Answer	Description
29	oto	(ear)
30	occipital	(posterior head)
31	acomial	(shoulder)
32	vertebral	(spinal column)
33	brachial	(arm)
34		(back)
35	cubital	(elbow)
36	lumbar	(loin)
37	sacral	(between hips)
38	gluteal	(buttocks)
39	perineal	(between anus and genitals)
40	femoral	(thigh)
41	popliteal	(posterior knee)
42	crural	(calf)
43	plantar surface	(sole)

(b)

Figure 2.7 *Continued.*

(a) (b) (c)

Figure 2.8 Three possible cuts of a banana: (*a*) cross section; (*b*) oblique section; (*c*) longitudinal section. Sections through an organ, as a body tube, frequently produce views similar to the cut banana.

Name _____

Date _____

Section _____

The ⬅ corresponds to the indicated outcome(s) found at the beginning of the laboratory exercise.

Body Organization and Terminology

Part A Assessments

Match the cavities in column A with the organs contained in the cavities in column B. Place the letter of your choice in the space provided. ⬅1

Column A		Column B
a. Abdominal cavity	a	**1.** Liver
b. Cranial cavity	d	**2.** Lungs
c. Pelvic cavity	a	**3.** Spleen
d. Thoracic cavity	a	**4.** Stomach
e. Vertebral canal (spinal cavity)	b	**5.** Brain
	d	**6.** Trachea and esophagus
	c	**7.** Urinary bladder
	a	**8.** Small intestine
	e	**9.** Spinal cord
	c	**10.** Internal reproductive organs
	d	**11.** Heart
	d	**12.** Mediastinum

Part B Assessments

Match the organ systems in column A with the general functions in column B. Place the letter of your choice in the space provided. ◂2

Column A		Column B
a. Cardiovascular system	C	**1.** Main system that secretes hormones
b. Digestive system	d	**2.** Provides an outer covering
c. Endocrine system		
d. Integumentary system	h	**3.** Produces gametes (eggs and sperm)
e. Lymphatic system	g	**4.** Stimulates muscles to contract and interprets information from sensory units
f. Muscular system		
g. Nervous system	j	**5.** Provides a framework for soft tissues and produces blood cells in red marrow
h. Reproductive system		
i. Respiratory system	i	**6.** Exchanges gases between air and blood
j. Skeletal system	e	**7.** Transports excess fluid from tissues to blood
k. Urinary system	f	**8.** Movement via contractions and creates most body heat
	K	**9.** Removes liquid wastes from blood and transports them to the outside of the body
	b	**10.** Converts food molecules into absorbable forms
	a	**11.** Transports nutrients, wastes, and gases throughout the body

Part C Assessments

Indicate whether each of the following sentences makes correct or incorrect usage of the word in boldface type (assume that the body is in the anatomical position as observed in fig. 2.4). If the sentence is incorrect, in the space provided supply a term that will make it correct. ◂4

1. The mouth is **superior** to the nose. IC – Inferior

2. The stomach is **inferior** to the diaphragm. C

3. The trachea is **anterior** to the spinal cord. C

4. The larynx is **posterior** to the esophagus. ___

5. The heart is **medial** to the lungs. C

6. The kidneys are **inferior** to the adrenal glands. C

7. The hand is **proximal** to the elbow. C

8. The knee is **proximal** to the ankle. C

9. Blood in **deep** blood vessels gives color to the skin. IC – Superficial

10. A **peripheral** nerve passes from the spinal cord into the limbs. C

11. The dermis is the **superficial** layer of the skin. IC

Part D Assessments

Match the body regions in column A with the body parts in column B. Place the letter of your choice in the space provided. 6

Column A		Column B
a. Antecubital	__e__	**1.** Wrist
b. Axillary	__k__	**2.** Reproductive organs
c. Brachial	__b__	**3.** Armpit
d. Buccal	__i__	**4.** Elbow
e. Carpal	__l__	**5.** Buttocks
f. Cephalic	__j__	**6.** Back
g. Cervical	__g__	**7.** Neck
h. Crural	__c__	**8.** Arm
i. Cubital	__d__	**9.** Cheek
j. Dorsal	__h__	**10.** Leg
k. Genital	__f__	**11.** Head
l. Gluteal	__a__	**12.** Front of elbow

Part E Assessments

Match the body regions in column A with the locations in column B. Place the letter of your choice in the space provided. 6

Column A		Column B
a. Inguinal	__h__	**1.** Pelvis
b. Lumbar	__c__	**2.** Breasts
c. Mammary	__i__	**3.** Between anus and reproductive organs
d. Occipital	__j__	**4.** Sole
e. Palmar	__l__	**5.** Middle of thorax
f. Pectoral	__f__	**6.** Chest
g. Pedal	__k__	**7.** Back of knee
h. Pelvic	__g__	**8.** Foot
i. Perineal	__d__	**9.** Lower posterior region of head
j. Plantar	__a__	**10.** Abdominal wall near thigh (groin)
k. Popliteal	__b__	**11.** Lower back
l. Sternal	__e__	**12.** Palm

Critical Thinking Application

State the quadrant of the abdominopelvic cavity where the pain or sound would be located for each of the six common conditions listed. In some cases, there may be more than one correct answer, and pain is sometimes referred to another region. This phenomenon, called *referred pain,* occurs when pain is interpreted as originating from some area other than the parts being stimulated. When referred pain is involved in the patient's interpretation of the pain location, the proper diagnosis of the ailment is more challenging. For the purpose of this exercise, assume the pain interpretation is originating from the organ involved. ◄**3**

1. Stomach ulcer _____
2. Appendicitis R ↓ quad.
3. Bowel sounds _____
4. Gallbladder attack R ↑ quad
5. Kidney stone in left ureter L ↓ quad
6. Ruptured spleen L ↑ quad

Laboratory Exercise 3

Chemistry of Life

Materials Needed

For pH Tests:
Chopped fresh red cabbage
Beaker (250 mL)
Distilled water
Tap water
Vinegar
Baking soda
Laboratory scoop for measuring
Pipets for measuring
Full-range pH test papers
7 assorted common liquids clearly labeled in closed bottles on a tray or in a tub
Droppers labeled for each liquid

For Organic Tests:
Test tubes
Test-tube rack
Test-tube clamps
China marker
Hot plate
Beaker for hot water bath (500 mL)
Pipets for measuring
Benedict's solution
Biuret reagent (or 10% NaOH and 1% $CuSO_4$)
Iodine-potassium-iodide (IKI) solution
Sudan IV dye
Egg albumin
10% glucose solution
Clear carbonated soft drink
10% starch solution
Potatoes for potato water
Distilled water
Vegetable oil
Brown paper
Numbered unknown organic samples

⚠ Safety

- Review all safety guidelines inside the front cover of your laboratory manual.
- Clean laboratory surfaces before and after laboratory procedures using soap and water.
- Use extreme caution when working with chemicals.
- Safety goggles must be worn at all times.
- Wear disposable gloves while working with the chemicals.
- Precautions should be taken to prevent chemicals from contacting your skin.
- Do not mix any of the chemicals together unless instructed to do so.
- Clean up any spills immediately and notify the instructor at once.
- Wash your hands before leaving the laboratory.

The complexities of the human body arise from the organization and interactions of chemicals. Organisms are made of matter, and the most basic unit of matter is the chemical element. The smallest unit of an element is an atom, and two or more of those can unite to form a molecule. All processes that occur within the body involve chemical reactions—interactions between atoms and molecules. We breathe to supply oxygen to our cells for energy. We eat and drink to bring chemicals into our bodies that our cells need. Water fills and bathes all of our cells and allows an amazing array of reactions to occur, all of which are designed to keep us alive. Chemistry forms the basis of life and thus forms the foundation of anatomy and physiology.

Purpose of the Exercise

To review the organization of atoms and molecules, types of chemical interactions, and basic categories of organic compounds, and to differentiate between types of organic compounds.

Learning Outcomes

After completing this exercise, you should be able to

1. Associate and illustrate the basic organization of atoms and molecules.
2. Measure pH values of various substances through testing methods.
3. Determine categories of organic compounds with basic colorimetric tests.
4. Discover the organic composition of an unknown solution.

Procedure A—Matter, Molecules, Bonding, and pH

1. Review the section entitled "Structure of Matter" in chapter 2 of the textbook.
2. Review figures 3.1, 3.2, and 3.3, which show molecular diagrams, formation of an ionic bond, and formation of a covalent bond, respectively.
3. Complete Part A of Laboratory Report 3.

Procedure B—The pH Scale

1. Review the section entitled "Acids and Bases" in chapter 2 of the textbook.
2. Review figure 3.4, which shows the pH scale. Note the range of the scale and the pH value considered neutral.
3. **Cabbage water tests.** Many tests can determine a pH value, but among the more interesting are colorimetric tests in which an indicator chemical changes color when it reacts. Many plant pigments, especially anthocyanins (which give plants color, from blue to red), can be used as colorimetric pH indicators. One that works well is the red pigment in red cabbage.
   a. Prepare cabbage water to be used as a general pH indicator. Fill a 250 mL beaker to the 100 mL level with chopped red cabbage. Add water to make 150 mL. Place the beaker on a hot plate and simmer the mixture until the pigments come out of the cabbage and the water turns deep purple. Allow the water to cool. (You may proceed to step 4 while you wait for this to finish.)

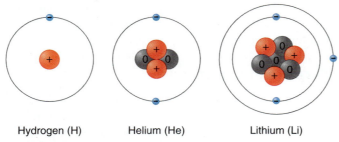

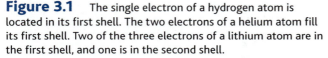

Figure 3.1 The single electron of a hydrogen atom is located in its first shell. The two electrons of a helium atom fill its first shell. Two of the three electrons of a lithium atom are in the first shell, and one is in the second shell.

b. Label three clean test tubes, one for water, one for vinegar, and one for baking soda.
c. Place 2 mL of cabbage water into each test tube.
d. To the first test tube, add 2 mL of distilled water and swirl the mixture. Record the color in Part B of the laboratory report.
e. Repeat this procedure for test tube 2, adding 2 mL of vinegar and swirling the mixture. Record the results.
f. Repeat this procedure for test tube 3, adding one laboratory scoop of baking soda. Swirl the mixture. Record the results.

4. **Testing with pH paper.** Many commercial pH indicators are available. A simple one to use is pH paper, which comes in small strips. Be cautious when handling the paper so your skin secretions do not contaminate the strip. Gloves are recommended for this procedure, as well as to protect you from the chemicals you are testing, some of which may be harmful.
   a. Test the pH of distilled water by dropping one or two drops of distilled water onto a strip of pH paper and note the color. Compare the color to the color guide on the container. Record the pH value in Part B of the laboratory report.

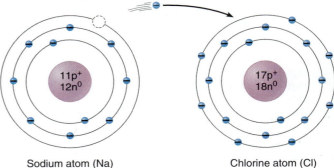

Sodium atom (Na)   Chlorine atom (Cl)

(a) Separate atoms
If a sodium atom loses an electron to a chlorine atom, the sodium atom becomes a sodium ion (Na^+), and the chlorine atom becomes a chloride ion (Cl^-).

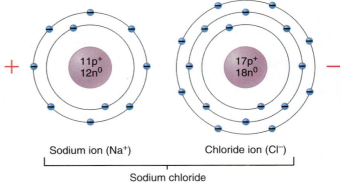

Sodium ion (Na^+)   Chloride ion (Cl^-)

Sodium chloride

(b) Bonded ions
These oppositely charged particles attract electrically and join by an ionic bond.

Figure 3.2 Formation of an ionic bond.

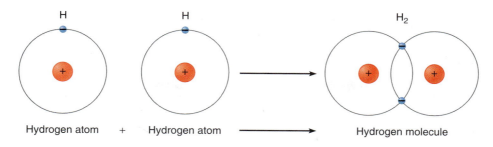

H   H   H₂

Hydrogen atom   +   Hydrogen atom   ⟶   Hydrogen molecule

Figure 3.3   A hydrogen molecule forms when two hydrogen atoms share a pair of electrons and join by a covalent bond.

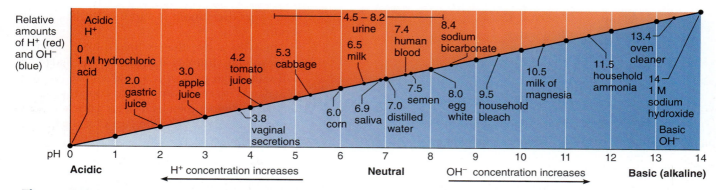

Figure 3.4   As the concentration of hydrogen ions (H⁺) increases, a solution becomes more acidic and the pH value decreases. As the concentration of ions that combine with hydrogen ions (such as hydroxide ions) increases, a solution becomes more basic (alkaline) and the pH value increases. The pH values of some common substances are shown.

b. Repeat this procedure for tap water. Record the results.

c. Individually test the various common substances found on your lab table. Record your results.

d. Complete Part B of the laboratory report.

Procedure C—Organic Molecules

In this section, you will perform some simple tests to check for the presence of the following categories of organic molecules (biomolecules): protein, sugar, starch, and lipid. Specific color changes will occur if the target compound is present. Please note the original color of the indicator being added so you can tell if the color really changed. For example, Biuret reagent and Benedict's solution are both initially blue, so an end color of blue would indicate no change. Use caution when working with these chemicals, and follow all directions. Carefully label the test tubes and avoid contaminating any of your samples. Only the Benedict's test requires heating and a time delay for accurate results. Do not heat any other tubes. To prepare for the Benedict's test, fill a 500 mL beaker half full with water and place it on the hot plate. Turn the hot plate on and bring the water to a boil. To save time, start the water bath before doing the Biuret test.

1. Biuret test for protein. In the presence of protein, Biuret reagent reacts with peptide bonds and changes to violet or purple. A pinkish color indicates that shorter polypeptides are present. The color intensity is proportional to the number of peptide bonds, thus the intensity reflects the length of polypeptides (amount of protein).

a. Label six test tubes as follows: 1W, 1E, 1G, 1D, 1S, and 1P.

b. To each test tube, add 2 mL of one of the samples as follows:

1W—2 mL distilled water

1E—2 mL egg albumin

1G—2 mL 10% glucose solution

1D—2 mL carbonated soft drink

1S—2 mL 10% starch solution

1P—2 mL potato juice

c. To each, add 2 mL of Biuret reagent, swirl the tube to mix it, and note the final color. [*Note:* If Biuret reagent is not available, add 2 mL of 10% NaOH (sodium hydroxide) and about 10 drops of 1% $CuSO_4$ (copper sulfate)]. **Be careful—NaOH is very caustic.**

d. Record your results in Part C of the laboratory report.

e. Mark a "+" on any tubes with positive results and retain them for comparison while testing your unknown sample. Put remaining test tubes aside so they will not get confused with future trials.

2. **Benedict's test for sugar (monosaccharides).** In the presence of sugar, Benedict's solution changes from its initial blue to a green, yellow, orange, or reddish color, depending on the amount of sugar present. Orange and red indicate greater amounts of sugar.

a. Label six test tubes as follows: 2W, 2E, 2G, 2D, 2S, and 2P.

b. To each test tube, add 2 mL of one of the samples as follows:

2W—2 mL distilled water

2E—2 mL egg albumin

2G—2 mL 10% glucose solution

2D—2 mL carbonated soft drink

2S—2 mL 10% starch solution

2P—2 mL potato juice

c. To each, add 2 mL of Benedict's solution, swirl the tube to mix it, and place all tubes into the boiling water bath for 3 to 5 minutes.

d. Note the final color of each tube after heating and record the results in Part C of the laboratory report.

e. Mark a "+" on any tubes with positive results and retain them for comparison while testing your unknown sample. Put remaining test tubes aside so they will not get confused with future trials.

3. **Iodine test for starch.** In the presence of starch, iodine turns a dark purple or blue-black color. Starch is a long chain formed by many glucose units linked together side by side. This regular organization traps the iodine molecules and produces the dark color.

a. Label six test tubes as follows: 3W, 3E, 3G, 3D, 3S, and 3P.

b. To each test tube, add 2 mL of one of the samples as follows:

3W—2 mL distilled water

3E—2 mL egg albumin

3G—2 mL 10% glucose solution

3D—2 mL carbonated soft drink

3S—2 mL 10% starch solution

3P—2 mL potato juice

c. To each, add 0.5 mL of IKI (iodine solution) and swirl the tube to mix it.

d. Record the final color of each tube in Part C of the laboratory report.

e. Mark a "+" on any tubes with positive results and retain them for comparison while testing your unknown sample. Put remaining test tubes aside so they will not get confused with future trials.

4. **Tests for lipids.**

a. Label two separate areas of a piece of brown paper as "water" or "oil."

b. Place a drop of water on the area marked "water" and a drop of vegetable oil on the area marked "oil."

c. Let the spots dry several minutes, then record your observations in Part C of the laboratory report. Upon drying, oil leaves a stain (grease spot) on brown paper; water does not leave such a spot.

d. In a test tube, add 2 mL of water and 2 mL of vegetable oil and observe. Shake the tube vigorously then let it sit for 5 minutes and observe again. Record your observations in Part C of the laboratory report.

e. Sudan IV is a dye that is lipid-soluble but not water-soluble. If lipids are present, the Sudan IV will stain them pink or red. Add a small amount of Sudan IV to the test tube that contains the oil and water. Swirl it then let it sit for a few minutes. Record your observations in Part C of the laboratory report.

Procedure D—Identifying Unknown Compounds

Now you will apply the information you gained with the tests in the previous section. You will retrieve an unknown sample that contains none, one, or any combination of the following types of organic compounds: protein, sugar, starch, or lipid. You will test your sample using each test from the previous section and record your results in Part D of the laboratory report.

1. Label four test tubes (1, 2, 3, 4).

2. Add 2 mL of your unknown sample to each test tube.

3. **Test for protein.** Add 2 mL of Biuret reagent to tube 1. Swirl the tube and observe the color. Record your observations in Part D of the laboratory report.

4. **Test for sugar.** Add 2 mL of Benedict's solution to tube 2. Swirl the tube and place it in a boiling water bath for 3 to 5 minutes. Record your observations in Part D of the laboratory report.

5. **Test for starch.** Add several drops of iodine solution to tube 3. Swirl the tube and observe the color. Record your observations in Part D of the laboratory report.

6. **Test for lipid.** Add 2 mL of water to tube 4. Swirl the tube and note if there is any separation.

7. Add a small amount of Sudan IV to test tube 4, swirl the tube and record your observations in Part D of the laboratory report.

8. Based on the results of these tests, determine if your unknown sample contains any organic compounds and, if so, what are they? Record and explain your identification in Part D of the laboratory report.

Chemistry of Life

Name _____

Date _____

Section _____

The ⬅ corresponds to the indicated outcome(s) found at the beginning of the laboratory exercise.

Part A Assessments

Match the terms in column A with the descriptions in column B. Place the letter of your choice in the space provided. **⬅1**

Column A	Column B
a. Atomic number	_b_ **1.** The number of protons plus the number of neutrons
b. Atomic weight	_e_ **2.** A combination of two or more atoms of different elements
c. Base	_h_ **3.** A small, negatively charged particle that orbits the nucleus
d. Catalyst (enzyme)	_j_ **4.** An atom that has gained or lost electrons, and thus carries an electrical charge
e. Compound	
f. Covalent bond	_f_ **5.** Atoms are held together by sharing electrons
g. Electrolyte	_i_ **6.** The most fundamental substances of matter
h. Electron	_c_ **7.** A substance that combines with hydrogen ions
i. Elements	_d_ **8.** A molecule that influences the rate of chemical reactions but is not consumed in the reaction
j. Ion	
k. Isotopes	_K_ **9.** Two atoms with the same atomic number but different atomic weights
l. Nucleus	_g_ **10.** A substance that releases ions in water
	_a_ **11.** The number of protons in an atom
	_l_ **12.** The area of an atom where protons and neutrons are located

Molecules And Bonding

After reviewing figures 3.1, 3.2, and 3.3, complete the following:

1. The atomic number of hydrogen is ___1___, and it has ___1___ electron(s) in its outer shell. **⬅1**

2. The atomic number of chlorine is ___17___, and it has ___17___ electron(s) in its outer shell. **⬅1**

3. Using the style shown in figure 3.2, draw an atom of hydrogen and one of chlorine, clearly indicating the numbers and positions of the protons, neutrons, and electrons. **⬅1**

Hydrogen

Chlorine

4. Is the hydrogen atom stable? **1** _____

   Is the chlorine atom stable? **1**_____

5. What type of bond is likely to form the compound HCl (hydrochloric acid)? **1** _____

6. Repeat your drawings here, but also show and explain how this bond of HCl would form. Clearly indicate the positions of protons, neutrons, and electrons. **1**

Part B—The pH Scale Assessments

1. Results from cabbage water test: **2**

Substance	Cabbage Water	Distilled Water	Vinegar	Baking Soda
Color				
Acid, base, or neutral?				

2. Results from pH paper tests: **2**

Substance Tested	Distilled Water	Tap Water	Sample 1:_____	Sample 2:_____	Sample 3:_____	Sample 4:_____	Sample 5:_____	Sample 6:_____	Sample 7:_____
pH Value									

3. Are the pH values the same for distilled water and tap water? **2** _____

4. If not, what might explain this difference? _____

_____

5. Draw the pH scale here, and indicate the following values: 0, 7 (neutral), and 14. Now label the scale with the names and indicate the location of the pH values for each substance you tested.

Part C—Testing For Organic Molecules (Biomolecules) Assessments

1. **Biuret test results for protein.** Enter your results from the Biuret test for protein. A color change to purple indicates that protein is present; pink indicates that short polypeptides are present. ◄3

Tube	Contents	Color	Protein Present (+) or Absent (–)
1W	Distilled water		
1E	Egg albumin		
1G	Glucose solution		
1D	Soft drink		
1S	Starch solution		
1P	Potato juice		

2. **Benedict's test results for most sugars (monosaccharides).** Enter your results from the Benedict's test for sugars. A color change to green, yellow, orange, or red indicates that sugar is present. Note the color after the mixture has been heated for 3 to 5 minutes. ◄3

Tube	Contents	Color	Sugar Present (+) or Absent (–)
2W	Distilled water		
2E	Egg albumin		
2G	Glucose solution		
2D	Soft drink		
2S	Starch solution		
2P	Potato juice		

3. **Iodine test results for starch.** Enter your results from the iodine test for starch. A color change to dark blue or black indicates that starch is present. ◄3

Tube	Contents	Color	Starch Present (+) or Absent (–)
3W	Distilled water		
3E	Egg albumin		
3G	Glucose solution		
3D	Soft drink		
3S	Starch solution		
3P	Potato juice		

4. **Lipid test results.** What did you observe when you allowed the drops of oil and water to dry on the brown paper? _____

   What did you observe when you mixed the oil and water together? _____

   _____

   What did you observe when you added the Sudan IV dye? ◄3 _____

Critical Thinking Application

If a person were on a low-carbohydrate, high-protein diet, which of the six substances tested would the person want to increase?_____

Explain your answer.

_____

_____

_____

_____

Part D—Testing For Unknown Organic Compounds Assessments

1. What is the number of your sample? _____

2. Record the results of your tests on the unknown here: ◄

Test Performed	Results
Biuret test for protein	
Benedict's test for sugars	
Iodine test for starch	
Water test for lipid	
Sudan IV test for lipid	

3. Based on these results, what organic compound(s) does your unknown sample contain? ◄ _____

_____

4. Explain your answer. ◄ _____

_____

_____

Laboratory Exercise  4

Care and Use of the Microscope

Materials Needed

Compound light microscope
Lens paper
Microscope slides
Coverslips
Transparent plastic millimeter ruler
Prepared slide of letter "e"
Slide of three colored threads
Medicine dropper
Dissecting needle (needle probe)
Specimen examples for wet mounts
Methylene blue (dilute) or iodine-potassium-iodide stain

For Demonstrations:
Micrometer scale
Stereomicroscope (dissecting microscope)

The human eye cannot perceive objects less than 0.1 mm in diameter, so a microscope is an essential tool for the study of small structures such as cells. The microscope usually used for this purpose is the *compound light microscope.* It is called compound because it uses two sets of lenses: an eyepiece or ocular lens system and an objective lens system. The eyepiece lens system magnifies, or compounds, the image reaching it after the image is magnified by the objective lens system. Such an instrument can magnify images of small objects up to about one thousand times.

Purpose of the Exercise

To become familiar with the major parts of a compound light microscope and their functions and to make use of the microscope to observe small objects.

Learning Outcomes

After completing this exercise, you should be able to
1. Locate and identify the major parts of a compound light microscope and differentiate the functions of these parts.
2. Calculate the total magnification produced by various combinations of eyepiece and objective lenses.

3. Demonstrate proper use of the microscope to observe and measure small objects.
4. Prepare a simple microscope slide and sketch the objects you observed.

Procedure A—Microscope Basics

1. Familiarize yourself with the following list of rules for care of the microscope:
   a. Keep the microscope under its *dustcover* and in a cabinet when it is not being used.
   b. Handle the microscope with great care. It is an expensive and delicate instrument. To move it or carry it, hold it by its *arm* with one hand and support its *base* with the other hand (fig. 4.1).
   c. Always store the microscope with the scanning or lowest power objective in place. Always start with this objective when using the microscope.
   d. To clean the lenses, rub them gently with *lens paper* or a high-quality cotton swab. If the lenses need additional cleaning, follow the directions in the "Lens-Cleaning Technique" section that follows.

Lens-Cleaning Technique

1. Moisten one end of a high-quality cotton swab with one drop of Kodak lens cleaner. Keep the other end dry.
2. Clean the optical surface with the wet end. Dry it with the other end, using a circular motion.
3. Use a hand aspirator to remove lingering dust particles.
4. Start with the scanning objective and work upward in magnification, using a new cotton swab for each objective.
5. When cleaning the eyepiece, do not open the lens unless it is absolutely necessary.
6. Use alcohol for difficult cleaning and only as a last resort use xylene. Regular use of xylene will destroy lens coatings.

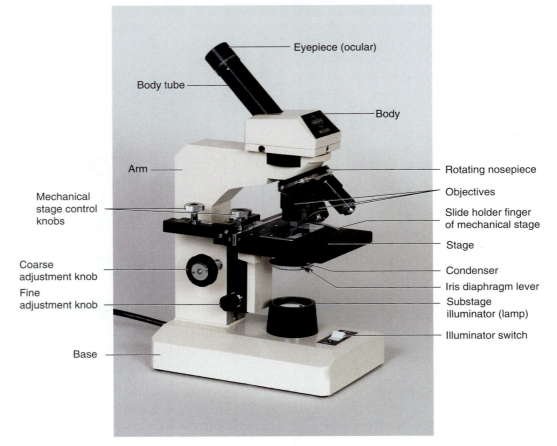

Figure 4.1 Major parts of a compound light microscope with a monocular body and a mechanical stage. Some compound microscopes are equipped with a binocular body.

Labels in figure:
- Eyepiece (ocular)
- Body tube
- Body
- Arm
- Rotating nosepiece
- Objectives
- Mechanical stage control knobs
- Slide holder finger of mechanical stage
- Stage
- Coarse adjustment knob
- Condenser
- Iris diaphragm lever
- Fine adjustment knob
- Substage illuminator (lamp)
- Illuminator switch
- Base

e. If the microscope has a substage lamp, be sure the electric cord does not hang off the laboratory table where someone might trip over it. The bulb life can be extended if the lamp is cool before the microscope is moved.

f. Never drag the microscope across the laboratory table after you have placed it.

g. Never remove parts of the microscope or try to disassemble the eyepiece or objective lenses.

h. If your microscope is not functioning properly, report the problem to your laboratory instructor immediately.

2. Observe a compound light microscope and study figure 4.1 to learn the names of its major parts. The lens system of a compound microscope includes three parts—the condenser, objective lens, and eyepiece or ocular.

Light enters this system from a *substage illuminator (lamp)* or *mirror* and is concentrated and focused by a *condenser* onto a microscope slide or specimen placed on the *stage.* The condenser, which contains a set of lenses, usually is kept in its highest position possible.

The *iris diaphragm,* located between the light source and the condenser, can be used to increase or decrease the intensity of the light entering the condenser. Locate the lever that operates the iris diaphragm beneath the stage and move it back and forth. Note how this movement causes the size of the opening in the diaphragm to change. (Some microscopes have a revolving plate called a disc diaphragm beneath the stage instead of an iris diaphragm. Disc diaphragms have different-sized holes to admit varying amounts of light.) Which way do you move the diaphragm to increase the light intensity? _____ Which way to decrease it? _____

After light passes through a specimen mounted on a microscope slide, it enters an *objective lens system.* This lens projects the light upward into the *body tube,* where it produces a magnified image of the object being viewed.

The *eyepiece (ocular) lens* system then magnifies this image to produce another image, seen by the eye. Typically, the eyepiece lens magnifies the image ten times (10×). Look for the number in the metal of the eyepiece that indicates its power (fig. 4.2). What is the eyepiece power of your microscope?

_____

The objective lenses are mounted in a *rotating nosepiece* so that different magnifications can be achieved by rotating any one of several objective lenses into position above the specimen. Commonly, this set of lenses includes a scanning objective (4×), a low-power objective (10×), and a high-power objective, also called a high-dry-power objective (about 40×). Sometimes an oil immersion objective (about 100×) is present. Look for the number printed on each objective that indicates its power. What are the objective lens powers of your microscope? _____

To calculate the *total magnification* achieved when using a particular objective, multiply the power of the eyepiece by the power of the objective used. Thus, the 10× eyepiece and the 40× objective produce a total magnification of 10 × 40, or 400×.

(a)

Microscope Lenses

Objective Lens Name	Common Objective Lens Magnification	Common Eyepiece Lens Magnification	Total Magnification
Scan	4×	10×	40×
Low power (LP)	10×	10×	100×
High power (HP)	40×	10×	400×
Oil immersion	100×	10×	1,000×

Note: If you wish to observe an object under LP, HP, or oil immersion, locate and then center and focus the object first under scan magnification.

3. Complete Part A of Laboratory Report 4.
4. Turn on the substage illuminator and look through the eyepiece. You will see a lighted circular area called the *field of view.*

You can measure the diameter of this field of view by focusing the lenses on the millimeter scale of a transparent plastic ruler. To do this, follow these steps:

   a. Place the ruler on the microscope stage in the spring clamp of a slide holder finger on a mechanical stage or under the stage (slide) clips. (*Note:* If

(b)

Figure 4.2 The powers of this 10× eyepiece (*a*) and this 40× objective (*b*) are marked in the metal. DIN is an international optical standard on quality optics. The 0.65 on the 40× objective is the numerical aperture, a measure of the light-gathering capabilities.

your microscope is equipped with a mechanical stage, it may be necessary to use a short section cut from a transparent plastic ruler. The section should be several millimeters long and can be mounted on a microscope slide for viewing.)

b. Center the millimeter scale in the beam of light coming up through the condenser, and rotate the scanning objective into position.

c. While you watch from the side to prevent the lens from touching anything, raise the stage until the objective is as close to the ruler as possible, using the coarse adjustment knob and then using the fine adjustment knob (fig. 4.3). (*Note:* The adjustment knobs on some microscopes move the body and objectives downward and upward for focusing.)

d. Look into the eyepiece and use the coarse adjustment knob to raise the objective lens until the lines of the millimeter scale come into sharp focus.

e. Adjust the light intensity by moving the *iris diaphragm lever* so that the field of view is brightly illuminated but comfortable to your eye. At the same time, take care not to overilluminate the field because transparent objects tend to disappear in bright light.

f. Position the millimeter ruler so that its scale crosses the greatest diameter of the field of view. Also, move the ruler so that one of the millimeter marks is against the edge of the field of view.

g. In millimeters, measure the distance across the field of view.

5. Complete Part B of the laboratory report.

6. Most microscopes are designed to be *parfocal.* This means that when a specimen is in focus with a lower-power objective, it will be in focus (or nearly so) when a higher-power objective is rotated into position. Always center the specimen in the field of view before changing to higher objectives.

   Rotate the low-power objective into position, and then look at the millimeter scale of the transparent plastic ruler. If you need to move the low-power objective to sharpen the focus, use the *fine adjustment knob.*

   Adjust the iris diaphragm so that the field of view is properly illuminated. Once again, adjust the millimeter ruler so that the scale crosses the field of view through its greatest diameter, and position the ruler so that a millimeter mark is against one edge of the field. Try to measure the distance across the field of view in millimeters.

7. Rotate the high-power objective into position while you watch from the side, and then observe the millimeter scale on the plastic ruler. All focusing using high-power magnification should be done only with the fine adjustment knob. If you use the coarse adjustment knob with the high-power objective,

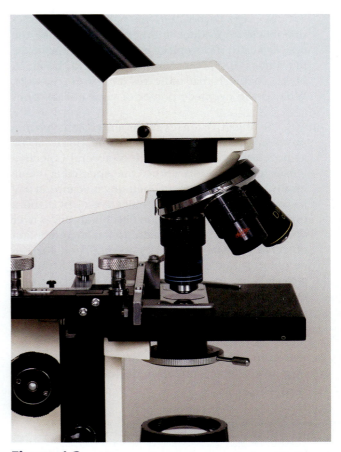

Figure 4.3 When you focus using a particular objective, you can prevent it from touching the specimen by watching from the side.

you can accidently force the objective into the coverslip and break the slide. This is because the *working distance* (the distance from the objective lens to the slide on the stage) is much shorter when using higher magnifications.

Adjust the iris diaphragm for proper illumination. Usually more illumination when using higher magnifications will help you view the objects more clearly. Try to measure the distance across the field of view in millimeters.

8. Locate the numeral 4 (or 9) on the plastic ruler and focus on it using the scanning objective. Note how the number appears in the field of view. Move the plastic ruler to the right and note which way the image moves. Slide the ruler away from you and again note how the image moves.

9. Observe a letter "e" slide in addition to the numerals on the plastic ruler. Note the orientation of the letter "e" using the scan, LP, and HP objectives. As you increase magnifications, note that the amount of the letter "e" is decreased. Move the slide to the

left, and then away from you, and note the direction the observed image moves.

10. Examine the slide of the three colored threads using the low-power objective and then the high-power objective. Focus on the location where the three threads cross. By using the fine adjustment knob, determine the order from top to bottom by noting which color is in focus at different depths. The other colored threads will still be visible, but they will be blurred. Be sure to notice whether the stage or the body tube moves up and down with the adjustment knobs of the microscope being used for this depth determination. The vertical depth of a specimen clearly in focus is called the *depth of field (focus)*. Whenever specimens are examined, continue to use the fine adjustment focusing knob to determine relative depths of structures clearly in focus within cells, giving a three-dimensional perspective. The depth of field is less at higher magnifications.

Critical Thinking Application

What was the sequence of the three colored threads from top to bottom? Explain how you came to that conclusion.

_____

_____

_____

11. Complete Parts C and D of the laboratory report.

Demonstration

A compound light microscope is sometimes equipped with a micrometer scale mounted in the eyepiece. Such a scale is subdivided into fifty to one hundred equal divisions (fig. 4.4). These arbitrary divisions can be calibrated against the known divisions of a micrometer slide placed on the microscope stage. Once the values of the divisions are known, the length and width of a microscopic object can be measured by superimposing the scale over the magnified image of the object.

Observe the micrometer scale in the eyepiece of the demonstration microscope. Focus the low-power objective on the millimeter scale of a micrometer slide (or a plastic ruler) and measure the distance between the divisions on the micrometer scale in the eyepiece. What is the distance between the finest divisions of the scale in micrometers?

_____

Procedure B—Slide Preparation

1. Prepare several temporary *wet mounts,* using any small, transparent objects of interest, and examine the specimens using the low-power objective and then a high-power objective to observe their details. To prepare a wet mount, follow these steps (fig. 4.5):
   a. Obtain a precleaned microscope slide.
   b. Place a tiny, thin piece of the specimen you want to observe in the center of the slide, and use a medicine dropper to put a drop of water over it. Consult with your instructor if a drop of stain might enhance the image of any cellular structures of your specimen. If the specimen is solid, you might want to tease some of it apart with dissecting needles. In any case, the specimen must be thin enough so that light can pass through it. Why is it necessary for the specimen to be so thin?

   _____

   c. Cover the specimen with a coverslip. Try to avoid trapping bubbles of air beneath the coverslip by slowly lowering it at an angle into the drop of water.
   d. Remove any excess water from the edge of the coverslip with absorbent paper. If your microscope has an inclination joint, do not tilt the microscope while observing wet mounts because the fluid will flow.
   e. Place the slide under the stage (slide) clips or in the slide holder on a mechanical stage and position the slide so that the specimen is centered in the light beam passing up through the condenser.
   f. Focus on the specimen using the scanning objective first. Next focus using the low-power objective, and then examine it with the high-power objective.

2. If an oil immersion objective is available, use it to examine the specimen. To use the oil immersion objective, follow these steps:
   a. Center the object you want to study under the high-power field of view.
   b. Rotate the high-power objective away from the microscope slide, place a small drop of immersion oil on the coverslip, and swing the oil immersion objective into position. To achieve sharp focus, use the fine adjustment knob only.
   c. You will need to open the iris diaphragm more fully for proper illumination. More light is needed because the oil immersion objective covers a very small lighted area of the microscope slide.
   d. The oil immersion objective must be very close to the coverslip to achieve sharp focus, so care must be taken to avoid breaking the coverslip or damaging the objective lens. For this reason, never lower the objective when you are looking into the eyepiece. Instead, always raise the objective

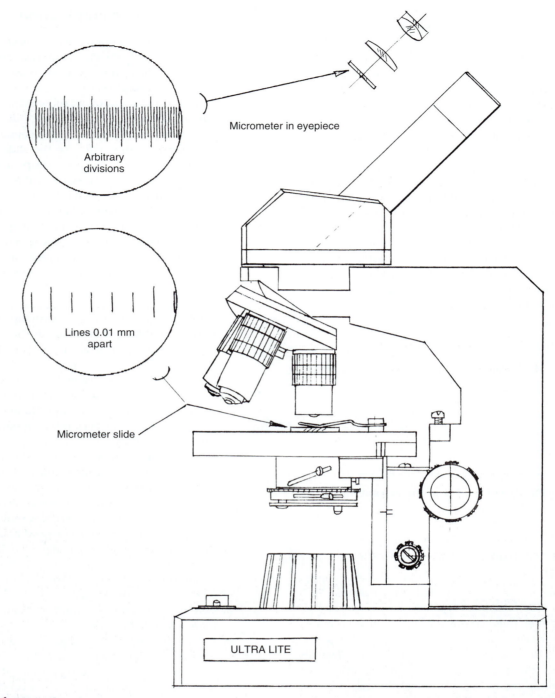

Arbitrary
divisions

Micrometer in eyepiece

Lines 0.01 mm
apart

Micrometer slide

ULTRA LITE

Figure 4.4 The divisions of a micrometer scale in an eyepiece can be calibrated against the known divisions of a micrometer slide. (Courtesy of Swift Instruments, Inc., San Jose, California)

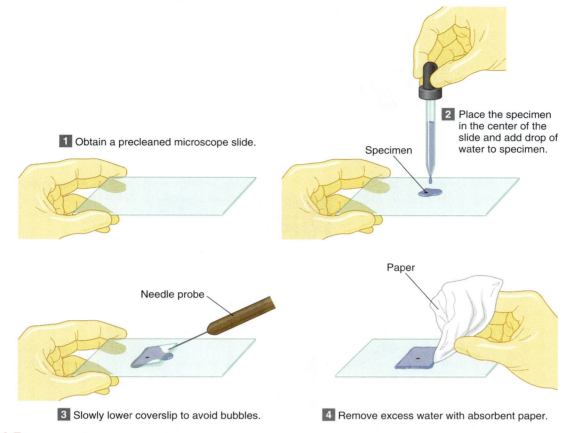

1 Obtain a precleaned microscope slide.

2 Place the specimen in the center of the slide and add drop of water to specimen.

Specimen

Needle probe

Paper

3 Slowly lower coverslip to avoid bubbles.

4 Remove excess water with absorbent paper.

Figure 4.5 Steps in the preparation of a wet mount.

to achieve focus, or prevent the objective from touching the coverslip by watching the microscope slide and coverslip from the side if the objective needs to be lowered. Usually when using the oil immersion objective, only the fine adjustment knob needs to be used for focusing.

3. When you have finished working with the microscope, remove the microscope slide from the stage and wipe any oil from the objective lens with lens paper or a high-quality cotton swab. Swing the scanning objective or the low-power objective into position. Wrap the electric cord around the base of the microscope and replace the dustcover.

4. Complete Part E of the laboratory report.

Demonstration

A stereomicroscope (dissecting microscope) (fig. 4.6) is useful for observing the details of relatively large, opaque specimens. Although this type of microscope achieves less magnification than a compound light microscope, it has the advantage of producing a three-dimensional image rather than the flat, two-dimensional image of the compound microscope. In addition, the image produced by the stereomicroscope is positioned in the same manner as the specimen, rather than being reversed and inverted as it is by the compound light microscope.

Observe the stereomicroscope. The eyepieces can be pushed apart or together to fit the distance between your eyes. Focus the microscope on the end of your finger. Which way does the image move when you move your finger to the right? _____ When you move it away?_____

If the instrument has more than one objective, change the magnification to higher power. Use the instrument to examine various small, opaque objects available in the laboratory.

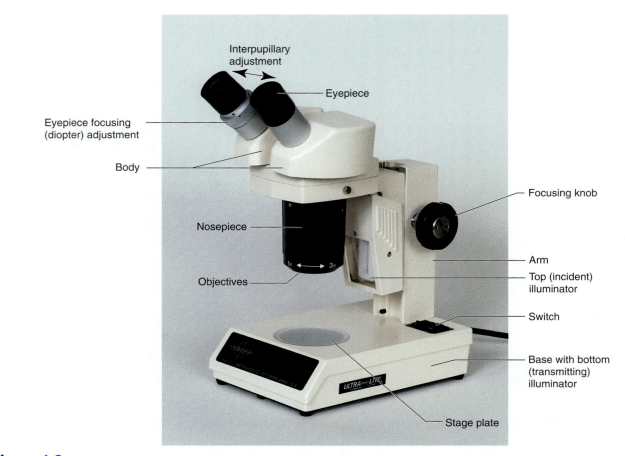

Figure 4.6   A stereomicroscope, also called a dissecting microscope.

Name _____

Date _____

Section _____

The ⬅ corresponds to the indicated outcome(s) found at the beginning of the laboratory exercise.

Care and Use of the Microscope

4

Part A Assessments

Revisit Procedure A, number 2, then complete the following:

1. What total magnification will be achieved if the 10× eyepiece and the 10× objective are used? ⬅**2** _____

2. What total magnification will be achieved if the 10× eyepiece and the 100× objective are used? ⬅**2** _____

Part B Assessments

Revisit Procedure A, number 2, then complete the following:

1. Sketch the millimeter scale as it appears under the scanning objective magnification. (The circle represents the field of view through the microscope.)

2. In millimeters, what is the diameter of the scanning field of view? ⬅**3** _____

3. Microscopic objects often are measured in *micrometers*. A micrometer equals 1/1,000 of a millimeter and is symbolized by μm. In micrometers, what is the diameter of the scanning power field of view? ⬅**3** _____

4. If a circular object or specimen extends halfway across the scanning field, what is its diameter in millimeters? ⬅**3** _____

5. In micrometers, what is its diameter? ⬅**3** _____

Part C Assessments

Complete the following:

1. Sketch the millimeter scale as it appears using the low-power objective.

2. What do you estimate the diameter of this field of view to be in millimeters? ⬅**3** _____

3. How does the diameter of the scanning power field of view compare with that of the low-power field? _____

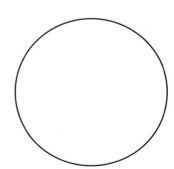

4. Why is it more difficult to measure the diameter of the high-power field of view than the low-power field? ◀**3**

_____

_____

5. What change occurred in the light intensity of the field of view when you exchanged the low-power objective

for the high-power objective? _____

6. Sketch the numeral 4 (or 9) as it appears through the scanning objective of the

compound microscope.

7. What has the lens system done to the image of the numeral? (Is it right side up,

upside down, or what?) _____

8. When you moved the ruler to the right, which way did the image move?_____

9. When you moved the ruler away from you, which way did the image move?_____

Part D Assessments

Match the names of the microscope parts in column A with the descriptions
in column B. Place the letter of your choice in the space provided. ◀**1**

Column A	Column B
a. Adjustment knob	_____ **1.** Increases or decreases the light intensity
b. Arm	
c. Condenser	_____ **2.** Platform that supports a microscope slide
d. Eyepiece (ocular)	_____ **3.** Concentrates light onto the specimen
e. Field of view	
f. Iris diaphragm	_____ **4.** Causes objective lens (or stage) to move upward or downward
g. Nosepiece	_____ **5.** After light passes through specimen, it next enters this lens system
h. Objective lens system	
i. Stage	_____ **6.** Holds a microscope slide in position
j. Stage (slide) clip	_____ **7.** Contains a lens at the top of the body tube
	_____ **8.** Serves as a handle for carrying the microscope
	_____ **9.** Part to which the objective lenses are attached
	_____ **10.** Circular area seen through the eyepiece

Part E Assessments

Prepare sketches of the objects you observed using the microscope. For each sketch, include the name of the object,
the magnification you used to observe it, and its estimated dimensions in millimeters and micrometers. ◀**3** ◀**4**

Laboratory Exercise 5

Cell Structure and Function

Cells are the "building blocks" from which all parts of the human body are formed. They account for the shape, organization, and construction of the body and are responsible for carrying on its life processes. Under the light microscope, with a properly applied stain to make structures visible, the **cell (plasma) membrane,** the **cytoplasm,** and a **nucleus** are easily seen. The cytoplasm is composed of a clear fluid, the *cytosol,* and numerous *cytoplasmic organelles* suspended in the cytosol.

The cell membrane, composed of lipids and proteins, composes the cell boundary and functions in various methods of membrane transport. The nucleus contains fine strands of DNA and protein called chromatin. Various cytoplasmic organelles, including mitochondria, endoplasmic reticulum, and Golgi apparatus, provide specialized metabolic functions.

Purpose of the Exercise

To review the structure and functions of major cellular components and to observe examples of human cells.

Learning Outcomes

After completing this exercise, you should be able to
1. Name and locate the components of a cell on a model or diagram.
2. Differentiate functions of cellular components.
3. Prepare a wet mount of cells lining the inside of the cheek; stain the cells; and identify the cell membrane, nucleus, and cytoplasm of these cells.
4. Locate cells on prepared slides of human tissues and identify their major components.

Procedure—Cell Structure and Function

1. Review the section entitled "Composite Cell" in chapter 3 of the textbook.
2. Observe the animal cell model and identify its major structures.
3. As a review activity, label figure 5.1 and study figure 5.2.
4. Complete Parts A and B of Laboratory Report 5.
5. Prepare a wet mount of cells lining the inside of the cheek. To do this, follow these steps:
   a. Gently scrape (force is not necessary and should be avoided) the inner lining of your cheek with the broad end of a flat toothpick.
   b. Stir the toothpick in a drop of water on a clean microscope slide and dispose of the toothpick as directed by your instructor.

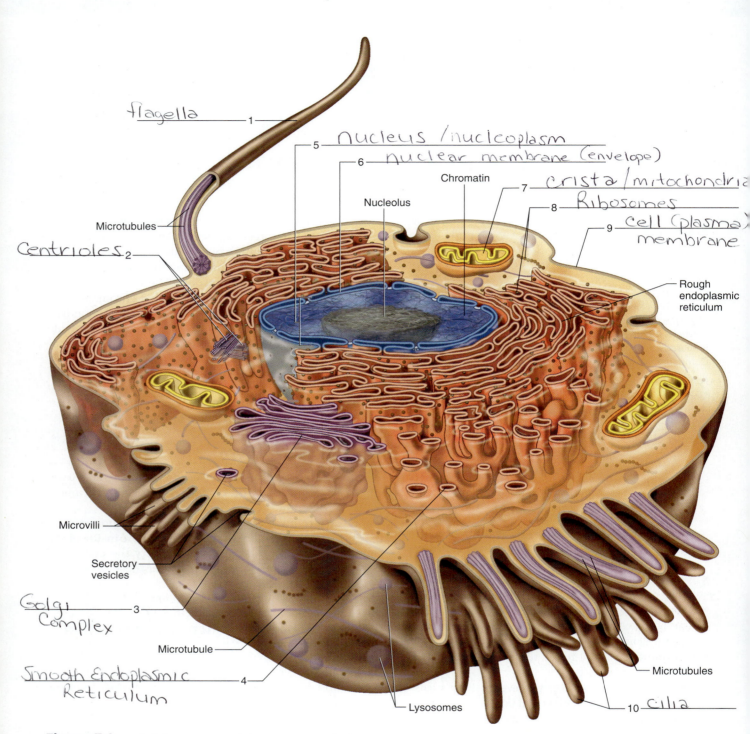

flagella —————1————

Nucleus / nucleoplasm

—————5

—————6 nuclear membrane (envelope)

Chromatin

crista / mitochondria

—————7

Nucleolus

—————8 Ribosomes

Microtubules

—————9 Cell (plasma) membrane

Centrioles ₂

Rough endoplasmic reticulum

Microvilli ———

Secretory vesicles ———

Golgi Complex ———3———

Microtubule ———

Smooth Endoplasmic Reticulum ———4———

Lysosomes

———10 cilia

Microtubules

Figure 5.1 Label the structures of this composite cell. The structures are not drawn to scale.

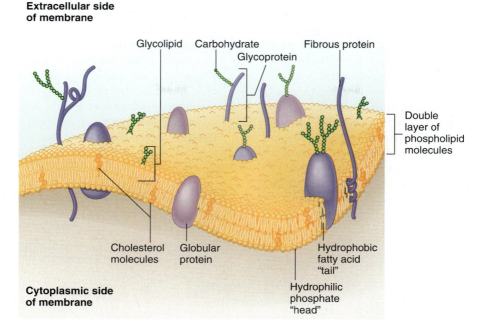

Extracellular side of membrane

Glycolipid · Carbohydrate · Glycoprotein · Fibrous protein

Double layer of phospholipid molecules

Cholesterol molecules · Globular protein · Hydrophobic fatty acid "tail"

Cytoplasmic side of membrane

Hydrophilic phosphate "head"

Figure 5.2   Structures of the cell membrane.

c. Cover the drop with a coverslip.

d. Observe the cheek cells by using the microscope. Compare your image with figure 5.3. To report what you observe, sketch a single cell in the space provided in Part C of the laboratory report.

6. Prepare a second wet mount of cheek cells, but this time, add a drop of dilute methylene blue or iodine-potassium-iodide stain to the cells. Cover the liquid with a coverslip and observe the cells with the microscope. Add to your sketch any additional structures you observe in the stained cells.

7. Complete Part C of the laboratory report.

8. Using the microscope, observe each of the prepared slides of human tissues. To report what you observe, sketch a single cell of each type in the space provided in Part D of the laboratory report.

9. Complete Part D of the laboratory report.

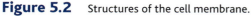 ## Critical Thinking Application

The cells lining the inside of the cheek are frequently removed for making observations of basic cell structure. The cells are from stratified squamous epithelium. Explain why these cells are used instead of outer body surface tissue.

_____

_____

_____

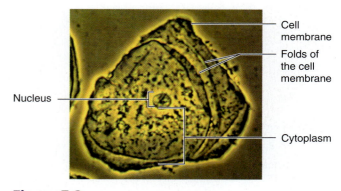

Cell membrane

Folds of the cell membrane

Nucleus

Cytoplasm

Figure 5.3   Stained cell lining the inside of the cheek as viewed through the compound light microscope using the high-power objective (400×).

Name _____

Date _____

Section _____

The ← corresponds to the indicated outcome(s) found at the beginning of the laboratory exercise.

Cell Structure and Function

Part A Assessments

Match the cellular components in column A with the descriptions in column B. Place the letter of your choice in the space provided. ◀1 ◀2

Column A		Column B
a. Chromatin	_a_	1. Loosely coiled fiber containing DNA and protein within nucleus
b. Cytoplasm	_g_	2. Location of the energy production from digested food molecules
c. Endoplasmic reticulum	_k_	3. Small RNA-containing particle for synthesis of proteins
d. Golgi apparatus	_l_	4. Membranous sac formed by pinching off of pieces of cell membrane
e. Lysosome		
f. Microtubule	_i_	5. Dense body of RNA within nucleus
g. Mitochondrion	_f_	6. Slender tubes that provide movement in cilia and flagella
h. Nuclear envelope	_c_	7. Organelle composed of membrane-bound canals
i. Nucleolus	_b_	8. Occupies space between cell membrane and nucleus
j. Nucleus	_d_	9. Flattened membranous sacs that package a secretion
k. Ribosome	_e_	10. Membranous sac that contains digestive enzymes
l. Vesicle (vacuole)	_h_	11. Separates nuclear contents from cytoplasm
	_j_	12. Spherical organelle that contains chromatin and nucleolus

Part B Assessments

Complete the following:

1. The cell membrane is composed mainly of _proteins + lipids · 98% lipids (75% of that is phospholipids, 20% cholesterol, 5% glycolipids) 2% protein_ . ◀1

2. The basic framework of a cell membrane can be described as _____

_____ . ◀1

3. A cell membrane is relatively impermeable to substances that are _____

_____ . ◀2

4. Molecules of _protein_ are responsible for special functions of cell membranes.

5. Chemically, membrane pores consist of _____ . ◀1

Part C Assessments

Complete the following:

1. Sketch a single cell from inside the cheek that has been stained. Label the cellular components you recognize. (The circle represents the field of view through the microscope.) ◀**3**

Magnification ____ ×

2. After comparing the wet mount and the stained cheek cells, describe the advantage gained by staining.

   _____

3. Are the stained cheek cells nearly the same size and shape? _____

   Explain your answer. _____

Part D Assessments

Complete the following:

1. Sketch a single cell of each type you observed in the prepared slides of human tissues. Name the tissue, indicate the magnification used, and label the cellular components you recognize. ◀**4**

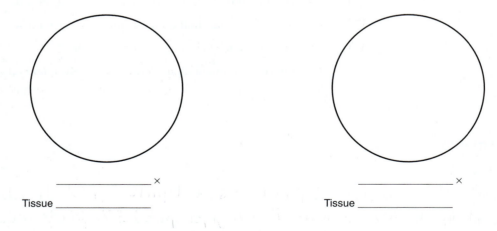

_____ ×   _____ ×

Tissue _____   Tissue _____

2. What do the various types of cells in these tissues have in common? _____

   _____

   _____

3. What are the main differences you observed among these cells? _____

   _____

   _____

Laboratory Exercise

Movements Through Cell Membranes

Materials Needed

For Procedure A—Diffusion:
Textbook
Petri dish
White paper
Forceps
Potassium permanganate crystals
Millimeter ruler (thin and transparent)

For Procedure B—Osmosis:
Thistle tube
Molasses (or Karo dark corn syrup)
Selectively permeable (semipermeable) membrane
    (presoaked dialysis tubing of 1⁵⁄₁₆" or greater diameter)
Ring stand and clamp
Beaker
Rubber band
Millimeter ruler

*For Procedure C—Hypertonic, Hypotonic,
    and Isotonic Solutions:*
Test tubes
Marking pen
Test-tube rack
10 mL graduated cylinder
Medicine dropper
Uncoagulated animal blood
Distilled water
0.9% NaCl (aqueous solution)
3.0% NaCl (aqueous solution)
Clean microscope slides
Coverslips
Compound light microscope

For Procedure D—Filtration:
Glass funnel
Filter paper
Ring stand and ring
Beaker
Powdered charcoal or ground black pepper
1% glucose (aqueous solution)
1% starch (aqueous solution)
Test tubes
10 mL graduated cylinder
Water bath (boiling water)
Benedict's solution
Iodine-potassium-iodide solution
Medicine dropper

For Alternative Osmosis Activity:
Fresh chicken egg
Beaker
Laboratory balance
Spoon
Vinegar
Corn syrup (Karo)

⚠ Safety

- Clean laboratory surfaces before and after laboratory procedures.
- Wear disposable gloves when handling chemicals and animal blood.
- Wear safety glasses when using chemicals.
- Dispose of laboratory gloves and blood-contaminated items as instructed.
- Wash your hands before leaving the laboratory.

A cell membrane functions as a gateway through which chemical substances and small particles may enter or leave a cell. These substances move through the membrane by physical processes such as diffusion, osmosis, and filtration, or by physiological processes such as active transport, phagocytosis, or pinocytosis.

Purpose of the Exercise

To demonstrate some of the physical processes by which substances move through cell membranes.

Learning Outcomes

After completing this exercise, you should be able to
1. Demonstrate *diffusion* and identify examples of diffusion.
2. Prepare and explain a graph to interpret diffusion.
3. Demonstrate *osmosis* and identify examples of osmosis.
4. Distinguish among hypertonic, hypotonic, and isotonic solutions and observe the effects of these solutions on animal cells.
5. Demonstrate *filtration* and identify examples of filtration.

Procedure A—Diffusion

1. Review the section entitled "Diffusion" in chapter 3 of the textbook.
2. To demonstrate diffusion, follow these steps:
   a. Place a petri dish, half filled with water, on a piece of white paper that has a millimeter ruler positioned on the paper. Wait until the water surface is still. Allow approximately 3 minutes. *Note:* The petri dish should remain level. A second millimeter ruler may be needed under the petri dish as a shim to obtain a level amount of the water inside the petri dish.
   b. Using forceps, place one crystal of potassium permanganate near the center of the petri dish and near the millimeter ruler (fig. 6.1).
   c. Measure the radius of the purple circle at 1-minute intervals for 10 minutes.
3. Complete Part A of Laboratory Report 6.

Learning Extension

Repeat the demonstration of diffusion using a petri dish filled with ice-cold water and a second dish filled with very hot water. At the same moment, add a crystal of potassium permanganate to each dish and observe the circle as before. What difference do you note in the rate of diffusion in the two dishes? How do you explain this difference? _____

_____

_____

_____

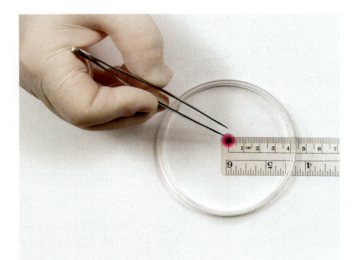

Figure 6.1 To demonstrate diffusion, place one crystal of potassium permanganate in the center of a petri dish containing water. Place the crystal near the millimeter ruler (positioned under the petri dish).

Procedure B—Osmosis

1. Review the section entitled "Osmosis" in chapter 3 of the textbook.
2. To demonstrate osmosis, refer to figure 6.2 as you follow these steps:
   a. One person plugs the tube end of a thistle tube with a finger.
   b. Another person then fills the bulb with molasses until it is about to overflow at the top of the bulb. Air remains trapped in the stem.
   c. Cover the bulb opening with a single thickness piece of moist selectively permeable (semipermeable) membrane. Dialysis tubing that has been soaked for 30 minutes can easily be cut open because it becomes pliable.
   d. Tightly secure the membrane in place with several wrappings of a rubber band.
   e. Immerse the bulb end of the tube in a beaker of water. If leaks are noted, repeat the procedures.
   f. Support the upright portion of the tube with a clamp on a ring stand. Folded paper under the clamp will protect the thistle tube stem from breakage.
   g. Mark the meniscus level of the molasses in the tube. *Note:* The best results will occur if the mark of the molasses is a short distance up the stem of the thistle tube when the experiment starts.
   h. Measure the level changes after 10 minutes and 30 minutes.
3. Complete Part B of the laboratory report.

Alternative Procedure

Eggshell membranes possess selectively permeable properties. To demonstrate osmosis using a natural membrane, soak a fresh chicken egg in vinegar for about 24 hours to remove the shell. Use a spoon to carefully handle the delicate egg. Place the egg in a hypertonic solution (corn syrup) for about 24 hours. Remove the egg, rinse it, and using a laboratory balance, weigh the egg to establish a baseline weight. Place the egg in a hypotonic solution (distilled water). Remove the egg and weigh it every 15 minutes for an elapsed time of 75 minutes. Explain any weight changes that were noted during this experiment.

_____

_____

_____

_____

_____

_____

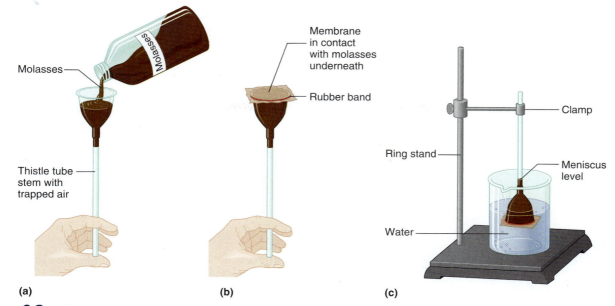

Figure 6.2 (*a*) Fill the bulb of the thistle tube with molasses; (*b*) tightly secure a piece of selectively permeable (semipermeable) membrane over the bulb opening; and (*c*) immerse the bulb in a beaker of water. *Note:* These procedures require the participation of two people.

Procedure C—Hypertonic, Hypotonic, and Isotonic Solutions

1. Review the section entitled "Osmosis" in chapter 3 of the textbook.
2. To demonstrate the effect of hypertonic, hypotonic, and isotonic solutions on animal cells, follow these steps:
   a. Place three test tubes in a rack and mark them as *tube 1, tube 2,* and *tube 3.* (*Note:* One set of tubes can be used to supply test samples for the entire class.)
   b. Using 10 mL graduated cylinders, add 3 mL of distilled water to tube 1; add 3 mL of 0.9% NaCl to tube 2; and add 3 mL of 3.0% NaCl to tube 3.
   c. Place three drops of fresh, uncoagulated animal blood into each of the tubes, and gently mix the blood with the solutions. Wait 5 minutes.
   d. Using three separate medicine droppers, remove a drop from each tube and place the drops on three separate microscope slides marked *1, 2,* and *3.*
   e. Cover the drops with coverslips and observe the blood cells, using the high power of the microscope.
3. Complete Part C of the laboratory report.

Alternative Procedure

Various substitutes for blood can be used for Procedure C. Onion, cucumber, or cells lining the inside of the cheek represent three possible options.

Procedure D—Filtration

1. Review the section entitled "Filtration" in chapter 3 of the textbook.
2. To demonstrate filtration, follow these steps:
   a. Place a glass funnel in the ring of a ring stand over an empty beaker. Fold a piece of filter paper in half and then in half again. Open one thickness of the filter paper to form a cone. Wet the cone, and place it in the funnel. The filter paper is used to demonstrate how movement across membranes is limited by size of the molecules, but it does not represent a working model of biological membranes.
   b. Prepare a mixture of 5 cc (approximately 1 teaspoon) powdered charcoal (or ground black pepper) and equal amounts of 1% glucose solution and 1% starch solution in a beaker. Pour some of the mixture into the funnel until it nearly reaches the top of the filter-paper cone. Care should be taken to prevent the mixture from spilling over the top of the filter paper. Collect the filtrate in the beaker below the funnel (fig. 6.3).
   c. Test some of the filtrate in the beaker for the presence of glucose. To do this, place 1 mL of filtrate in a clean test tube and add 1 mL of Benedict's solution. Place the test tube in a water bath of boiling water for 2 minutes and then allow the liquid to cool slowly. If the color of the solution changes to green, yellow, or red, glucose is present (fig. 6.4).

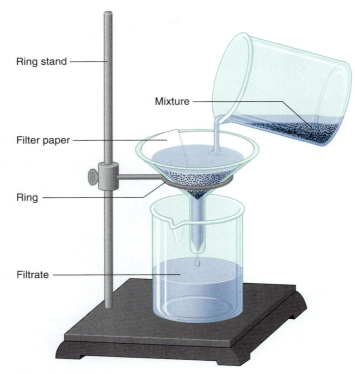

Figure 6.3 Apparatus used to illustrate filtration.

Ring stand

Mixture

Filter paper

Ring

Filtrate

Water bath

Filtrate and Benedict s solution

Hot plate

Figure 6.4 Heat the filtrate and Benedict's solution in a boiling water bath for 2 minutes.

    d. Test some of the filtrate in the beaker for the presence of starch. To do this, place a few drops of filtrate in a test tube and add one drop of iodine-potassium-iodide solution. If the color of the solution changes to blue-black, starch is present.

    e. Observe any charcoal in the filtrate.

3. Complete Part D of the laboratory report.

Name _____

Date _____

Section _____

The ◀ corresponds to the indicated outcome(s) found at the beginning of the laboratory exercise.

Movements Through Cell Membranes

Part A Assessments

Complete the following:

1. Enter data for changes in the movement of the potassium permanganate. ◀ **1**

Elapsed Time	Radius of Purple Circle in Millimeters
Initial	_____
1 minute	_____
2 minutes	_____
3 minutes	_____
4 minutes	_____
5 minutes	_____
6 minutes	_____
7 minutes	_____
8 minutes	_____
9 minutes	_____
10 minutes	_____

2. Prepare a graph that illustrates the diffusion distance of potassium permanganate in 10 minutes. ◀ **2**

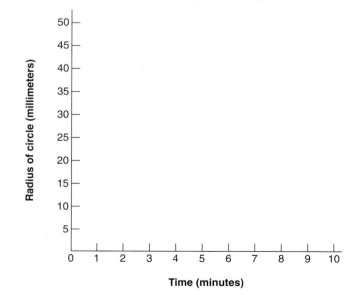

3. Explain your graph. ◀**2** _____

_____

4. Briefly define *diffusion*. _____

_____

 ## Critical Thinking Application

By answering yes or no, indicate which of the following provides an example of diffusion. ◀**1**

1. A perfume bottle is opened, and soon the odor can be sensed in all parts of the room._____

2. A sugar cube is dropped into a cup of hot water, and, without being stirred, all of the liquid becomes sweet tasting._____

3. Water molecules move from a faucet through a garden hose when the faucet is turned on._____

4. A person blows air molecules into a balloon by forcefully exhaling._____

5. A crystal of blue copper sulfate is placed in a test tube of water. The next day, the solid is gone, but the water is evenly colored._____

Part B Assessments

Complete the following:

1. What was the change in the level of molasses in 10 minutes? ◀**3** _____

2. What was the change in the level of molasses in 30 minutes? ◀**3** _____

3. How do you explain this change? ◀**3** _____

_____

4. Briefly define *osmosis*. _____

_____

Critical Thinking Application

By answering yes or no, indicate which of the following involves osmosis. ◀**3**

1. A fresh potato is peeled, weighed, and soaked in a strong salt solution. The next day, it is discovered that the potato has lost weight._____

2. Garden grass wilts after being exposed to dry chemical fertilizer._____

3. Air molecules escape from a punctured tire as a result of high pressure inside._____

4. Plant seeds soaked in water swell and become several times as large as before soaking._____

5. When the bulb of a thistle tube filled with water is sealed by a selectively permeable membrane and submerged in a beaker of molasses, the water level in the tube falls._____

Part C Assessments

Complete the following:

1. In the spaces, sketch a few blood cells from each of the test tubes, and indicate the magnification. ◄4

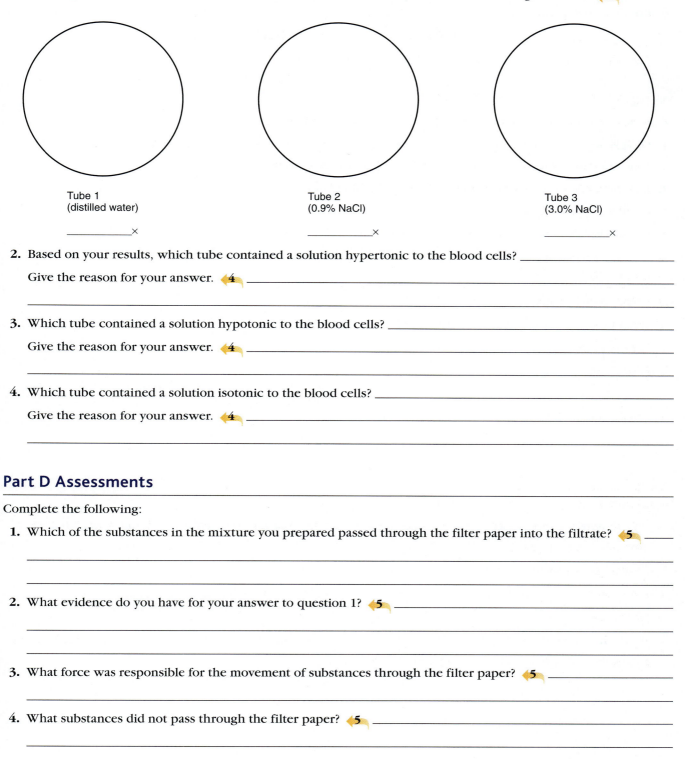

Tube 1
(distilled water)

_____×

Tube 2
(0.9% NaCl)

_____×

Tube 3
(3.0% NaCl)

_____×

2. Based on your results, which tube contained a solution hypertonic to the blood cells? _____

   Give the reason for your answer. ◄4 _____

   _____

3. Which tube contained a solution hypotonic to the blood cells? _____

   Give the reason for your answer. ◄4 _____

   _____

4. Which tube contained a solution isotonic to the blood cells? _____

   Give the reason for your answer. ◄4 _____

   _____

Part D Assessments

Complete the following:

1. Which of the substances in the mixture you prepared passed through the filter paper into the filtrate? ◄5 _____

   _____

   _____

2. What evidence do you have for your answer to question 1? ◄5 _____

   _____

   _____

3. What force was responsible for the movement of substances through the filter paper? ◄5 _____

   _____

4. What substances did not pass through the filter paper? ◄5 _____

   _____

5. What factor prevented these substances from passing through? ◄**5** _____

_____

_____

6. Briefly define *filtration*. _____

_____

_____

Critical Thinking Application

By answering yes or no, indicate which of the following involves filtration. ◄**5**

1. Oxygen molecules move into a cell and carbon dioxide molecules leave a cell because of differences in the concentrations of these substances on either side of the cell membrane._____

2. Blood pressure forces water molecules from the blood outward through the thin wall of a blood capillary._____

3. Urine is forced from the urinary bladder through the tubular urethra by muscular contractions._____

4. Air molecules enter the lungs through the airways when air pressure is greater outside these organs than inside._____

5. Coffee is made using a coffeemaker (not instant)._____

7

Cell Cycle

The cell cycle consists of the series of changes a cell undergoes from the time it is formed until it divides. Typically, a newly formed cell grows to a certain size and then divides to form two new cells (*daughter cells*). This cell division process involves two major steps: (1) division of the cell's nuclear parts, *mitosis,* and (2) division of the cell's cytoplasm, *cytokinesis.* Before the cell divides, it must synthesize biochemicals and other contents. This period of preparation is called *interphase.* The extensive period of interphase is divided into three phases. The S phase, when DNA synthesis occurs, is between two gap phases (G_1 and G_2), when cell growth occurs and cytoplasmic organelles duplicate. Eventually, some specialized cells, such as skeletal muscle cells and most nerve cells, cease further cell division, but remain alive. A special type of cell division, called *meiosis,* occurs in the reproductive system to produce gametes and is not included in this laboratory exercise.

Purpose of the Exercise

To review the stages in the cell cycle and to observe cells in various stages of their life cycles.

Learning Outcomes

After completing this exercise, you should be able to

1. Describe the cell cycle, and locate structures involved with the process.
2. Identify and sketch the stages in the life cycle of a particular cell.
3. Arrange into a correct sequence a set of models or drawings of cells in various stages of their life cycles.

Procedure—Cell Cycle

1. Review the section entitled "The Cell Cycle" in chapter 3 of the textbook.
2. As a review activity, study the various stages of the cell's life cycle represented in figure 7.1 and label the structures indicated in figure 7.2.
3. Observe the animal mitosis models, and review the major events in a cell's life cycle represented by each of them. Be sure you can arrange these models in correct sequence if their positions are changed. The acronym IPMAT can help you arrange the correct order of phases in the cell cycle. This includes interphase followed by the four phases of mitosis. Cytokinesis overlaps anaphase and telophase.
4. Complete Part A of Laboratory Report 7.
5. Obtain a slide of the whitefish mitosis (blastula).
   a. Examine the slide using the high-power objective of a microscope. The tissue on this slide was obtained from a developing embryo (blastula) of a fish, and many of the embryonic cells are undergoing mitosis. The chromosomes of these dividing cells are darkly stained (fig. 7.3).
   b. Search the tissue for cells in various stages of cell division. There are several sections on the slide. If you cannot locate different stages in one section, examine the cells of another section because the stages occur randomly on the slide.
   c. Each time you locate a cell in a different stage, sketch it in an appropriate circle in Part B of the laboratory report.

Critical Thinking Application

Which stage (phase) of the cell cycle was the most numerous in the blastula?_____
Explain your answer._____
_____
_____

6. Complete Parts C, D, and E of the laboratory report.

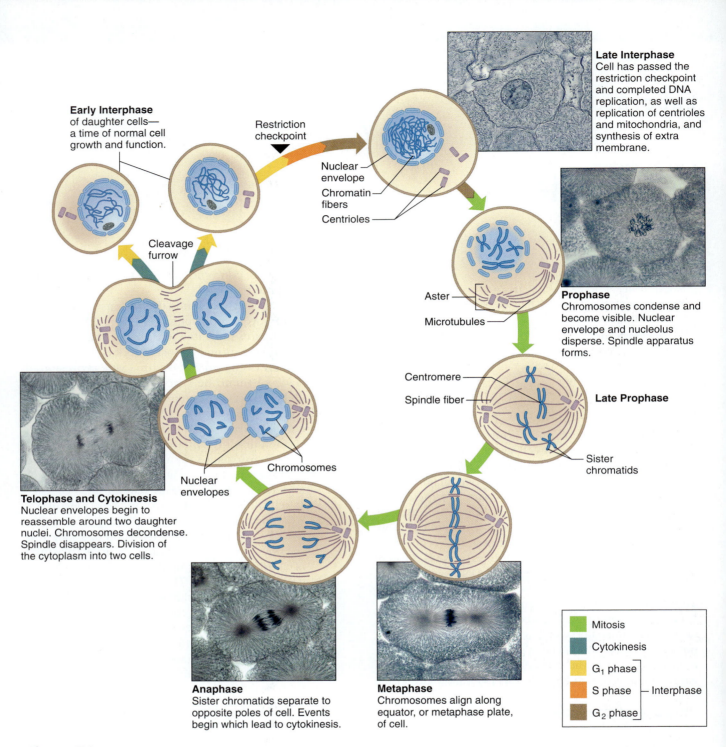

Early Interphase of daughter cells—a time of normal cell growth and function.

Restriction checkpoint

Late Interphase
Cell has passed the restriction checkpoint and completed DNA replication, as well as replication of centrioles and mitochondria, and synthesis of extra membrane.

Nuclear envelope
Chromatin fibers
Centrioles

Cleavage furrow

Aster
Microtubules

Prophase
Chromosomes condense and become visible. Nuclear envelope and nucleolus disperse. Spindle apparatus forms.

Centromere
Spindle fiber

Late Prophase

Sister chromatids

Telophase and Cytokinesis
Nuclear envelopes begin to reassemble around two daughter nuclei. Chromosomes decondense. Spindle disappears. Division of the cytoplasm into two cells.

Nuclear envelopes

Chromosomes

Anaphase
Sister chromatids separate to opposite poles of cell. Events begin which lead to cytokinesis.

Metaphase
Chromosomes align along equator, or metaphase plate, of cell.

🟩	Mitosis
🟦	Cytokinesis
🟨	G₁ phase
🟧	S phase — Interphase
🟫	G₂ phase

Figure 7.1 The cell cycle: interphase, mitosis, and cytokinesis.

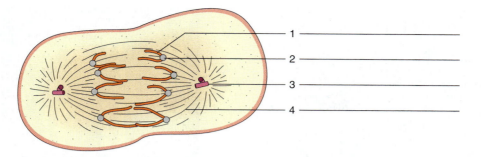

Figure 7.2 Label the structures indicated in the dividing cell. ◀**1**

1 _____

2 _____

3 _____

4 _____

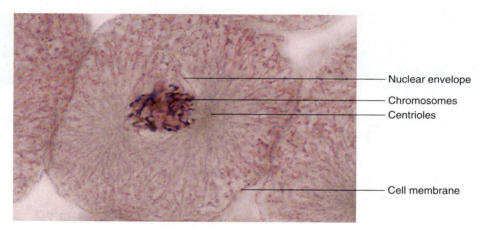

Nuclear envelope

Chromosomes
Centrioles

Cell membrane

Figure 7.3 Cell in prophase (250× micrograph enlarged to 1,000×).

Demonstration

Using the oil immersion objective of a microscope, see if you can locate some human chromosomes by examining a prepared slide of human chromosomes from leukocytes. The cells on this slide were cultured in a special medium and were stimulated to undergo mitosis. The mitotic process was arrested in metaphase by exposing the cells to a chemical called colchicine, and the cells were caused to swell osmotically. As a result of this treatment, the chromosomes were spread apart. A complement of human chromosomes should be visible when they are magnified about 1,000×. Each chromosome is double-stranded and consists of two chromatids joined by a common centromere (fig. 7.4).

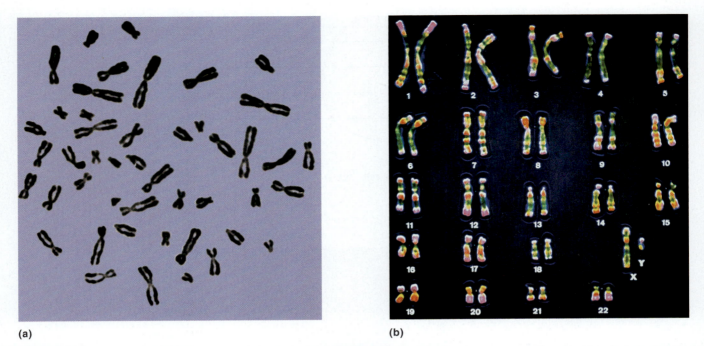

(a)   (b)

Figure 7.4   (*a*) A complement of human chromosomes (2,700×). (*b*) A *karyotype* can be constructed by arranging the homologous chromosome pairs together in a chart. The completed karyotype indicates a normal male.

The ⬅ corresponds to the indicated outcome(s) found at the beginning of the laboratory exercise.

Cell Cycle

Part A Assessments

Complete the table. ⬅1

Stage	Major Events Occurring
Interphase (G₁, S, and G₂)	
Mitosis Prophase	
Metaphase	
Anaphase	
Telophase	
Cytokinesis	

Part B Assessments

Sketch an interphase cell and cells in different stages of mitosis to illustrate the whitefish cell's life cycle. Label the major cellular structures represented in the sketches and indicate cytokinesis locations. (The circles represent fields of view through the microscope.) 2

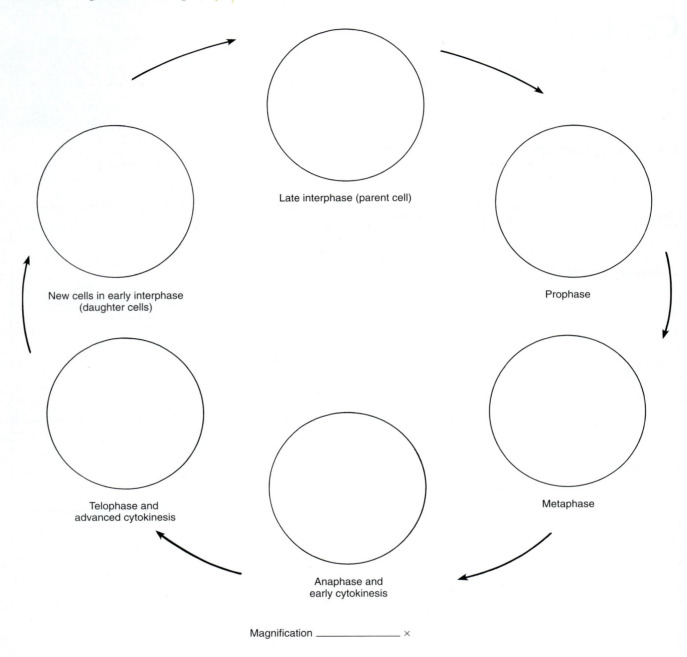

Late interphase (parent cell)

New cells in early interphase
(daughter cells)

Prophase

Telophase and
advanced cytokinesis

Metaphase

Anaphase and
early cytokinesis

Magnification _____ ×

Part C Assessments

Complete the following:

1. In what ways are the new cells (daughter cells), which result from a cell cycle, similar?_____

_____

_____

2. How do the new cells slightly differ? _____

_____

_____

3. Distinguish between mitosis and cytokinesis. ◀1 _____

_____

_____

Part D Assessments

Identify the mitotic stage represented by each of the micrographs in figure 7.5(*a-d*). ◀3

a. _____

b. _____

c. _____

d. _____

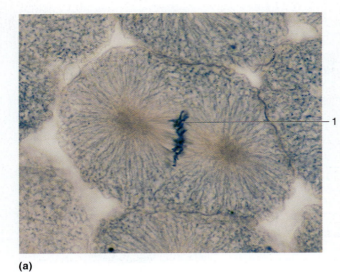

(a)

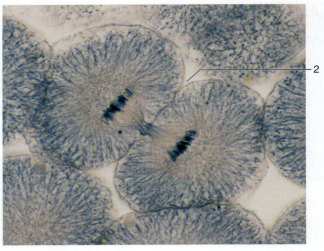

(b)

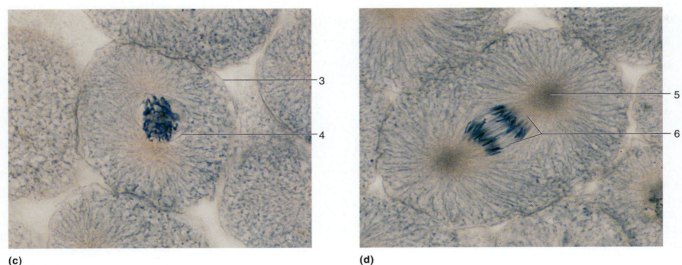

(c)

(d)

Figure 7.5   Identify the mitotic stage of the cell in each of these micrographs of the whitefish blastula (250×).

Part E Assessments

Identify the structures indicated by numbers in figure 7.5.   **1**

   1._____

   2. _____

   3. _____

   4. _____

   5. _____

   6. _____

Laboratory Exercise

8

Epithelial Tissues

A tissue is composed of a layer or group of cells similar in size, shape, and function. Within the human body, there are four major types of tissues: (1) epithelial, which cover the body's external and internal surfaces and most glands; (2) connective, which bind and support parts; (3) muscle, which make movement possible; and (4) nervous, which conduct impulses from one part of the body to another and help to control and coordinate body activities.

Epithelial tissues are tightly packed single (simple) to multiple (stratified) layers of cells that provide protective barriers. The underside of this tissue layer contains a basement membrane layer of adhesives and collagen to which the epithelial cells anchor to an underlying connective tissue. The cells readily divide and lack blood vessels. Epithelial cells always have a free (apical) surface exposed to the outside or to an open space internally and a basal surface that attaches to the basement membrane. Epithelial cell functions include protection, filtration, secretion, and absorption. Many shapes of the cells exist that are used to name and identify the variations. Many of the prepared slides contain more than the tissue to be studied, so be certain that your view matches the correct tissue. Also be aware that stained colors of all tissues might vary.

Purpose of the Exercise

To review the characteristics of epithelial tissues and to observe examples.

Learning Outcomes

After completing this exercise, you should be able to

1. Differentiate the special characteristics of each type of epithelial tissue.
2. Sketch and label the characteristics of epithelial tissues that you were able to observe.
3. Indicate a location and function of each type of epithelial tissue.
4. Identify the major types of epithelial tissues on microscope slides.

Procedure—Epithelial Tissues

1. Review the section entitled "Epithelial Tissues" in chapter 5 of the textbook.
2. Complete Part A of Laboratory Report 8.
3. Use the microscope to observe the prepared slides of types of epithelial tissues. As you observe each tissue, look for its special distinguishing features as described in the textbook, such as cell size, shape, and arrangement. Compare your prepared slides of epithelial tissues to the micrographs in figure 8.1. As you observe each type of epithelial tissue, prepare a labeled sketch of a representative portion of the tissue in Part B of the laboratory report.
4. Complete Part B of the laboratory report.
5. Test your ability to recognize each type of epithelial tissue. To do this, have a laboratory partner select one of the prepared slides, cover its label, and focus the microscope on the tissue. Then see if you can correctly identify the tissue. 4

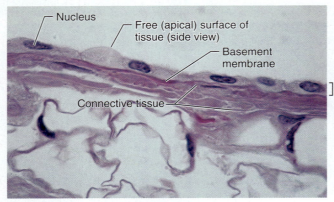

(a) Simple squamous epithelium (side view) (1,100×)

Labels: Nucleus — Free (apical) surface of tissue (side view) — Basement membrane — Connective tissue

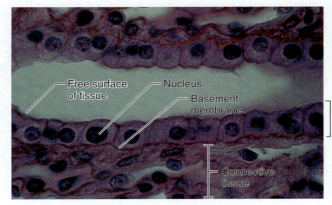

(b) Simple cuboidal epithelium (800×)

Labels: Free surface of tissue — Nucleus — Basement membrane — Connective tissue

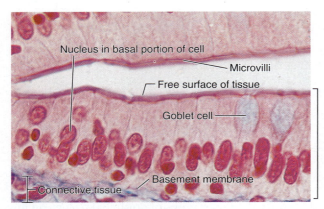

(c) Simple columnar epithelium (650×)

Labels: Nucleus in basal portion of cell — Microvilli — Free surface of tissue — Goblet cell — Basement membrane — Connective tissue

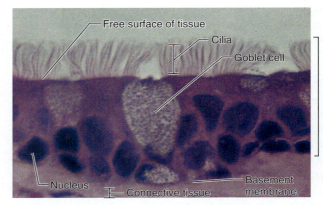

(d) Pseudostratified columnar epithelium with cilia (1,500×)

Labels: Free surface of tissue — Cilia — Goblet cell — Nucleus — Connective tissue — Basement membrane

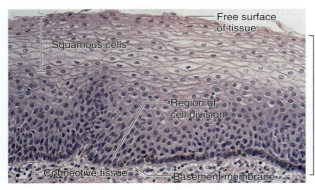

(e) Stratified squamous epithelium (nonkeratinized) (370×)

Labels: Free surface of tissue — Squamous cells — Region of cell division — Connective tissue — Basement membrane

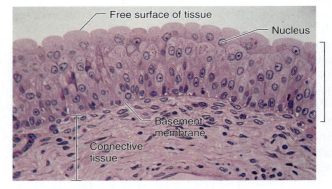

(f) Transitional epithelium (unstretched) (190×)

Labels: Free surface of tissue — Nucleus — Basement membrane — Connective tissue

Figure 8.1 Micrographs of epithelial tissues. *Note:* The brackets to the right of each micrograph indicate the tissue.

Name _____

Date _____

Section _____

The ⬅ corresponds to the indicated outcome(s) found at the beginning of the laboratory exercise.

Epithelial Tissues

Part A Assessments

Match the tissues in column A with the characteristics in column B. Place the letter of your choice in the space provided. (Some answers may be used more than once.) ⬅1

Column A	Column B
a. Simple columnar epithelium	_____ **1.** Consists of several layers of cube-shaped, elongated, and irregular cells
b. Simple cuboidal epithelium	
c. Simple squamous epithelium	_____ **2.** Commonly possesses cilia that move dust and mucus out of the airways
d. Pseudostratified columnar epithelium	
e. Stratified squamous epithelium	_____ **3.** Single layer of flattened cells
f. Transitional epithelium	_____ **4.** Nuclei located at different levels within cells
	_____ **5.** Forms walls of capillaries and air sacs of lungs
	_____ **6.** Forms linings of trachea and bronchi
	_____ **7.** Younger cells cuboidal, older cells flattened
	_____ **8.** Forms inner lining of urinary bladder
	_____ **9.** Lines kidney tubules and ducts of salivary glands
	_____ **10.** Forms lining of stomach and intestines
	_____ **11.** Nuclei located near basement membrane
	_____ **12.** Forms lining of oral cavity, anal canal, and vagina

Part B Assessments

In the space that follows, sketch a few cells of each type of epithelium you observed. For each sketch, label the major characteristics, indicate the magnification used, write an example of a location in the body, and provide a function. ◀2 ◀3

Simple squamous epithelium (_____ ×)
Location example: _____
Function: _____

Simple cuboidal epithelium (_____ ×)
Location example: _____
Function: _____

Simple columnar epithelium (_____ ×)
Location example: _____
Function: _____

Pseudostratified columnar epithelium with cilia (_____ ×)
Location example: _____
Function: _____

Stratified squamous epithelium (_____ ×)
Location example: _____
Function: _____

Transitional epithelium (_____ ×)
Location example: _____
Function: _____

Learning Extension

Use colored pencils to differentiate various cellular structures in Part B. Select a different color for a nucleus, cytoplasm, cell membrane, basement membrane, goblet cell, and cilia whenever visible.

Critical Thinking Application

As a result of your observations of epithelial tissues, which one(s) provide(s) the best protection? Explain your answer._____

_____

Laboratory Exercise 9

Connective Tissues

Materials Needed

Textbook
Compound light microscope
Prepared slides of the following:
    Areolar tissue
    Adipose tissue
    Dense connective tissue (regular type)
    Hyaline cartilage
    Elastic cartilage
    Fibrocartilage
    Bone (compact, ground, cross section)
    Blood (human smear)

For Learning Extension:
Colored pencils

Connective tissues contain a variety of cell types and occur in all regions of the body. They bind structures together, provide support and protection, fill spaces, store fat, and produce blood cells.

Connective tissue cells are often widely scattered in an abundance of extracellular matrix. The matrix consists of fibers and a ground substance of various densities and consistencies. Many of the prepared slides contain more than the tissue to be studied, so be certain that your view matches the correct tissue. Additional study of bone and blood will be found in Laboratory Exercises 12 and 34.

Purpose of the Exercise

To review the characteristics of connective tissues and to observe examples of the major types.

Learning Outcomes

After completing this exercise, you should be able to

1. Differentiate the special characteristics of each of the major types of connective tissue.
2. Sketch and label the characteristics of connective tissues that you were able to observe.
3. Indicate a location and function of each type of connective tissue.
4. Identify the major types of connective tissues on microscope slides.

Procedure—Connective Tissues

1. Review the section entitled "Connective Tissues" in chapter 5 of the textbook.
2. Complete Part A of Laboratory Report 9.
3. Use a microscope to observe the prepared slides of various connective tissues. As you observe each tissue, look for its special distinguishing features as described in the textbook. Compare your prepared slides of connective tissues to the micrographs in figure 9.1. As you observe each type of connective tissue, prepare a labeled sketch of a representative portion of the tissue in Part B of the laboratory report.
4. Complete Part B of the laboratory report.
5. Test your ability to recognize each of these connective tissues by having a laboratory partner select a slide, cover its label, and focus the microscope on this tissue. Then see if you correctly identify the tissue. 4

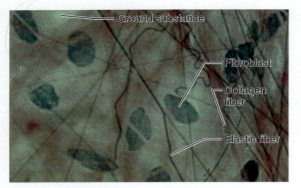

(a) Areolar tissue (400×)

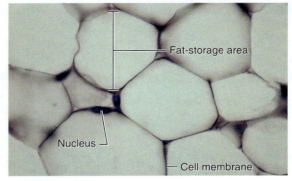

(b) Adipose tissue (265×)

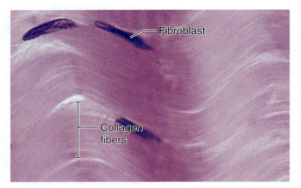

(c) Dense connective tissue (regular) (1,000×)

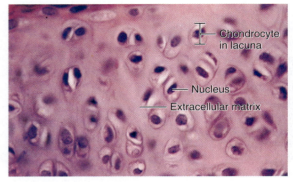

(d) Hyaline cartilage (500×)

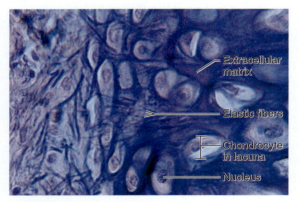

(e) Elastic cartilage (1,200×)

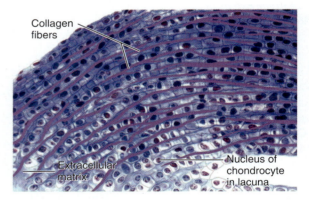

(f) Fibrocartilage (1,800×)

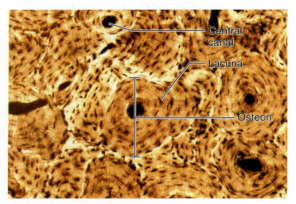

(g) Bone (compact) (100×)

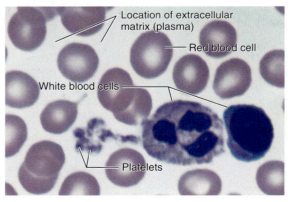

(h) Blood (1,200×)

Figure 9.1   Micrographs of connective tissues.

Name _____

Date _____

Section _____

The ⬅ corresponds to the indicated outcome(s) found at the beginning of the laboratory exercise.

Connective Tissues

Part A Assessments

Match the tissues in column A with the characteristics in column B. Place the letter of your choice in the space provided. (Some answers may be used more than once.) ⬅**1**

Column A	Column B
a. Adipose tissue	_____ **1.** Forms framework of outer ear
b. Areolar tissue	_____ **2.** Functions as heat insulator beneath skin
c. Blood	_____ **3.** Contains large amounts of fluid and lacks fibers
d. Bone (compact)	_____ **4.** Cells arranged around central canal
e. Dense connective tissue (regular)	_____ **5.** Binds skin to underlying organs
f. Elastic cartilage	_____ **6.** Main tissue of tendons and ligaments
g. Fibrocartilage	_____ **7.** Provides stored energy supply in fat vacuoles
h. Hyaline cartilage	_____ **8.** Forms the flexible soft part of the nasal septum
	_____ **9.** Pads between vertebrae that are shock absorbers
	_____ **10.** Forms supporting rings of respiratory passages
	_____ **11.** Cells greatly enlarged with nuclei pushed to sides
	_____ **12.** Forms delicate, thin layers between muscles

Part B Assessments

In the space provided, sketch a small section of each of the types of connective tissues you observed. For each sketch, label the major characteristics, indicate the magnification used, write an example of a location in the body, and provide a function. ◄2 ◄3

Areolar tissue (_____ ×)
Location example: _____
Function: _____

Adipose tissue (_____ ×)
Location example: _____
Function: _____

Dense connective tissue (regular type) (_____ ×)
Location example: _____
Function: _____

Hyaline cartilage (_____ ×)
Location example: _____
Function: _____

Elastic cartilage (_____ ×)
Location example: _____
Function: _____

Fibrocartilage (_____ ×)
Location example: _____
Function: _____

Bone (compact) (_____ ×)
Location example: _____
Function: _____

Blood (_____ ×)
Location example: _____
Function: _____

Learning Extension

Use colored pencils to differentiate various cellular structures in Part B. Select a different color for the cells, fibers, and ground substance whenever visible.

10

Muscle and Nervous Tissues

Materials Needed

Textbook
Compound light microscope
Prepared slides of the following:
    Skeletal muscle tissue
    Smooth muscle tissue
    Cardiac muscle tissue
    Nervous tissue (spinal cord smear and/or cerebellum)

For Learning Extension:
Colored pencils

Muscle tissues are characterized by the presence of elongated cells or muscle fibers that can contract in response to specific stimuli. As they shorten, these fibers pull at their attached ends and cause body parts to move. The three types of muscle tissues are skeletal, smooth, and cardiac.

Nervous tissues occur in the brain, spinal cord, and peripheral nerves. They consist of neurons (nerve cells), the impulse-conducting cells of the nervous system, and neuroglial cells, which perform supportive and protective functions.

Purpose of the Exercise

To review the characteristics of muscle and nervous tissues and to observe examples of these tissues.

Learning Outcomes

After completing this exercise, you should be able to

1. Differentiate the special characteristics of each type of muscle tissue and nervous tissue.
2. Sketch and label the characteristics of muscle tissues and nervous tissues that you observed.
3. Indicate a location and function of each type of muscle tissue and nervous tissue.
4. Identify three types of muscle tissues and nervous tissue on microscope slides.

Procedure—Muscle and Nervous Tissues

1. Review the sections entitled "Muscle Tissues" and "Nervous Tissues" in chapter 5 of the textbook.
2. Complete Part A of Laboratory Report 10.
3. Using the microscope, observe each of the types of muscle tissues on the prepared slides. Look for the special features of each type, as described in the textbook. Compare your prepared slides of muscle tissues to the micrographs in figure 10.1 and the muscle tissue characteristics in table 10.1. As you observe each type of muscle tissue, prepare a labeled sketch of a representative portion of the tissue in Part B of the laboratory report.
4. Observe the prepared slide of nervous tissue and identify neurons (nerve cells), neuron cellular processes, and neuroglial cells. Compare your prepared slide of nervous tissue to the micrograph in figure 10.1. Prepare a labeled sketch of a representative portion of the tissue in Part B of the laboratory report.
5. Complete Part B of the laboratory report.
6. Test your ability to recognize each of these muscle and nervous tissues by having a laboratory partner select a slide, cover its label, and focus the microscope on this tissue. Then see if you correctly identify the tissue. 4

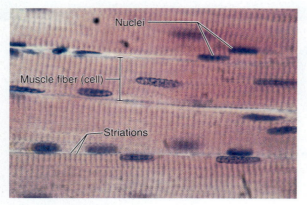

(a) Skeletal muscle (700×)

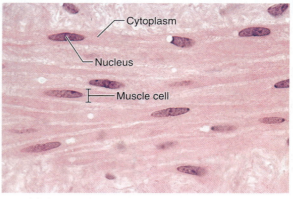

(b) Smooth muscle (900×)

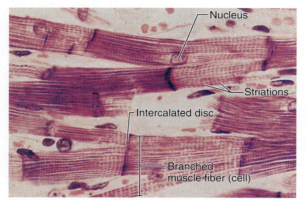

(c) Cardiac muscle (400×)

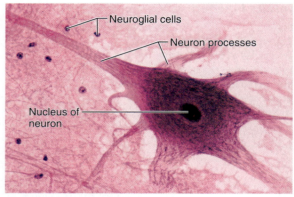

(d) Nervous tissue (300×)

Figure 10.1 Micrographs of muscle and nervous tissues.

Table 10.1 **Muscle Tissue Characteristics**

Characteristic	Skeletal Muscle	Smooth Muscle	Cardiac Muscle
Appearance of cells	Unbranched and relatively parallel	Spindle-shaped	Branched and connected in complex networks
Striations	Present and obvious	Absent	Present but faint
Nucleus	Multinucleated	Uninucleated	Uninucleated (usually)
Intercalated discs	Absent	Absent	Present
Control	Voluntary	Involuntary	Involuntary

Name _____

Date _____

Section _____

The ◄━ corresponds to the indicated outcome(s) found at the beginning of the laboratory exercise.

Muscle and Nervous Tissues

Part A Assessments

Match the tissues in column A with the characteristics in column B. Place the letter of your choice in the space provided. (Some answers may be used more than once.) ◄1

Column A	Column B
a. Cardiac muscle	_____ **1.** Coordinates, regulates, and integrates body functions
b. Nervous tissue	_____ **2.** Contains intercalated discs
c. Skeletal muscle	_____ **3.** Muscle that lacks striations
d. Smooth muscle	_____ **4.** Striated and involuntary
	_____ **5.** Striated and voluntary
	_____ **6.** Contains neurons and neuroglial cells
	_____ **7.** Muscle attached to bones
	_____ **8.** Muscle that composes heart
	_____ **9.** Moves food through the digestive tract
	_____ **10.** Transmits impulses along cellular processes

Part B Assessments

In the space that follows, sketch a few cells or fibers of each of the three types of muscle tissues and of nervous tissue as they appear through the microscope. For each sketch, label the major structures of the cells or fibers, indicate the magnification used, write an example of a location in the body, and provide a function. ◀2 ◀3

Skeletal muscle tissue (_____ ×) Location example: _____ Function: _____	Smooth muscle tissue (_____ ×) Location example: _____ Function: _____
Cardiac muscle tissue (_____ ×) Location example: _____ Function: _____	Nervous tissue (_____ ×) Location example: _____ Function: Spinal _____

Learning Extension

Use colored pencils to differentiate various cellular structures in Part B.

72

Integumentary System

The integumentary system includes the skin, hair, nails, sebaceous glands, and sweat glands. These structures provide a protective covering for deeper tissues, aid in regulating body temperature, retard water loss, house sensory receptors, synthesize various chemicals, and excrete small quantities of wastes.

The skin consists of two distinct layers. The outer layer, the epidermis, consists of stratified squamous epithelium. The inner layer, the dermis, consists of a thicker layer of dense connective tissue. Beneath the dermis is the subcutaneous layer (not considered a true layer of the skin), composed of adipose and areolar connective tissues.

Purpose of the Exercise

To observe the structures and tissues of the integumentary system and to review the functions of these parts.

Learning Outcomes

After completing this exercise, you should be able to

1. Locate the structures of the integumentary system.
2. Describe the major functions of these structures.
3. Distinguish the locations and tissues among epidermis, dermis, and the subcutaneous layer.
4. Sketch the layers of the skin and associated structures that you observed on the prepared slide.

Procedure—Inegumentary System

1. Review the sections entitled "Skin and Its Tissues" and "Accessory Structures of the Skin" in chapter 6 of your textbook.
2. As a review activity, label figures 11.1 and 11.2. Locate as many of these structures as possible on a skin model.
3. Complete Part A of Laboratory Report 11.
4. Use the hand magnifier or dissecting microscope to do the following:
   a. Observe the skin, hair, and nails of your hand.
   b. Compare the type and distribution of hairs on the front and back of your forearm.
5. Pull out a single hair with the forceps and mount it on a microscope slide under a coverslip. Use the hand magnifier or dissecting microscope to observe the root and shaft of the hair. Note the scalelike parts that make up the shaft.
6. Complete Part B of the laboratory report.
7. As vertical sections of human skin are observed, remember that the lenses of the microscope invert and reverse images. It is important to orient the position of the epidermis, dermis, and subcutaneous (hypodermis) layers using scan magnification before continuing with additional observations. Compare all of your skin observations to figure 11.3. Use low-power magnification of the compound light microscope and proceed as follows:
   a. Observe the prepared slide of human scalp or axilla.
   b. Locate the epidermis, dermis, and subcutaneous layer; a hair follicle; an arrector pili muscle; a sebaceous gland; and a sweat gland.
   c. Focus on the epidermis with high power and locate the stratum corneum and stratum basale. Note how the shapes of the cells in these two layers differ.
   d. Observe the dense connective tissue (irregular type) that makes up the bulk of the dermis.
   e. Observe the adipose tissue that composes most of the subcutaneous layer.

8. Complete Part C of the laboratory report.

9. Using low-power magnification, locate a hair follicle sectioned longitudinally through its bulblike base, a sebaceous gland close to the follicle, and a sweat gland (fig. 11.3). Observe the detailed structure of these parts with high-power magnification.

10. Complete Parts D and E of the laboratory report.

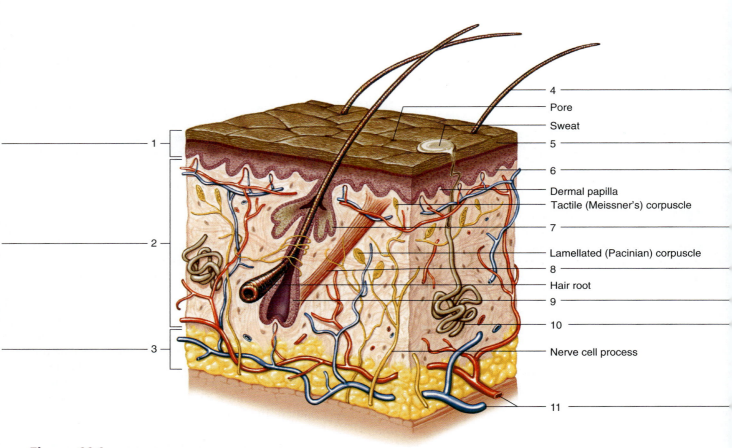

Labels on figure:
4
Pore
Sweat
5
6
Dermal papilla
Tactile (Meissner's) corpuscle
7
Lamellated (Pacinian) corpuscle
8
Hair root
9
10
Nerve cell process
11
1
2
3

Figure 11.1 Label this vertical section of the skin.

Critical Thinking Application

Explain the advantage for melanin granules being located in the deep layer of the epidermis.

_____

_____

_____

_____

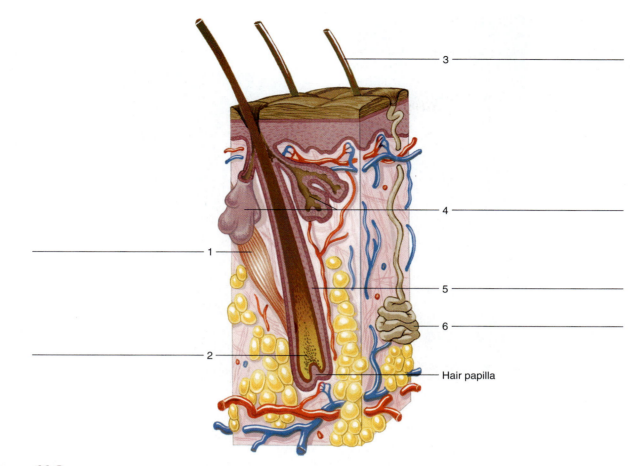

Figure 11.2 Label the features associated with this hair follicle. 1

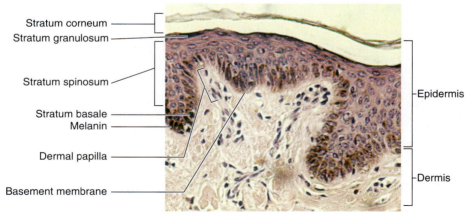

Stratum corneum
Stratum granulosum
Stratum spinosum
Stratum basale
Melanin
Dermal papilla
Basement membrane

Epidermis

Dermis

(a)

Figure 11.3 Features of human skin are indicated in these micrographs. Magnifications: (*a*) 290×; (*b*) 30× micrograph enlarged to 280×; (*c*) 45×; (*d*) 110×.

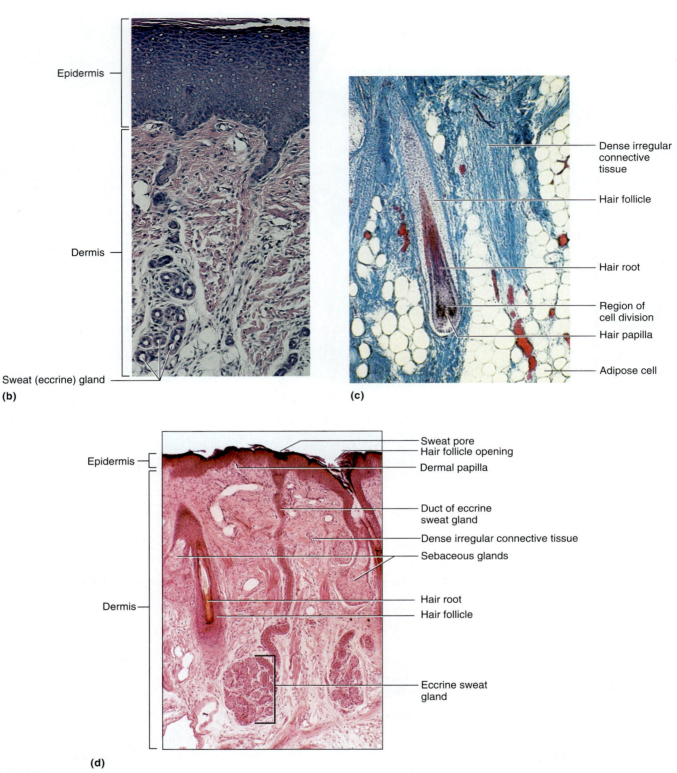

(b)

Epidermis

Dermis

Sweat (eccrine) gland

(c)

Dense irregular connective tissue

Hair follicle

Hair root

Region of cell division

Hair papilla

Adipose cell

(d)

Epidermis

Dermis

Sweat pore
Hair follicle opening
Dermal papilla

Duct of eccrine sweat gland

Dense irregular connective tissue

Sebaceous glands

Hair root
Hair follicle

Eccrine sweat gland

Figure 11.3 *Continued.*

Name _____

Date _____

Section _____

The ⬅ corresponds to the indicated outcome(s) found at the beginning of the laboratory exercise.

Integumentary System

Part A Assessments

Match the structures in column A with the description and functions in column B. Place the letter of your choice in the space provided. ◄1 ◄2

Column A	Column B

Column A

a. Apocrine sweat gland
b. Arrector pili muscle
c. Dermis
d. Eccrine sweat gland
e. Epidermis
f. Hair follicle
g. Keratin
h. Melanin
i. Sebaceous gland
j. Sebum
k. Stratum basale
l. Stratum corneum

Column B

_____ 1. An oily secretion that helps to waterproof body surface

_____ 2. Outermost layer of epidermis

_____ 3. Become active at puberty

_____ 4. Epidermal pigment

_____ 5. Inner layer of skin

_____ 6. Responds to elevated body temperature

_____ 7. General name of entire superficial layer of the skin

_____ 8. Gland that secretes an oily substance

_____ 9. Hard protein of nails and hair

_____ 10. Cell division and deepest layer of epidermis

_____ 11. Tubelike part that contains the root of the hair

_____ 12. Causes hair to stand erect and goose bumps to appear

Part B Assessments

Complete the following:

1. How does the skin on your palm differ from that on the back (posterior) of your hand?

_____

_____

2. Describe the differences you observed in the type and distribution of hair on the front (anterior) and back (posterior) of your forearm._____

_____

_____

3. Explain how the hair is formed. ◀**2** _____

_____

_____

4. What cells produce the pigment in hair? ◀**2** _____

Part C Assessments

Complete the following:

1. Distinguish the locations and tissues among the epidermis, dermis, and subcutaneous layers. ◀**3** _____

_____

_____

2. How do the cells of stratum corneum and stratum basale differ? ◀**3** _____

_____

_____

3. What special qualities does the connective tissue of the dermis have? ◀**3** _____

_____

_____

_____

Part D Assessments

Complete the following:

1. In which layer of skin are follicles usually found? ◀**1** _____

2. How are sebaceous glands associated with hair follicles? ◀**1** _____

_____

3. In which layer of skin are sweat glands usually located? ◀**1** _____

Part E Assessments

Using the scanning objective, sketch a vertical section of human skin. Label the skin layers and a hair follicle, a sebaceous gland, and a sweat gland. ◀**4**

Laboratory Exercise 12

Bone Structure

A bone represents an organ of the skeleton system. As such, it is composed of a variety of tissues including bone tissue, cartilage, dense connective tissue, blood, and nervous tissue. Bones are not only alive, but also multifunctional. They support and protect softer tissues, provide points of attachment for muscles, house blood-producing cells, and store inorganic salts.

Although bones of the skeleton vary greatly in size and shape, they have much in common structurally and functionally.

Purpose of the Exercise

To examine the structure of a long bone.

Learning Outcomes

After completing this exercise, you should be able to

1. Locate the major structures of a long bone.
2. Distinguish between compact and spongy bone.
3. Differentiate the special characteristics of compact bone tissue.
4. Describe the functions of various structures of a bone.

Procedure—Bone Structure

1. Review the section entitled "Bone Structure" in chapter 7 of the textbook.
2. As a review activity, label figures 12.1 and 12.2.
3. Examine the sectioned bones and locate the following:

epiphysis (proximal and distal)	**compact bone**
epiphyseal plate	**spongy bone**
articular cartilage	**medullary cavity**
diaphysis	**endosteum**
periosteum	**red marrow**
	yellow marrow

4. Use the dissecting microscope to observe the compact bone and spongy bone of the sectioned specimens. Also examine the marrow in the medullary cavity and the spaces within the spongy bone of the fresh specimen.
5. Reexamine the microscopic structure of bone tissue by observing a prepared microscope slide of ground compact bone. Use figure 12.3 of bone tissue to locate the following features:

   osteon (Haversian system)—cylinder-shaped unit

   central canal (Haversian canal)—contains blood vessels and nerves

   lacuna—small chamber for an osteocyte

   bone extracellular matrix—collagen and calcium phosphate

   lamella—concentric ring of matrix around central canal

   canaliculus—minute tube containing cellular process

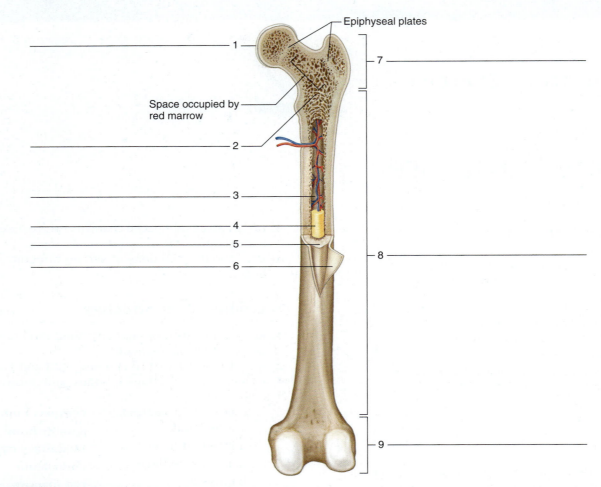

Epiphyseal plates

1

7

Space occupied by
red marrow

2

3

4

5

6

8

9

Figure 12.1   Label the major structures of this long bone (femur).   ◀1

Critical Thinking Application

Explain how bone cells embedded in a solid ground substance obtain nutrients and eliminate wastes.   ◀3

_____

_____

_____

_____

  6. Complete Parts A and B of Laboratory Report 12.

Demonstration

Examine a fresh chicken bone and a chicken bone that has
been soaked for several days in vinegar or exposed over-
night in dilute hydrochloric acid. Wear disposable gloves
for handling these bones. This acid treatment removes the
inorganic salts from the bone extracellular matrix. Rinse
the bones in water and note the texture and flexibility of
each (fig. 12.4a). Based on your observations, what quality
of the fresh bone seems to be due to the inorganic salts
removed by the acid treatment?   ◀4

_____

_____

  Examine the specimen of chicken bone that has been
exposed to high temperature (baked at 121°C [250°F] for
2 hours). This treatment removes the protein and other
organic substances from the bone extracellular matrix
(fig. 12.4b). What quality of the fresh bone seems to be
due to these organic materials?   ◀4

_____

_____

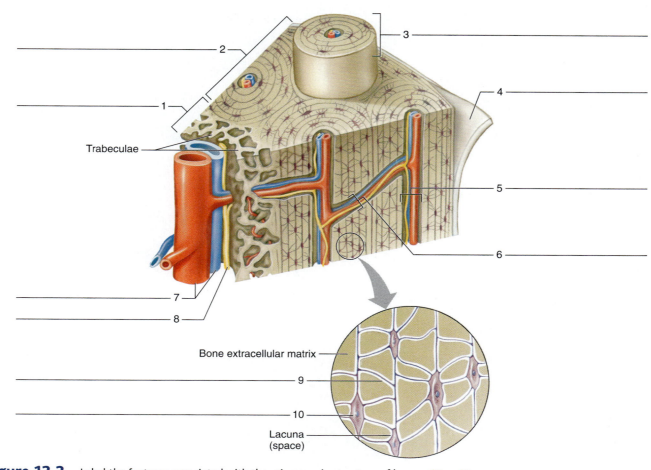

Trabeculae

Bone extracellular matrix

Lacuna
(space)

Figure 12.2   Label the features associated with the microscopic structure of bone.   **2**   **3**

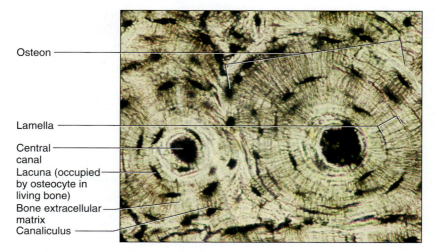

Osteon

Lamella

Central
canal

Lacuna (occupied
by osteocyte in
living bone)

Bone extracellular
matrix

Canaliculus

Figure 12.3   Micrograph of ground compact bone tissue (160×).

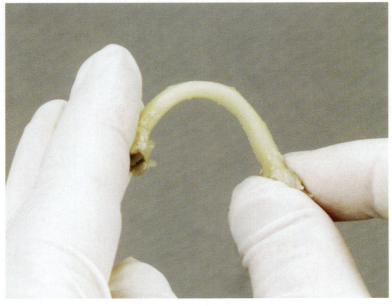

(a)

(b)

Figure 12.4 Results of fresh chicken bone demonstration: (*a*) soaked in vinegar; (*b*) baked in oven.

Name _____

Date _____

Section _____

The ◀ corresponds to the indicated outcome(s) found at the beginning of the laboratory exercise.

Bone Structure

Part A Assessments

Complete the following:

1. Where in the human skeleton are long bones found? ◀**1**

_____

2. Distinguish between the epiphysis and the diaphysis of a long bone. ◀**1**

_____

_____

3. Where is cartilage found on the surface of a long bone? ◀**1**

_____

_____

4. Where is dense connective tissue found on the surface of a long bone? ◀**1**

_____

5. In general, what is the function of bony processes? ◀**4**

_____

_____

6. Distinguish the locations and tissues between the periosteum and the endosteum. ◀**1**

_____

_____

_____

7. What structural differences did you note between the compact bone and the spongy bone? ◀**2**

_____

_____

_____

8. How are these structural differences related to the locations and functions of these two types of bone? ◀**2**

_____

_____

_____

9. From your observations, how does the marrow in the medullary cavity compare with the marrow in the spaces of the spongy bone? ◀**4**

_____

_____

_____

Part B Assessments

Identify the structures indicated in figure 12.5. ◀**1** ◀**2**

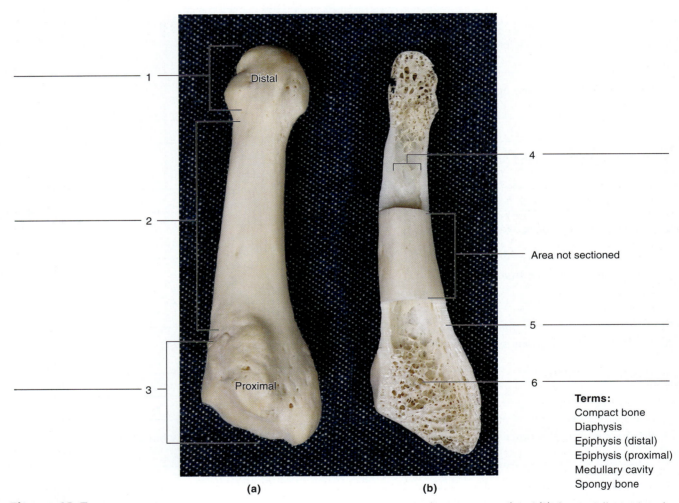

(a)   (b)

Figure 12.5   Identify the structures indicated in (a) the unsectioned long bone (fifth metatarsal) and (b) the partially sectioned long bone, using the terms provided.

Terms:
Compact bone
Diaphysis
Epiphysis (distal)
Epiphysis (proximal)
Medullary cavity
Spongy bone

Organization of the Skeleton

Materials Needed

Textbook
Human skeleton, articulated

For Learning Extension:
Colored pencils

For Demonstration:
Radiographs (X rays) of skeletal structures

The skeleton can be divided into two major portions: (1) the axial skeleton, which consists of the bones and cartilages of the head, neck, and trunk; and (2) the appendicular skeleton, which consists of the bones of the limbs and those that anchor the limbs to the axial skeleton. The bones that anchor the limbs include the pectoral and pelvic girdles.

Purpose of the Exercise

To review the organization of the skeleton, the major bones of the skeleton, and the terms used to describe skeletal structures.

Learning Outcomes

After completing this exercise, you should be able to

1. Distinguish between the axial skeleton and the appendicular skeleton.
2. Locate and label the major bones of the human skeleton.
3. Associate the terms used to describe skeletal structures and locate examples of such structures on the human skeleton.

Procedure—Organization of the Skeleton

1. Review the section entitled "Skeletal Organization" in chapter 7 of the textbook. (Pronunciations of the names for major skeletal structures are included within the narrative of chapter 7.)
2. As a review activity, label figure 13.1.

3. Examine the human skeleton and locate the following parts. As you locate the following bones, note the number of each in the skeleton. Palpate as many of the corresponding bones in your skeleton as possible.

axial skeleton

skull	
cranium	(8)
face	(14)
middle ear bone	(6)
hyoid bone	(1)
vertebral column	
vertebra	(24)
sacrum	(1)
coccyx	(1)
thoracic cage	
rib	(24)
sternum	(1)

appendicular skeleton

pectoral girdle	
scapula	(2)
clavicle	(2)
upper limbs	
humerus	(2)
radius	(2)
ulna	(2)
carpal	(16)
metacarpal	(10)
phalanx	(28)
pelvic girdle	
hip bone (coxal bone; pelvic bone; innominate bone)	(2)
lower limbs	
femur	(2)
tibia	(2)
fibula	(2)
patella	(2)
tarsal	(14)
metatarsal	(10)
phalanx	(28)
Total	**206 bones**

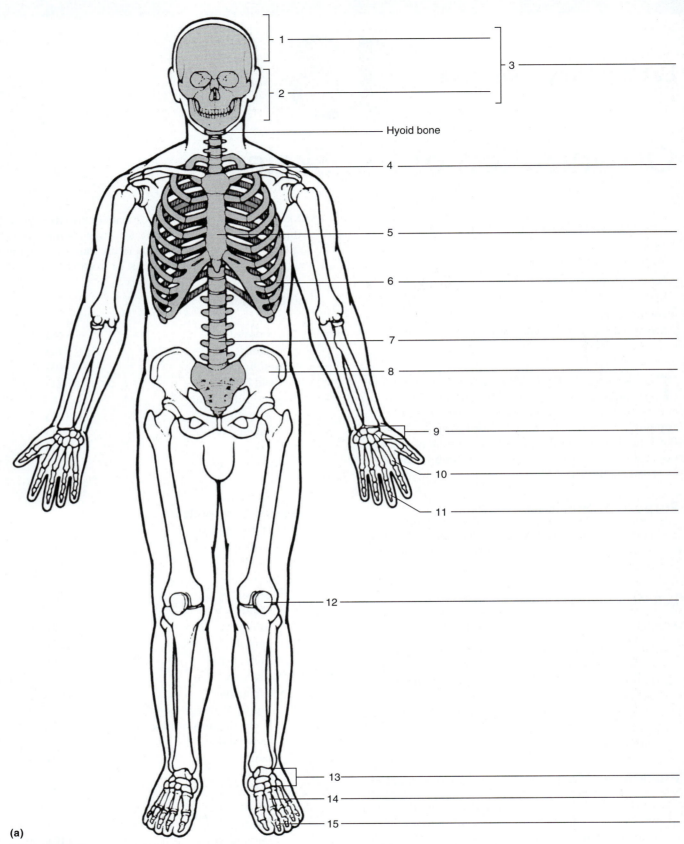

Hyoid bone

Figure 13.1 Label the major bones of the skeleton: (*a*) anterior view; (*b*) posterior view. 2

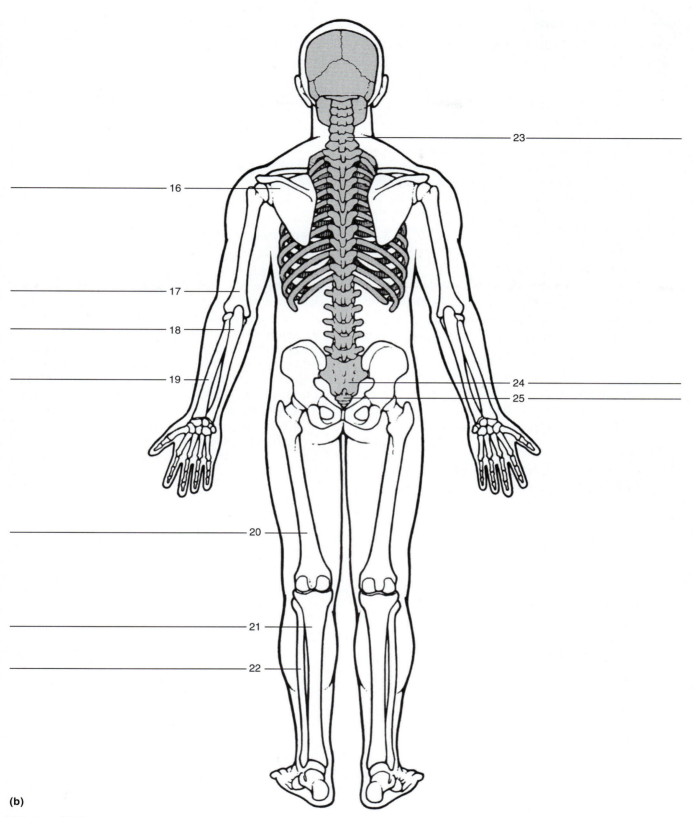

Figure 13.1 *Continued.*

4. Study table 7.2 in chapter 7 of the textbook. Bone feaures (bone markings) can be grouped together in a category of projections, articulations, depressions, or openings. Within each category more specific examples occur. The bones listed in this section only represent an example of a location in the human body. Locate each of the following features on the example bone listed, noting the size, shape, and location in the human skeleton:

Projections:

crest (ridgelike)—hip bone

epicondyle (superior to condyle)—femur

line (linea) (slightly raised ridge)—femur

process (prominent)—vertebra

protuberance (outgrowth)—skull (occipital)

ramus (extension)—hip bone

spine (thornlike)—scapula

trochanter (large)—femur

tubercle (knoblike)—humerus

tuberosity (rough elevation)—tibia

Articulations:

condyle (rounded process)—skull (occipital)

facet (nearly flat)—vertebra

head (expanded end)—femur

Depressions:

fossa (shallow basin)—humerus

fovea (tiny pit)—femur

Openings:

canal (tubular passage)—skull (occipital)

fissure (slit)—skull (orbit)

foramen (hole)—vertebra

meatus (tubelike)—skull (temporal)

sinus (cavity)—skull (maxilla)

Critical Thinking Application

Locate and name the largest foramen in the skull.

_____

_____

Locate and name the largest foramen in the skeleton.

_____

_____

5. Complete Parts A, B, C, and D of Laboratory Report 13.

Demonstration

Images on radiographs (X rays) are produced by allowing X rays from an X-ray tube to pass through a body part and to expose photographic film positioned on the opposite side of the part. The image that appears on the film after it is developed reveals the presence of parts with different densities. Bone, for example, is very dense tissue and is a good absorber of X rays. Thus, bone generally appears light on the film. Air-filled spaces, on the other hand, absorb almost no X rays and appear as dark areas on the film. Liquids and soft tissues absorb intermediate quantities of X rays, so they usually appear in various shades of gray.

Examine the available radiographs (X rays) of skeletal structures by holding each film in front of a light source. Identify as many of the bones and features as you can.

The ⬅ corresponds to the indicated outcome(s) found at the beginning of the laboratory exercise.

Organization of the Skeleton

Part A Assessments

Complete the following statements:

1. The two divisions of the skeleton are the _____ skeleton and the appendicular skeleton. ⬅**1**

2. The _____ bone supports the tongue. ⬅**2**

3. The _____ at the inferior end of the sacrum is composed of several fused vertebrae. ⬅**2**

4. The ribs are attached posteriorly to the _____. ⬅**2**

5. The thoracic cage is composed of _____ pairs of ribs. ⬅**2**

6. The scapulae and clavicles together form the _____. ⬅**2**

7. The humerus, radius, and _____ articulate to form the elbow joint. ⬅**2**

8. The wrist is composed of eight bones called _____. ⬅**2**

9. The hip bones are attached posteriorly to the _____. ⬅**2**

10. The pelvic girdle (hip bones), sacrum, and coccyx together form the _____. ⬅**2**

11. The _____ covers the anterior surface of the knee. ⬅**2**

12. The bones that articulate with the distal ends of the tibia and fibula are called _____. ⬅**2**

13. All finger and toe bones are called _____. ⬅**2**

Part B Assessments

Match the terms in column A with the definitions in column B. Place the letter of your choice in the space provided. ⬅**3**

Column A		Column B
a. Condyle	_____	1. Small, nearly flat articular surface
b. Crest		
c. Facet	_____	2. Deep depression or shallow basin
d. Fontanel	_____	3. Rounded process
e. Foramen	_____	4. Opening or passageway
f. Fossa		
g. Suture	_____	5. Interlocking line of union
	_____	6. Narrow, ridgelike projection
	_____	7. Soft region between bones of skull

Part C Assessments

Match the terms in column A with the definitions in column B. Place the letter of your choice in the space provided. ◂ **3**

Column A

a. Fovea
b. Head
c. Meatus
d. Sinus
e. Spine
f. Trochanter
g. Tubercle

Column B

_____ **1.** Tubelike passageway

_____ **2.** Tiny pit or depression

_____ **3.** Small, knoblike process

_____ **4.** Rounded enlargement at end of bone

_____ **5.** Air-filled cavity within bone

_____ **6.** Relatively large process

_____ **7.** Thornlike projection

Part D Assessments

Identify the bones indicated in figure 13.2. ◂ **2**

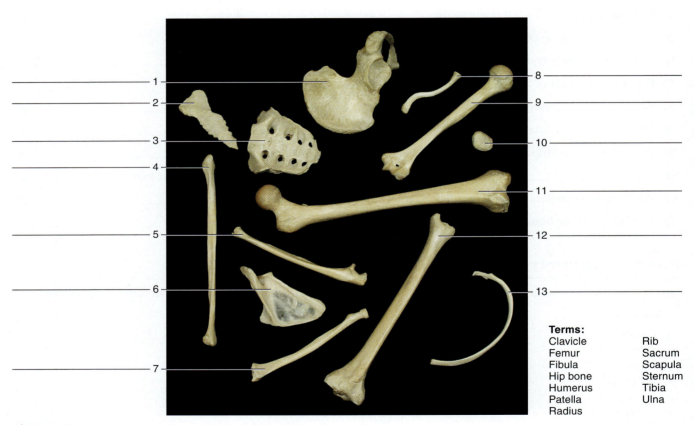

1
2
3
4
5
6
7

8
9
10
11
12
13

Terms:

Clavicle	Rib
Femur	Sacrum
Fibula	Scapula
Hip bone	Sternum
Humerus	Tibia
Patella	Ulna
Radius	

Figure 13.2   Using the terms provided, identify the bones in this random arrangement.

Laboratory Exercise 14

Skull

Materials Needed

Textbook
Human skull, articulated
Human skull, disarticulated (Beauchene)
Human skull, sagittal section

For Learning Extension:
Colored pencils

For Demonstration:
Fetal skull

A human skull consists of twenty-two bones that, except for the lower jaw, are firmly interlocked along sutures. Eight of these immovable bones make up the braincase, or cranium, and thirteen more immovable bones and mandible form the facial skeleton.

Purpose of the Exercise

To examine the structure of the human skull and to identify the bones and major features of the skull.

Learning Outcomes

After completing this exercise, you should be able to

1. Distinguish between the cranium and the facial skeleton.
2. Locate and label the bones of the skull and their major features.
3. Locate and label the major sutures of the cranium.
4. Locate and label the sinuses of the skull.

Procedure—Skull

1. Review the section entitled "Skull" in chapter 7 of the textbook.
2. As a review activity, label figures 14.1, 14.2, 14.3, 14.4, and 14.5.

3. Examine the **cranial bones** of the articulated human skull and the sectioned skull. Also observe the corresponding disarticulated bones. Locate the following bones and features in the laboratory specimens and, at the same time, palpate as many of these bones and features in your skull as possible.

frontal bone (1)
supraorbital foramen
frontal sinus

parietal bone (2)
sagittal suture
coronal suture

occipital bone (1)
lambdoid suture
foramen magnum
occipital condyle

temporal bone (2)
squamous suture
external acoustic meatus
mandibular fossa
mastoid process
styloid process
zygomatic process

sphenoid bone (1)
sella turcica
sphenoidal sinus

ethmoid bone (1)
cribriform plate
perpendicular plate
superior nasal concha
middle nasal concha
ethmoidal sinus
crista galli

4. Complete Parts A and B of Laboratory Report 14.

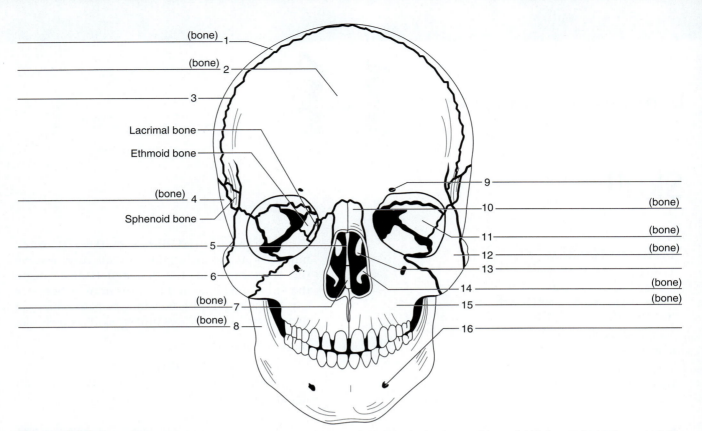

(bone) 1
(bone) 2
3
Lacrimal bone
Ethmoid bone
(bone) 4
Sphenoid bone
5
6
(bone) 7
(bone) 8

9
10 (bone)
11 (bone)
12 (bone)
13
14 (bone)
15 (bone)
16

Figure 14.1 Label the anterior bones and features of the skull. (If the line lacks the word *bone*, label the particular feature of that bone.) ◀2

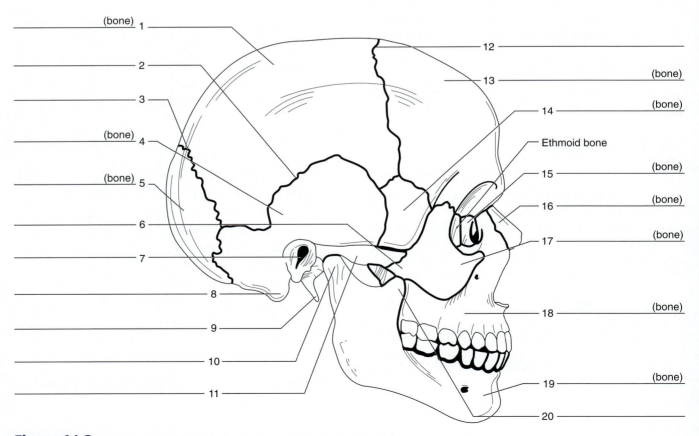

(bone) 1
2
3
(bone) 4
(bone) 5
6
7
8
9
10
11

12
13 (bone)
14 (bone)
Ethmoid bone
15 (bone)
16 (bone)
17 (bone)
18 (bone)
19 (bone)
20

Figure 14.2 Label the lateral bones and features of the skull.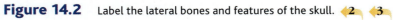

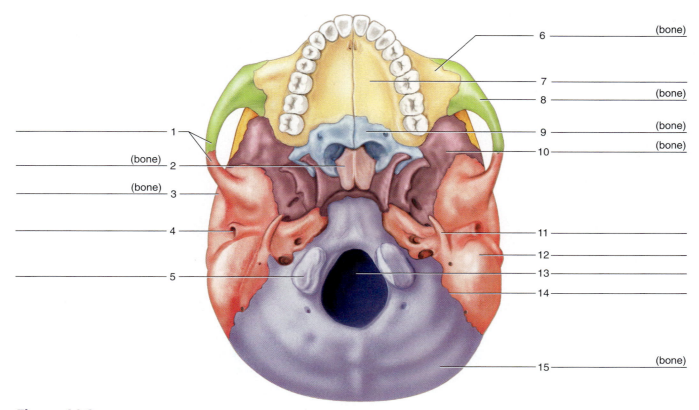

Figure 14.3 Label the inferior bones and features of the skull. ◀2

Anterior

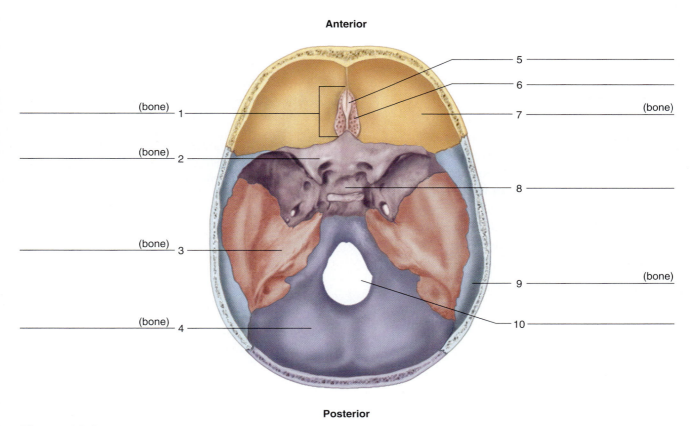

Posterior

Figure 14.4 Label the bones and features of the floor of the cranial cavity as viewed from above. ◀2

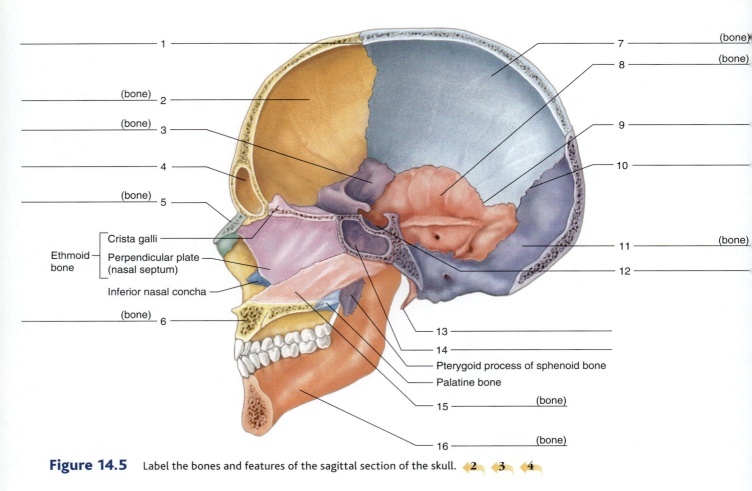

1 _____

(bone) 2 _____

(bone) 3 _____

4 _____

(bone) 5 _____

Crista galli _____

Ethmoid — Perpendicular plate
bone — (nasal septum) _____

Inferior nasal concha _____

(bone) 6 _____

7 _____ (bone)

8 _____ (bone)

9 _____

10 _____

11 _____ (bone)

12 _____

13 _____

14 _____

Pterygoid process of sphenoid bone

Palatine bone

15 _____ (bone)

16 _____ (bone)

Figure 14.5   Label the bones and features of the sagittal section of the skull. ◄2 ◄3 ◄4

5. Examine the **facial bones** of the articulated and sectioned skulls and the corresponding disarticulated bones. Locate the following:

maxilla (2)
  maxillary sinus
  palatine process
  alveolar process
  alveolar arch
palatine bone (2)
zygomatic bone (2)
  temporal process
  zygomatic arch

lacrimal bone (2)
nasal bone (2)
vomer bone (1)
inferior nasal concha (2)
mandible (1)
  mandibular condyle
  coronoid process
  alveolar arch

6. Complete Parts C and D of the laboratory report.

Learning Extension

Use colored pencils to color the bones illustrated in figures 14.1 and 14.2. Select a different color for each bone in the series. This activity should help you locate various bones shown in different views in the figures. You can check your work by referring to the corresponding figures in the textbook, presented in full color.

Demonstration

Examine the fetal skull (fig. 14.6). The skull is incompletely developed and the cranial bones are separated by fibrous membranes. These membranous areas are called *fontanels,* or "soft spots." The fontanels close as the cranial bones grow together. The posterior fontanel usually closes within a few months after birth, whereas the anterior fontanel may not close until the middle or end of the second year. What other features characterize the fetal skull?_____

_____

_____

_____

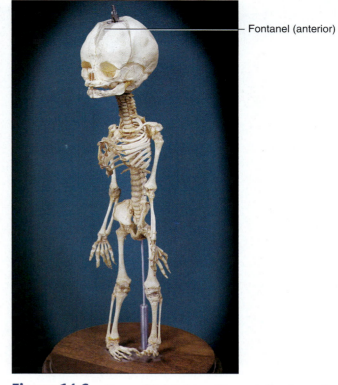

Fontanel (anterior)

Figure 14.6   Human fetal skeleton. The gestational age of this skeleton is 7–8 months.

Notes

The ◀ corresponds to the indicated outcome(s) found at the beginning of the laboratory exercise.

Skull

Part A Assessments

Match the bones in column A with the features in column B. Place the letter of your choice in the space provided. (Some answers are used more than once.) ◀2

Column A	Column B
a. Ethmoid bone	_____ **1.** Forms sagittal, coronal, squamous, and lambdoid sutures
b. Frontal bone	_____ **2.** Cribriform plate
c. Occipital bone	_____ **3.** Crista galli
d. Parietal bone	_____ **4.** External acoustic meatus
e. Sphenoid bone	_____ **5.** Foramen magnum
f. Temporal bone	_____ **6.** Mandibular fossa
	_____ **7.** Mastoid process
	_____ **8.** Middle nasal concha
	_____ **9.** Occipital condyle
	_____ **10.** Sella turcica
	_____ **11.** Styloid process
	_____ **12.** Supraorbital foramen

Part B Assessments

Complete the following statements:

1. The _____ suture joins the frontal bone to the parietal bones. ◀3

2. The parietal bones are fused along the midline by the _____ suture. ◀3

3. The _____ suture joins the parietal bones to the occipital bone. ◀3

4. The temporal bones are joined to the parietal bones along the _____ sutures. ◀3

5. Name the three cranial bones that contain sinuses. ◀1 ◀4 _____

_____

6. Name a facial bone that contains a sinus. ◀1 ◀4 _____

Part C Assessments

Match the bones in column A with the characteristics in column B. Place the letter of your choice in the space provided. ◀2

Column A	Column B
a. Inferior nasal concha	_____ **1.** Forms bridge of nose
b. Lacrimal bone	_____ **2.** Only movable bone in facial skeleton
c. Mandible	_____ **3.** Contains coronoid process
d. Maxilla	_____ **4.** Creates prominence of cheek inferior and lateral to the eye
e. Nasal bone	_____ **5.** Contains sockets of upper teeth
f. Palatine bone	_____ **6.** Forms inferior portion of nasal septum in back
g. Vomer bone	_____ **7.** Forms anterior portion of zygomatic arch
h. Zygomatic bone	_____ **8.** Scroll-shaped bone
	_____ **9.** Forms anterior roof of mouth
	_____ **10.** Forms posterior roof of mouth
	_____ **11.** Scalelike part in medial wall of orbit

Part D Assessments

Identify the numbered bones and features of the skulls in figures 14.7, 14.8, 14.9, 14.10, and 14.11. ◀2 ◀3

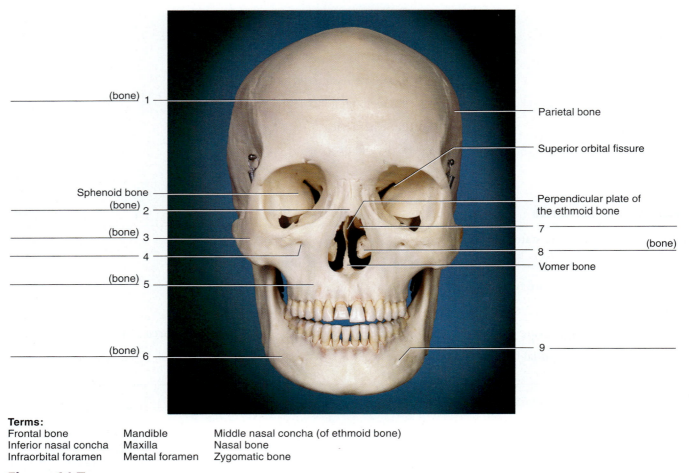

Terms:

Frontal bone	Mandible	Middle nasal concha (of ethmoid bone)
Inferior nasal concha	Maxilla	Nasal bone
Infraorbital foramen	Mental foramen	Zygomatic bone

Figure 14.7 Using the terms provided, identify the bones and features indicated on this anterior view of the skull.

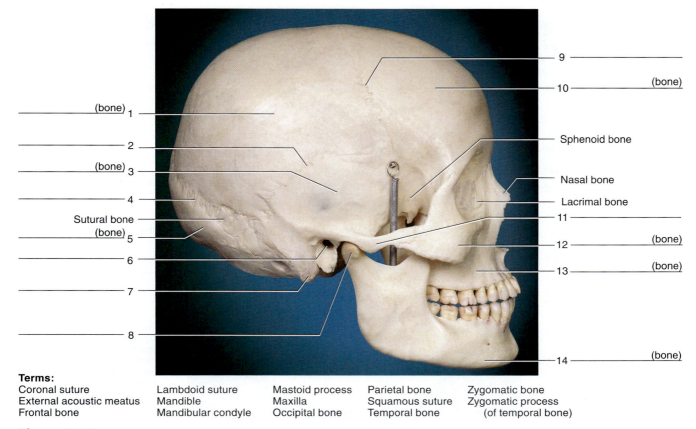

9
10 _____ (bone)

Sphenoid bone

(bone) 1

2

(bone) 3

4

Sutural bone
(bone) 5

6

7

8

Nasal bone

Lacrimal bone

11 _____

12 _____ (bone)

13 _____ (bone)

14 _____ (bone)

Terms:

Coronal suture
External acoustic meatus
Frontal bone

Lambdoid suture
Mandible
Mandibular condyle

Mastoid process
Maxilla
Occipital bone

Parietal bone
Squamous suture
Temporal bone

Zygomatic bone
Zygomatic process
(of temporal bone)

Figure 14.8 Using the terms provided, identify the bones and features indicated on this lateral view of the skull.

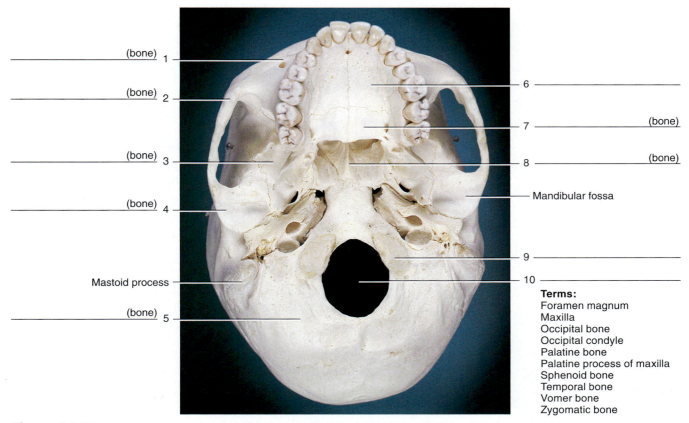

(bone) 1

(bone) 2

(bone) 3

(bone) 4

Mastoid process

(bone) 5

6

7 _____ (bone)

8 _____ (bone)

Mandibular fossa

9

10

Terms:

Foramen magnum
Maxilla
Occipital bone
Occipital condyle
Palatine bone
Palatine process of maxilla
Sphenoid bone
Temporal bone
Vomer bone
Zygomatic bone

Figure 14.9 Using the terms provided, identify the bones and features indicated on this inferior view of the skull.

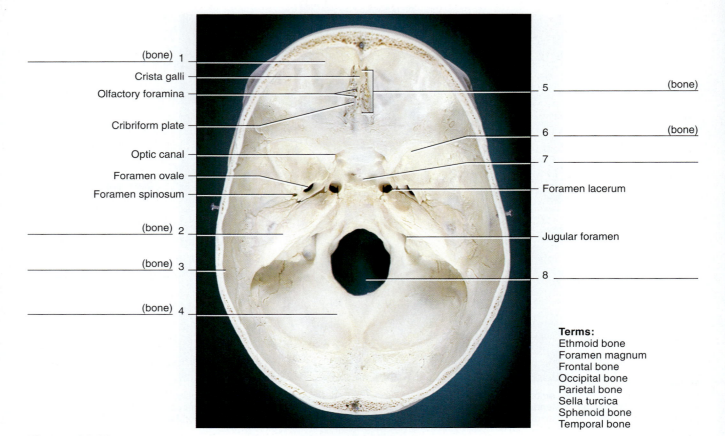

_____ (bone) 1 ____
                  Crista galli _____
            Olfactory foramina _____
                Cribriform plate _____
                   Optic canal _____
                 Foramen ovale _____
            Foramen spinosum _____

_____ (bone) 2 ____

_____ (bone) 3 ____

_____ (bone) 4 ____

5 _____ (bone)

6 _____ (bone)

7 _____

_____ Foramen lacerum

_____ Jugular foramen

8 _____

Terms:
Ethmoid bone
Foramen magnum
Frontal bone
Occipital bone
Parietal bone
Sella turcica
Sphenoid bone
Temporal bone

Figure 14.10   Using the terms provided, identify the bones and features on this floor of the cranial cavity of a skull.

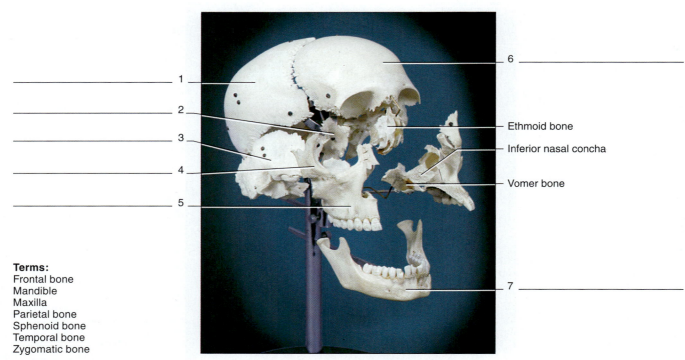

_____ 1 ____
_____ 2 ____
_____ 3 ____
_____ 4 ____
_____ 5 ____

6 _____

_____ Ethmoid bone

_____ Inferior nasal concha

_____ Vomer bone

7 _____

Terms:
Frontal bone
Mandible
Maxilla
Parietal bone
Sphenoid bone
Temporal bone
Zygomatic bone

Figure 14.11   Using the terms provided, identify the bones on this disarticulated skull.

100

Laboratory Exercise  15

Vertebral Column and Thoracic Cage

Materials Needed

Textbook
Human skeleton, articulated
Samples of cervical, thoracic, and lumbar vertebrae
Human skeleton, disarticulated

The vertebral column, consisting of twenty-six bones, extends from the skull to the pelvis and forms the vertical axis of the human skeleton. The vertebral column includes seven cervical vertebrae, twelve thoracic vertebrae, five lumbar vertebrae, one sacrum of five fused vertebrae, and one coccyx of usually four fused vertebrae. To help remember the number of cervical, thoracic, and lumbar vertebrae from superior to inferior consider this saying: breakfast at 7, lunch at 12, and dinner at 5. These vertebrae are separated from one another by cartilaginous intervertebral discs and are held together by ligaments.

The thoracic cage surrounds the thoracic and upper abdominal cavities. It includes the ribs, the thoracic vertebrae, the sternum, and the costal cartilages. Men and women, although variations can exist, possess the same total bone number of 206.

Purpose of the Exercise

To examine the vertebral column and the thoracic cage of the human skeleton and to identify the bones and major features of these parts.

Learning Outcomes

After completing this exercise, you should be able to

1. Identify the major features of the vertebral column.
2. Locate the features of a typical vertebra.
3. Distinguish among a cervical, thoracic, and lumbar vertebra and locate the sacrum and coccyx.
4. Identify the structures and functions of the thoracic cage.
5. Distinguish between true and false ribs.

Procedure A—The Vertebral Column

1. Review the section entitled "Vertebral Column" in chapter 7 of the textbook.
2. As a review activity, label figures 15.1, 15.2, 15.3, and 15.4.
3. Examine the vertebral column of the human skeleton and locate the following bones and features. At the same time, locate as many of the corresponding bones and features in your skeleton as possible.

atlas (C1)	(1)
axis (C2)	(1)
vertebra prominens (C7)	(1)
cervical vertebrae	(7)
(includes atlas, axis, and vertebra prominens)	
thoracic vertebrae	(12)
lumbar vertebrae	(5)
intervertebral discs	
vertebral canal	
sacrum	(1)
coccyx	(1)
cervical curvature	
thoracic curvature	
lumbar curvature	
sacral curvature	
intervertebral foramina	

Critical Thinking Application

Note the four curvatures of the vertebral column. What functional advantages exist with curvatures for skeletal structure instead of a straight vertebral column?

_____
_____
_____
_____
_____

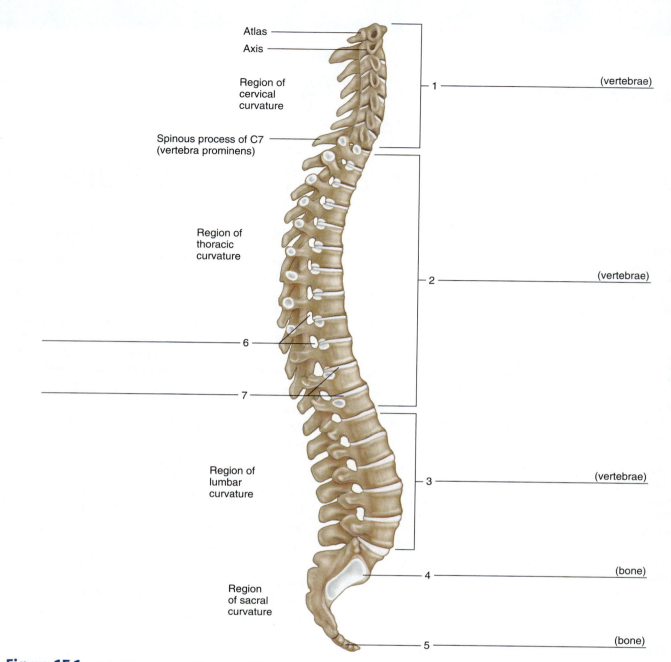

Atlas

Axis

Region of
cervical
curvature

Spinous process of C7
(vertebra prominens)

Region of
thoracic
curvature

Region of
lumbar
curvature

Region
of sacral
curvature

1 ————————————— (vertebrae)

2 ————————————— (vertebrae)

6

7

3 ————————————— (vertebrae)

4 ————————————— (bone)

5 ————————————— (bone)

Figure 15.1   Label the bones and features of the vertebral column (right lateral view).   ◀**1**

(a) Atlas (C1)

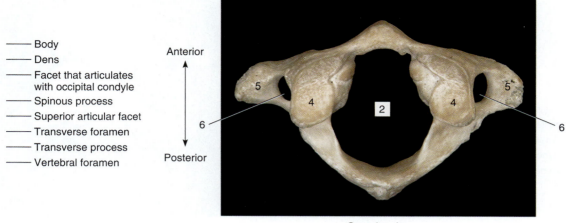

Anterior

Posterior

Superior view

—— Body
—— Dens
—— Facet that articulates
       with occipital condyle
—— Spinous process
—— Superior articular facet
—— Transverse foramen
—— Transverse process
—— Vertebral foramen

(b) Axis (C2)

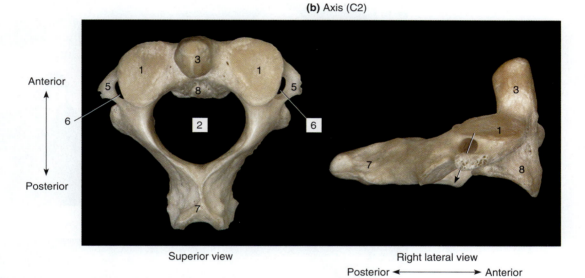

Anterior

Posterior

Superior view

Right lateral view

Posterior ◄——————► Anterior

Figure 15.2 Label the superior features of (a) the atlas and the superior and right lateral features of (b) the axis by placing the correct numbers in the spaces provided. (The broken arrow indicates a transverse foramen.) ◄2

Superior views　　　　　　Right lateral views

——— Body
——— Inferior notch
——— Lamina
——— Pedicle
——— Spinous process
——— Superior articular process
——— Transverse foramen
——— Transverse process
——— Vertebral foramen

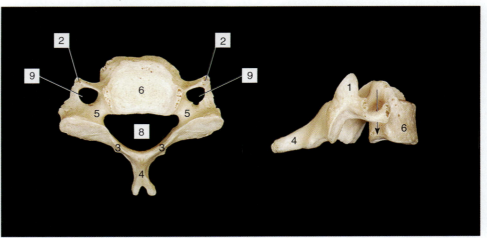

(a) Cervical vertebra

Anterior

Posterior

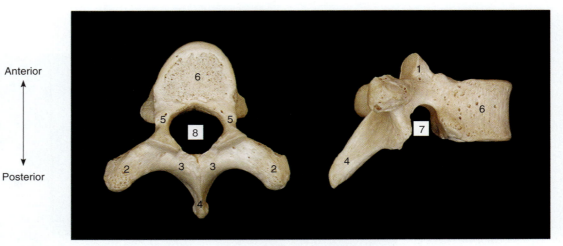

(b) Thoracic vertebra

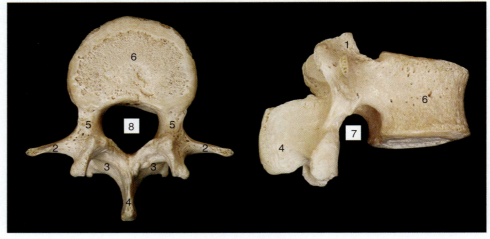

(c) Lumbar vertebra

Posterior ◄————► Anterior

Figure 15.3　　Label the superior and right lateral features of the (a) cervical, (b) thoracic, and (c) lumbar vertebrae by placing the correct numbers in the spaces provided. (The broken arrow indicates a transverse foramen.) ◄**2**

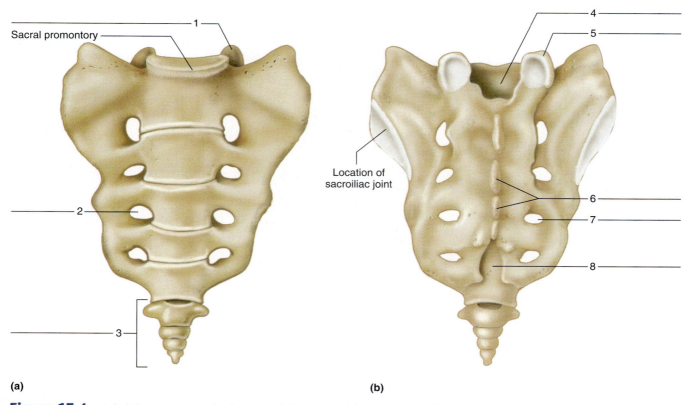

Sacral promontory

Location of
sacroiliac joint

(a)

(b)

Figure 15.4 Label the coccyx and the features of the sacrum: (a) anterior view; (b) posterior view. ◄3▸

4. Compare the available samples of cervical, thoracic, and lumbar vertebrae by noting differences in size and shape and by locating the following features:

vertebral foramen	superior articular
body	processes
pedicles	inferior articular
laminae	processes
spinous process	transverse foramina
vertebral arch	facets
transverse processes	dens of axis

5. Examine the sacrum and coccyx. Locate the following features:

sacrum
  superior articular process
  posterior sacral foramen
  anterior sacral foramen
  sacral promontory
  sacral canal
  tubercles
  median sacral crest
  sacral hiatus
coccyx

6. Complete Parts A and B of Laboratory Report 15.

Procedure B—The Thoracic Cage

1. Review the section entitled "Thoracic Cage" in chapter 7 of the textbook.
2. As a review activity, label figure 15.5.
3. Examine the thoracic cage of the human skeleton and locate the following bones and features:

rib
  head
  tubercle
  facets
  true ribs (pairs 1–7)
  false ribs (pairs 8–12; includes floating ribs)
    floating ribs (pairs 11–12)
costal cartilages
sternum
  jugular (suprasternal) notch
  manubrium
  sternal angle
  body
  xiphoid process

4. Complete Parts C and D of the laboratory report.

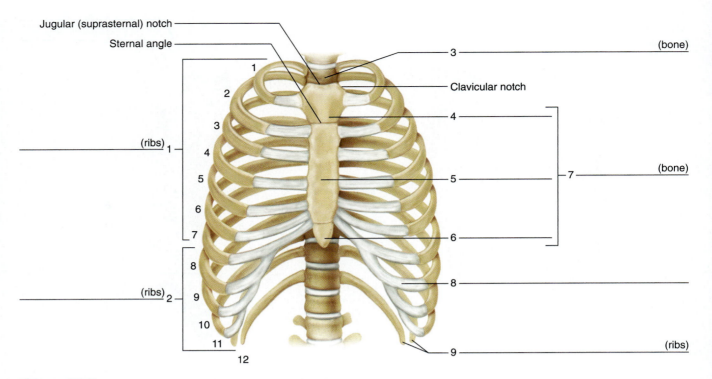

Jugular (suprasternal) notch

Sternal angle

(ribs) 1

(ribs) 2

3 ———————————— (bone)

Clavicular notch

4

5

6

7 ———————————— (bone)

8

9 ———————————— (ribs)

1
2
3
4
5
6
7
8
9
10
11
12

Figure 15.5   Label the bones and features of the thoracic cage (anterior view).

The ◀ corresponds to the indicated outcome(s) found at the beginning of the laboratory exercise.

Vertebral Column and Thoracic Cage

Part A Assessments

Complete the following statements:

1. The vertebral column encloses and protects the _____ . ◀**1**

2. The _____ of the vertebrae support the weight of the head and trunk. ◀**1**

3. The _____ separate adjacent vertebrae. ◀**1**

4. The pedicles, laminae, and spinous process of a vertebra form the _____ . ◀**2**

5. The intervertebral foramina provide passageways for _____ . ◀**1**

6. Transverse foramina of cervical vertebrae serve as passageways for _____ leading to the brain. ◀**2**

7. The first vertebra is also called the _____ . ◀**1**

8. When the head is moved from side to side, the first vertebra pivots around the _____ of the second vertebra. ◀**2**

9. The _____ vertebrae have the largest and strongest bodies. ◀**2**

10. The number of vertebrae that fuse to form the sacrum is _____ . ◀**1**

11. An opening called the _____ exists at the tip of the sacral canal. ◀**1**

Part B Assessments

Based on your observations, compare typical cervical, thoracic, and lumbar vertebrae in relation to the characteristics indicated in the table. For your responses, consider characteristics such as size, shape, presence or absence, and unique features. ◀**3**

Vertebra	Number	Size	Body	Spinous Process	Transverse Foramina
Cervical					
Thoracic					
Lumbar					

Part C Assessments

Complete the following statements:

1. The adult skeleton of most men and women contains a total of _____ (number) bones.

2. The last two pairs of ribs that have no cartilaginous attachments to the sternum are sometimes called _____ ribs. **5**

3. The tubercles of the ribs articulate with facets on the _____ processes of the thoracic vertebrae. **5**

4. The manubrium articulates with the _____ on its superior border. **4**

5. List three general functions of the thoracic cage. **4**

   a. _____

   b. _____

   c. _____

Part D Assessments

Identify the bones and features indicated in the radiograph of the neck in figure 15.6. **2**

Terms:
Atlas
Axis
Body
Intervertebral disc
Spinous process
Transverse process

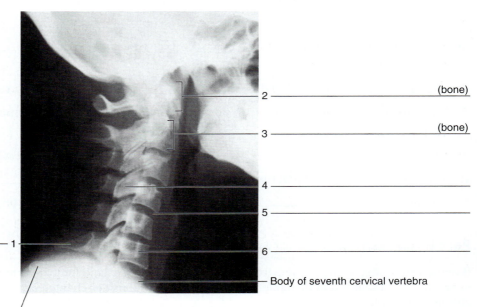

2 _____ (bone)

3 _____ (bone)

4 _____

5 _____

6 _____

Body of seventh cervical vertebra

1

Spinous process of C7 (vertebra prominens)

Figure 15.6   Using the terms provided, identify the bones and features indicated in this radiograph of the neck (lateral view).

Laboratory Exercise 16

Pectoral Girdle and Upper Limb

Materials Needed

Textbook
Human skeleton, articulated
Human skeleton, disarticulated

For Learning Extension:
Colored pencils

The pectoral girdle (shoulder girdle) consists of two clavicles and two scapulae. These parts support the upper limbs and serve as attachments for various muscles that move these limbs.

Each upper limb includes a humerus, a radius, an ulna, eight carpals, five metacarpals, and fourteen phalanges. These bones form the framework of the arm, forearm, and hand. They also function as parts of levers when the limbs are moved.

Purpose of the Exercise

To examine the bones of the pectoral girdle and upper limb and to identify the major features of these bones.

Learning Outcomes

After completing this exercise, you should be able to

1. Locate and identify the bones of the pectoral girdle and their major features.
2. Locate and identify the bones of the upper limb and their major features.

Procedure A—The Pectoral Girdle

1. Review the section entitled "Pectoral Girdle" in chapter 7 of the textbook.
2. As a review activity, label figures 16.1 and 16.2.
3. Examine the bones of the pectoral girdle and locate the following features. At the same time, locate as many of the corresponding surface bones and features of your own skeleton as possible.

   clavicle
   scapula
       spine
       acromion process
       coracoid process
       glenoid cavity
       borders
           superior border
           medial (vertebral) border
           lateral (axillary) border
       fossae
           supraspinous fossa
           infraspinous fossa

 **Critical Thinking Application**

Why is the clavicle a bone that can easily fracture?

_____

_____

_____

_____

4. Complete Part A of Laboratory Report 16.

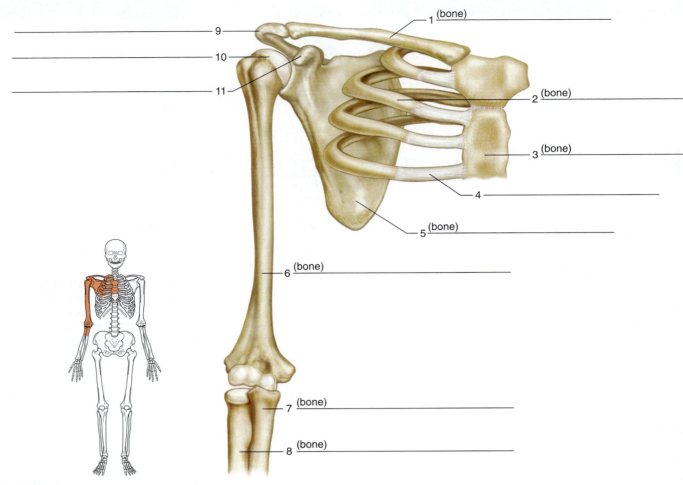

9 _____

10 _____

11 _____

1 (bone) _____

2 (bone) _____

3 (bone) _____

4 _____

5 (bone) _____

6 (bone) _____

7 (bone) _____

8 (bone) _____

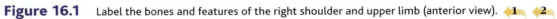

Figure 16.1 Label the bones and features of the right shoulder and upper limb (anterior view). ◀**1** ◀**2**

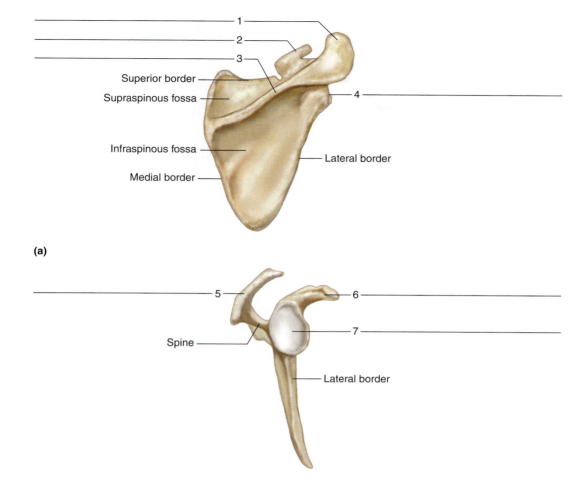

(a)

Superior border

Supraspinous fossa

Infraspinous fossa

Medial border

Lateral border

(b)

Spine

Lateral border

Figure 16.2   Label (a) the posterior surface of the right scapula and (b) the lateral aspect of the right scapula.   1

Procedure B—The Upper Limb

1. Review the section entitled "Upper Limb" in chapter 7 of the textbook.
2. As a review activity, label figures 16.3, 16.4, and 16.5.
3. Examine the following bones and features of the upper limb:

humerus
  proximal features
    head
    greater tubercle
    lesser tubercle
    anatomical neck
    surgical neck
    intertubercular groove (sulcus)
  shaft
    deltoid tuberosity

  distal features
    capitulum
    trochlea
    medial epicondyle
    lateral epicondyle
    coronoid fossa
    olecranon fossa

radius
    head
    radial tuberosity
    styloid process

ulna
    trochlear notch (semilunar notch)
    olecranon process
    coronoid process
    styloid process
    head

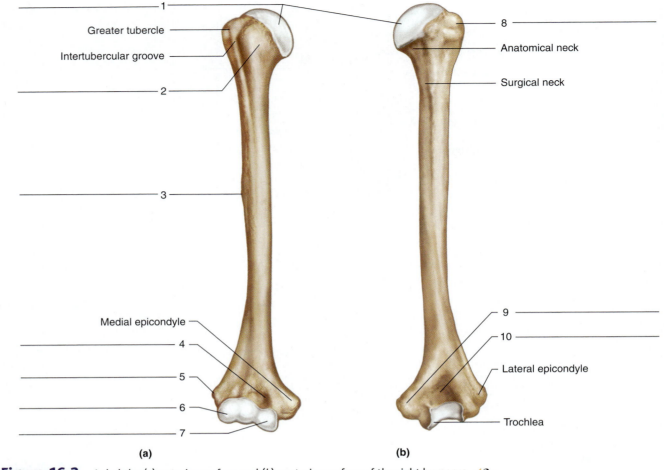

Greater tubercle

Intertubercular groove

1

2

3

Medial epicondyle

4

5

6

7

8

Anatomical neck

Surgical neck

9

10

Lateral epicondyle

Trochlea

(a)

(b)

Figure 16.3   Label the (a) anterior surface and (b) posterior surface of the right humerus. ◀2

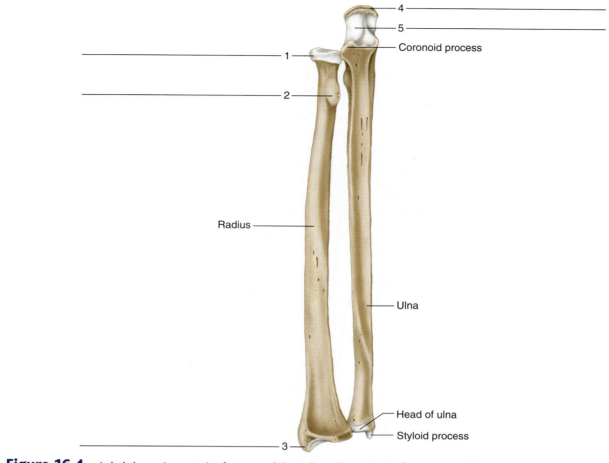

Coronoid process

Radius

Ulna

Head of ulna

Styloid process

1

2

3

4

5

Figure 16.4   Label the major anterior features of the right radius and ulna (anterior view). ◀**2**

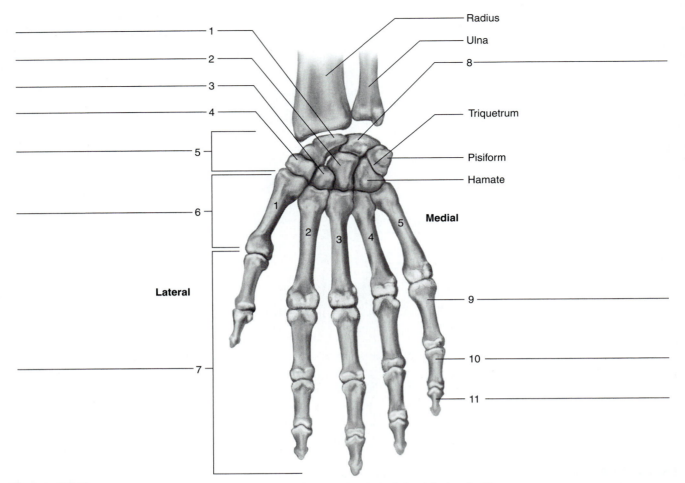

Figure 16.5 Label the bones and groups of bones in this anterior view of the right hand. ◀2

carpal bones

    proximal row (listed lateral to medial)

        scaphoid

        lunate

        triquetrum

        pisiform

    distal row (listed medial to lateral)

        hamate

        capitate

        trapezoid

        trapezium

The following mnemonic device will help you learn the eight carpals:

So Long Top Part

Here Comes The Thumb

The first letter of each word corresponds to the first letter of a carpal. This device arranges the carpals in order for the proximal, transverse row of four bones from lateral to medial, followed by the distal, transverse row from medial to lateral, which ends nearest the thumb. This arrangement assumes the anatomical position of the hand.

    metacarpal bones

    phalanges

        proximal phalanx

        middle phalanx

        distal phalanx

4. Complete Parts B, C, and D of the laboratory report.

Learning Extension

Use different colored pencils to distinguish the individual bones in figure 16.5.

The ⬅ corresponds to the indicated outcome(s) found at the beginning of the laboratory exercise.

Pectoral Girdle and Upper Limb

Part A Assessments

Complete the following statements:

1. The pectoral girdle is an incomplete ring because it is open in the back between the _____ . ⬅1

2. The medial ends of the clavicles articulate with the _____ of the sternum. ⬅1

3. The lateral ends of the clavicles articulate with the _____ of the scapula. ⬅1

4. The _____ divides the posterior side of the scapula into unequal portions. ⬅1

5. The tip of the shoulder is due to the _____ of the scapula. ⬅1

6. At the lateral end of the scapula, the _____ curves anteriorly and inferiorly from the clavicle. ⬅1

7. The glenoid cavity of the scapula articulates with the _____ of the humerus. ⬅1

Part B Assessments

Match the bones in column A with the features in column B. Place the letter of your choice in the space provided. ⬅2

Column A	Column B
a. Carpals	_____ 1. Capitate
b. Humerus	_____ 2. Coronoid fossa
c. Metacarpals	_____ 3. Deltoid tuberosity
d. Phalanges	_____ 4. Greater tubercle
e. Radius	_____ 5. Intertubercular groove
f. Ulna	_____ 6. Lunate
	_____ 7. Olecranon fossa
	_____ 8. Five palmar bones
	_____ 9. Radial tuberosity
	_____ 10. Trapezium
	_____ 11. Trochlear notch
	_____ 12. Fourteen bones in digits

Part C Assessments

Identify the bones and features indicated in the radiographs (X rays) of figures 16.6, 16.7, and 16.8. ◀1 ◀2

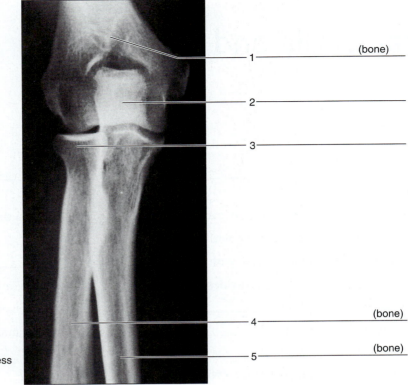

Terms:
Head of radius
Humerus
Olecranon process
Radius
Ulna

1 _____ (bone)

2 _____

3 _____

4 _____ (bone)

5 _____ (bone)

Figure 16.6   Using the terms provided, identify the bones and features indicated on this radiograph of the right elbow (anterior view).

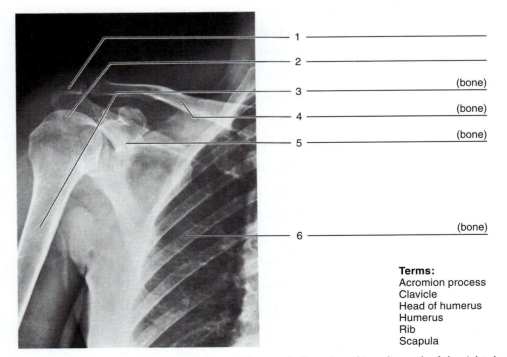

1 _____

2 _____

3 _____ (bone)

4 _____ (bone)

5 _____ (bone)

6 _____ (bone)

Terms:
Acromion process
Clavicle
Head of humerus
Humerus
Rib
Scapula

Figure 16.7   Using the terms provided, identify the bones and features indicated on this radiograph of the right shoulder (anterior view).

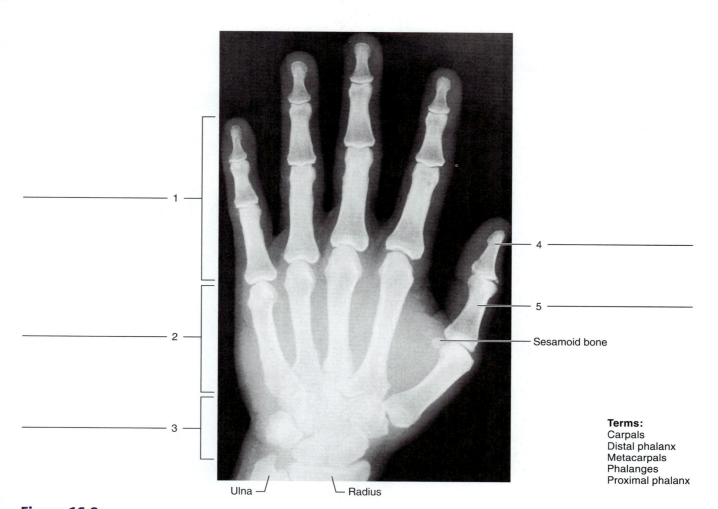

1 _____

2 _____

3 _____

4 _____

5 _____

Sesamoid bone

Ulna ⌐   ⌐ Radius

Terms:
Carpals
Distal phalanx
Metacarpals
Phalanges
Proximal phalanx

Figure 16.8   Using the terms provided, identify the bones indicated on this radiograph of the right hand (anterior view).

Part D Assessments

Identify the bones of the hand in figure 16.9.

_____	Capitate
_____	Distal phalanges
_____	Hamate
_____	Lunate
_____	Metacarpals
_____	Middle phalanges
_____	Pisiform
_____	Proximal phalanges
_____	Scaphoid
_____	Trapezium
_____	Trapezoid
_____	Triquetrum

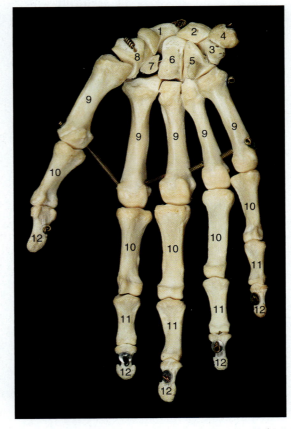

Figure 16.9   Label the bones numbered on this anterior view of the right hand by placing the correct numbers in the spaces provided.

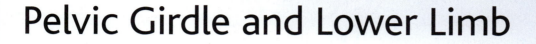

Laboratory Exercise 17

Pelvic Girdle and Lower Limb

Materials Needed

Textbook
Human skeleton, articulated
Human skeleton, disarticulated
Male and female pelves

For Learning Extension:
Colored pencils

The pelvic girdle includes two hip bones that articulate with each other anteriorly at the symphysis pubis and posteriorly with the sacrum. Together, the pelvic girdle, sacrum, and coccyx comprise the pelvis. The pelvis, in turn, provides support for the trunk of the body and provides attachments for the lower limbs.

The bones of the lower limb form the framework of the thigh, leg, and foot. Each limb includes a femur, a patella, a tibia, a fibula, seven tarsals, five metatarsals, and fourteen phalanges.

Purpose of the Exercise

To examine the bones of the pelvic girdle and lower limb and to identify the major features of these bones.

Learning Outcomes

After completing this exercise, you should be able to

1. Locate and identify the bones of the pelvic girdle and their major features.
2. Locate and identify the bones of the lower limb and their major features.

Procedure A—The Pelvic Girdle

1. Review the section entitled "Pelvic Girdle" in chapter 7 of the textbook.
2. As a review activity, label figures 17.1 and 17.2.

3. Examine the bones of the pelvic girdle and locate the following:

   hip bone (coxal bone; pelvic bone; innominate bone)
   acetabulum
   ilium
   iliac crest
   sacroiliac joint
   anterior superior iliac spine
   greater sciatic notch
   ischium
   ischial tuberosity
   ischial spine
   pubis
   symphysis pubis
   pubic arch
   obturator foramen

4. Complete Part A of Laboratory Report 17.

Critical Thinking Application

Examine the male and female pelves. Look for major differences between them. Note especially the flare of the iliac bones, the angle of the pubic arch, the distance between the ischial spines and ischial tuberosities, and the curve and width of the sacrum. In what ways are the differences you observed related to the function of the female pelvis as a birth canal?

_____

_____

_____

_____

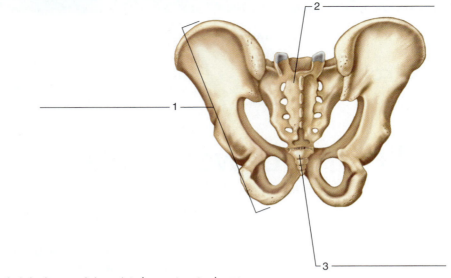

Figure 17.1   Label the bones of the pelvis (posterior view). ◀1

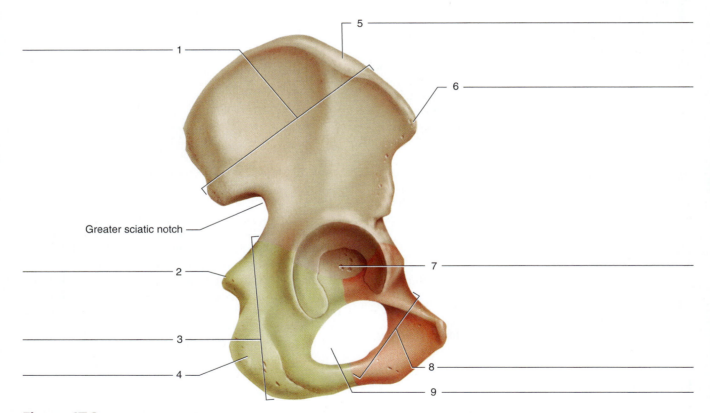

Greater sciatic notch

Figure 17.2   Label the lateral features of the right hip bone. ◀1

Procedure B—The Lower Limb

1. Review the section entitled "Lower Limb" in chapter 7 of the textbook.
2. As a review activity, label figures 17.3, 17.4, and 17.5.
3. Examine the bones of the lower limb and locate each of the following:

femur
    proximal features
        head
        fovea capitis
        neck
        greater trochanter
        lesser trochanter
    shaft
        gluteal tuberosity
        linea aspera
    distal features
        lateral epicondyle
        medial epicondyle
        lateral condyle
        medial condyle

patella

tibia
    medial condyle
    lateral condyle
    tibial tuberosity
    anterior crest
    medial malleolus

fibula
    head
    lateral malleolus

tarsal bones
    talus
    calcaneus
    navicular
    cuboid
    lateral cuneiform
    intermediate cuneiform
    medial cuneiform

metatarsal bones

phalanges
    proximal phalanx
    middle phalanx
    distal phalanx

4. Complete Parts B, C, and D of the laboratory report.

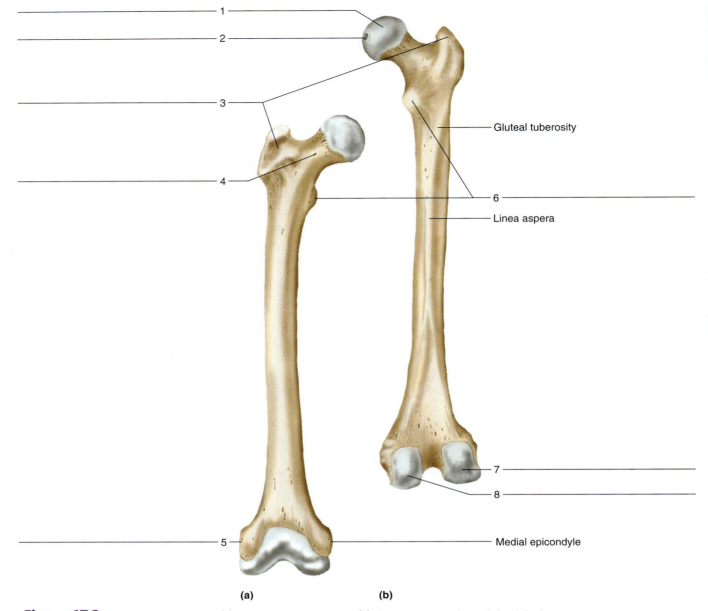

1 _____

2 _____

3 _____

4 _____

Gluteal tuberosity

6 _____

Linea aspera

7 _____

8 _____

5 _____

Medial epicondyle

(a)   (b)

Figure 17.3   Label the features of (a) the anterior surface and (b) the posterior surface of the right femur. 2

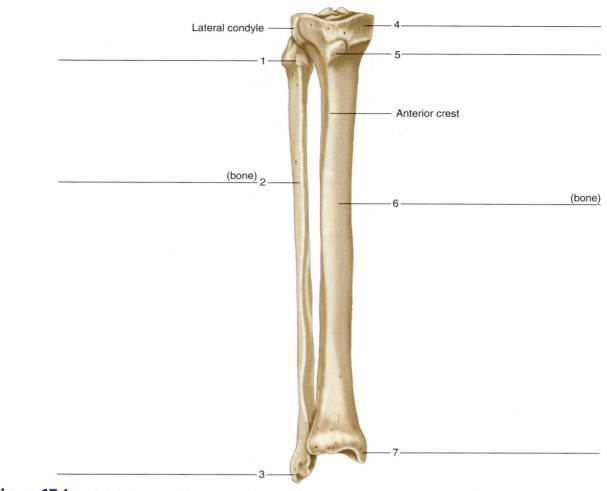

Lateral condyle

Anterior crest

(bone)

(bone)

Figure 17.4 Label the bones and features of the right tibia and fibula in this anterior view.

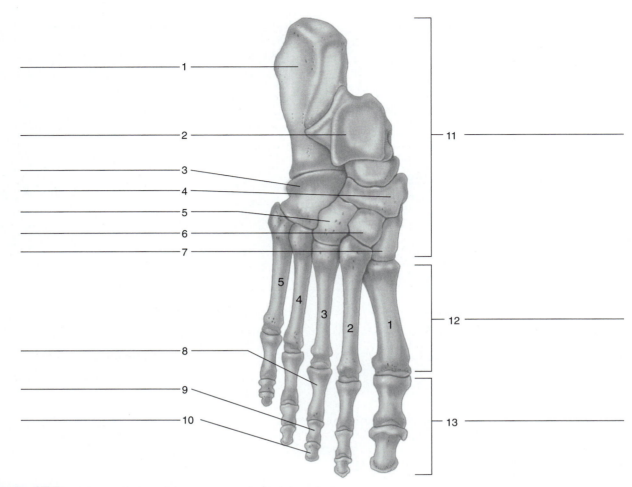

Figure 17.5 Label the bones of the superior surface of the right foot. ◀**2**

Name _____

Date _____

Section _____

The ⬅ corresponds to the indicated outcome(s) found at the beginning of the laboratory exercise.

Pelvic Girdle and Lower Limb

Part A Assessments

Complete the following statements:

1. The pelvic girdle consists of two _____ . ⬅**1**

2. The head of the femur articulates with the _____ of the hip bone. ⬅**1**

3. The _____ is the largest portion of the hip bone. ⬅**1**

4. The pubic bones come together anteriorly to form the joint called the _____ . ⬅**1**

5. The _____ is the portion of the ilium that causes the prominence of the hip. ⬅**1**

6. When a person sits, the _____ of the ischium supports the weight of the body. ⬅**1**

7. The angle formed by the pubic bones below the symphysis pubis is called the _____ . ⬅**1**

8. The _____ is the largest foramen in the skeleton. ⬅**1**

9. The ilium joins the sacrum at the _____ joint. ⬅**1**

Part B Assessments

Match the bones in column A with the features in column B. Place the letter of your choice in the space provided. ⬅**2**

Column A	Column B
a. Femur	_____ 1. Middle phalanx
b. Fibula	_____ 2. Lesser trochanter
c. Metatarsals	_____ 3. Medial malleolus
d. Patella	_____ 4. Fovea capitis
e. Phalanges	_____ 5. Calcaneus
f. Tarsals	_____ 6. Lateral cuneiform
g. Tibia	_____ 7. Tibial tuberosity
	_____ 8. Talus
	_____ 9. Cuboid
	_____ 10. Lateral malleolus
	_____ 11. Located in a tendon over the knee
	_____ 12. Five bones that form the instep

Part C Assessments

Identify the bones and features indicated in the radiographs of figures 17.6, 17.7, and 17.8. ◀**1** ◀**2**

Terms:
Head of femur
Ilium
Obturator foramen
Pubis
Sacrum
Symphysis pubis

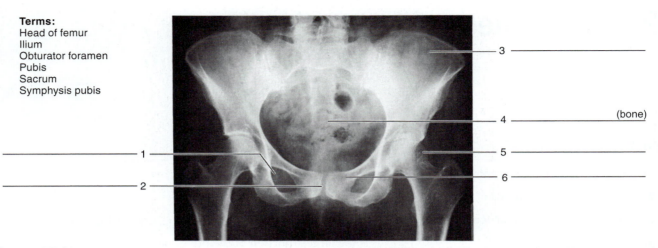

3 _____

4 _____ (bone)

1 _____

5 _____

2 _____

6 _____

Figure 17.6 Using the terms provided, identify the bones and features indicated on this radiograph of the anterior view of the pelvic region.

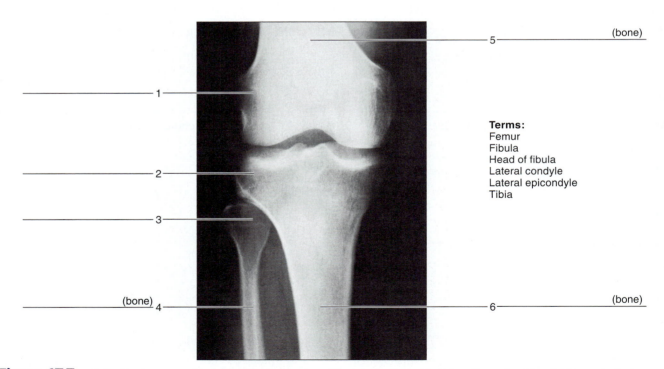

5 _____ (bone)

1 _____

Terms:
Femur
Fibula
Head of fibula
Lateral condyle
Lateral epicondyle
Tibia

2 _____

3 _____

(bone) 4 _____

6 _____ (bone)

Figure 17.7 Using the terms provided, identify the bones and features indicated in this radiograph of the right knee (anterior view).

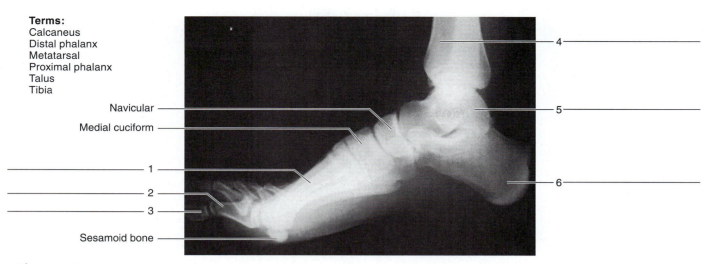

Terms:
Calcaneus
Distal phalanx
Metatarsal
Proximal phalanx
Talus
Tibia

Navicular —————
Medial cuciform —————

————— 1
————— 2
————— 3

Sesamoid bone —————

4
5
6

Figure 17.8   Using the terms provided, identify the bones indicated on this radiograph of the right foot (medial side).

Part D Assessments

Identify the bones of the foot in figure 17.9. ◀**1**

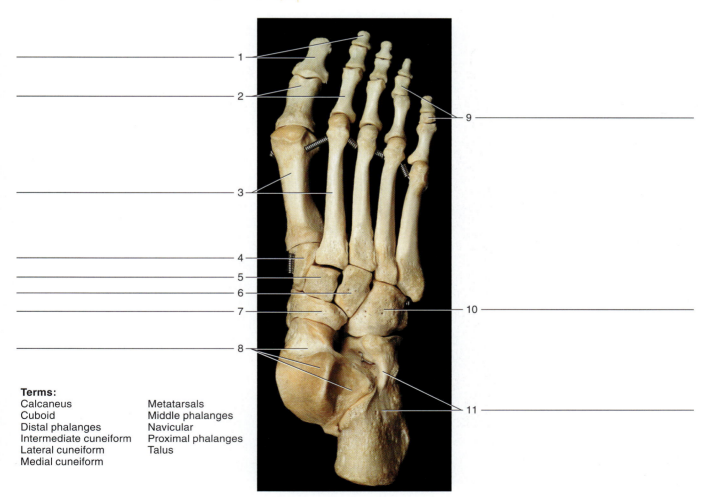

————— 1
————— 2

————— 3

————— 4
————— 5
————— 6
————— 7

————— 8

9

10

11

Terms:
Calcaneus   Metatarsals
Cuboid   Middle phalanges
Distal phalanges   Navicular
Intermediate cuneiform   Proximal phalanges
Lateral cuneiform Talus
Medial cuneiform

Figure 17.9   Using the terms provided, identify the bones indicated on this superior view of the right foot.

Laboratory Exercise 18

Joint Structure and Movements

Materials Needed

Textbook
Human skull
Human skeleton, articulated
Model of knee joint

For Demonstration:
Fresh animal joint (knee joint preferred)
Radiographs of major joints

⚠ Safety

- Wear disposable gloves when handling the fresh animal joint.
- Wash your hands before leaving the laboratory.

Joints are junctions between bones. Although they vary considerably in structure, they can be classified according to the type of tissue that binds the bones together. Thus, the three groups of joints can be identified as fibrous joints, cartilaginous joints, and synovial joints. Fibrous joints are filled with dense fibrous connective tissue, cartilaginous joints are filled with a type of cartilage, and synovial joints contain synovial fluid inside a joint cavity. Joints can also be classified by the degree of movement allowed: synarthroses are immovable, amphiarthroses allow slight movement, and diarthroses allow free movement.

Movements occurring at freely movable synovial joints are due to the contractions of skeletal muscles. In each case, the type of movement depends on the type of joint involved and the way in which the muscles are attached to the bones on either side of the joint.

Purpose of the Exercise

To examine examples of the three types of joints, to identify the major features of these joints, and to review the types of movements produced at synovial joints.

Learning Outcomes

After completing this exercise, you should be able to

1. Distinguish structural and functional characteristics among fibrous, cartilaginous, and synovial joints.
2. Locate examples of each type of joint.
3. Locate examples of each of the six types of synovial joints.
4. Demonstrate the types of movements that occur at synovial joints.
5. Examine the structure of the knee joint.

Procedure A—Types of Joints

1. Review the section entitled "Joints" in chapter 7 of the textbook.
2. Study the knee joint model and locate the structures indicated in figure 18.1.

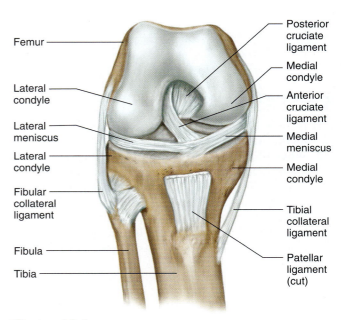

Femur
Lateral condyle
Lateral meniscus
Lateral condyle
Fibular collateral ligament
Fibula
Tibia

Posterior cruciate ligament
Medial condyle
Anterior cruciate ligament
Medial meniscus
Medial condyle
Tibial collateral ligament
Patellar ligament (cut)

Figure 18.1 Anterior view of right knee (patella removed).

3. Complete Part A of Laboratory Report 18.
4. Examine the skull and locate examples of *fibrous joints* as described in the textbook.
5. Locate and examine examples of *cartilaginous joints* in the skeleton.
6. Locate examples of the following types of *synovial joints* in the skeleton and in your body. Experiment with each joint to experience its range of movements.

   ball-and-socket joint

   condyloid (ellipsoidal) joint

   gliding (plane) joint

   hinge joint

   pivot joint

   saddle joint

7. Complete Parts B and C of the laboratory report.

Demonstration

Examine a longitudinal section of a fresh synovial animal joint. Locate the dense connective tissue that forms the joint capsule and the hyaline cartilage that forms the articular cartilage on the ends of the bones. Locate the synovial membrane on the inside of the joint capsule. Does the joint have any semilunar cartilages (menisci)?

_____

_____

What is the function of such cartilages? _____

_____

_____

Procedure B—Joint Movements

1. Review the section entitled "Types of Joint Movements" in chapter 7 of the textbook.
2. When the body is in anatomical position, most joints are extended and/or adducted. Skeletal muscle action involves the movable end (*insertion*) being pulled toward the stationary end (*origin*). In the limbs, the origin is usually proximal to the insertion; in the trunk, the origin is usually medial to the insertion. Use these concepts as reference points as

you move joints. Move various parts of your body to demonstrate the following joint movements:

flexion

extension

   hyperextension

dorsiflexion

plantar flexion

abduction

adduction

rotation

   medial (internal)

   lateral (external)

circumduction

supination

pronation

eversion (called pronation in some health professions)

inversion (called supination in some health professions)

protraction

retraction

elevation

depression

3. Have your laboratory partner do some of the preceding movements and see if you can correctly identify the movements. ◂4

Critical Thinking Application

Describe a body position that can exist when all major body parts are flexed.

_____

_____

_____

4. Complete Part D of the laboratory report.

Demonstration

Study the available radiographs of joints by holding them in front of a light source. Identify the type of joint and the bones incorporated in the joint. Also identify other major visible features.

The ◄ corresponds to the indicated outcome(s) found at the beginning of the laboratory exercise.

Joint Structure and Movements

Part A Assessments

Complete the following statements:

1. A _____ of the skull is an example of an immovable fibrous joint. ◄2

2. The joint between the first rib and the sternum is _____ . ◄2

3. The joint at the distal tibia and fibula is _____ . ◄2

4. Intervertebral discs are composed of _____ . ◄1

5. _____ are the most common type of joint in the human body. ◄1

6. _____ acts as a joint lubricant in synovial joints. ◄1

7. The discs of fibrocartilage between the articulating surfaces of the knee are called _____ . ◄5

8. The fluid-filled sacs sometimes associated with synovial joints are called _____ .

9. The symphysis pubis is an example of a(n) _____ joint. ◄2

10. Articular cartilage is composed of _____ tissue. ◄1

Part B Assessments

Match the types of synovial joints in column A with the examples in column B. Place the letter of your choice in the space provided. ◄3

Column A	Column B
a. Ball-and-socket	_____ 1. Hip joint
b. Condyloid (ellipsoidal)	_____ 2. Metacarpal-phalanx
c. Gliding (plane)	_____ 3. Proximal radius-ulna
d. Hinge	_____ 4. Humerus-ulna of the elbow joint
e. Pivot	_____ 5. Phalanx-phalanx
f. Saddle	_____ 6. Shoulder joint
	_____ 7. Tarsal-tarsal
	_____ 8. Carpal-metacarpal of the thumb
	_____ 9. Carpal-carpal

Part C Assessments

Identify the types of joints numbered in figure 18.2. ◀2 ◀3

1. _____
2. _____
3. _____
4. _____
5. _____

6. _____
7. _____
8. _____
9. _____

Part D Assessments

Identify the types of joint movements numbered in figure 18.3. ◀4

1. _____ (of head)
2. _____ (of shoulder)
3. _____ (of shoulder)
4. _____ (of hand)
5. _____ (of hand)
6. _____ (of arm at shoulder)
7. _____ (of arm at shoulder)
8. _____ (of hand at wrist)
9. _____ (of hand at wrist)
10. _____ (of thigh at hip)
11. _____ (of thigh at hip)
12. _____ (of lower limb at hip)
13. _____ (of chin/mandible)
14. _____ (of chin/mandible)

15. _____ (of vertebral column/trunk)
16. _____ (of vertebral column/trunk)
17. _____ (of head and neck)
18. _____ (of head and neck)
19. _____ (of arm at shoulder)
20. _____ (of arm at shoulder)
21. _____ (of forearm at elbow)
22. _____ (of forearm at elbow)
23. _____ (of thigh at hip)
24. _____ (of thigh at hip)
25. _____ (of leg at knee)
26. _____ (of leg at knee)
27. _____ (of foot at ankle)
28. _____ (of foot at ankle)

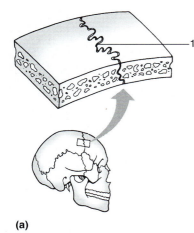

(a)

1

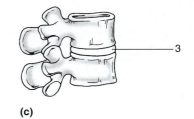

2

(b)

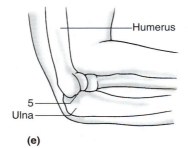

3

(c)

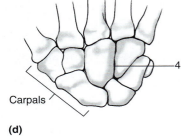

Carpals

4

(d)

Humerus

5

Ulna

(e)

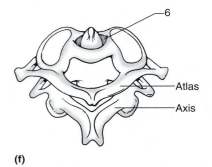

6

Atlas

Axis

(f)

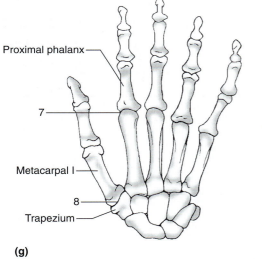

Proximal phalanx

7

Metacarpal I

8

Trapezium

(g)

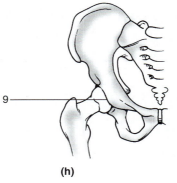

9

(h)

Figure 18.2   Identify the types of joints numbered in these illustrations.

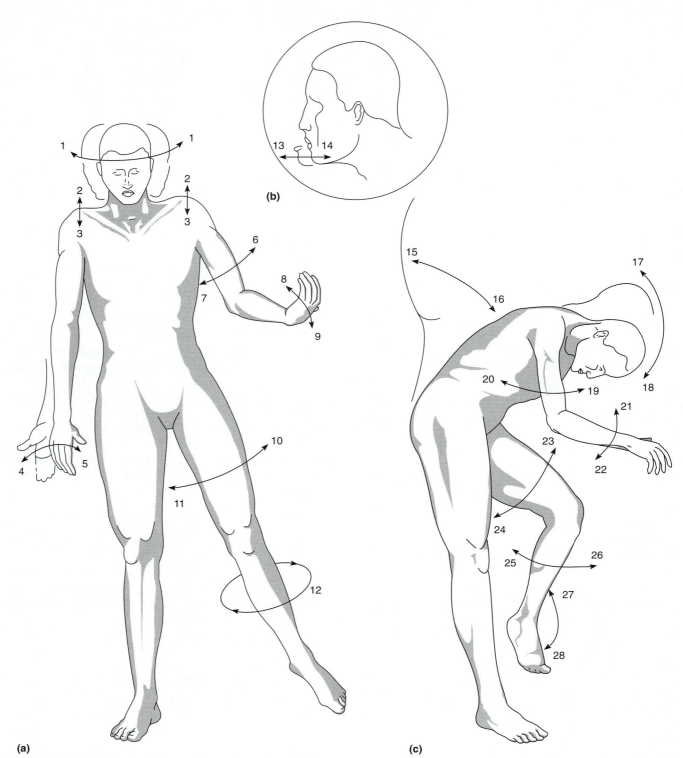

(a)

(b)

(c)

Figure 18.3　Identify each of the types of movements numbered and illustrated: (*a*) anterior view; (*b*) lateral view of head; (*c*) lateral view. (Original illustration drawn by Ross Martin)

Laboratory Exercise 19

Skeletal Muscle Structure

Materials Needed

Textbook
Compound light microscope
Prepared microscope slide of skeletal muscle tissue
Torso with musculature
Model of skeletal muscle fiber

For Demonstration:
Fresh round beefsteak

⚠ Safety

- Wear disposable gloves when handling the fresh beefsteak.
- Wash your hands before leaving the laboratory.

A skeletal muscle represents an organ of the muscular system and is composed of several types of tissues. These tissues include skeletal muscle tissue, nervous tissue, blood, and various connective tissues.

Each skeletal muscle is surrounded by connective tissue called fascia. The connective tissue often extends beyond the end of a muscle, providing an attachment to other muscles or to bones. Connective tissue also extends into the structure of a muscle and separates it into compartments. The outer layer, called epimysium, extends deeper into the muscle as perimysium, containing bundles of cells (fascicles), and farther inward extends around each muscle cell (fiber) as a thin endomysium. Some of the collagen fibers are continuous with the tendon, periosteum, and inward extensions as bone fibers, allowing a strong structural continuity.

Muscles are named according to their location, size, action, shape, attachments, or the direction of the fibers. Examples of how muscles are named include: gluteus maximus (location and size); adductor longus (action and shape); sternocleidomastoid (attachments); and orbicularis oculi (direction of fibers and location).

Purpose of the Exercise

To review the structure of a skeletal muscle.

Learning Outcomes

After completing this exercise, you should be able to

1. Locate the major structures of a skeletal muscle fiber on a microscope slide of skeletal muscle tissue and on a model.
2. Arrange how connective tissue is associated with muscle tissue within a skeletal muscle.
3. Distinguish between the origin and insertion of a muscle.
4. Describe and demonstrate the general actions of prime movers (agonists), synergists, and antagonists.

Procedure—Skeletal Muscle Structure

1. Review the section entitled "Skeletal Muscle Tissue" in chapter 5 of the textbook.
2. Reexamine the microscopic structure of skeletal muscle (fig. 19.1) by observing a prepared microscope slide of this tissue. Locate the following features:

   skeletal muscle fiber (cell)
   nuclei
   striations (alternating light and dark)

3. Review the section entitled "Structure of a Skeletal Muscle" in chapter 8 of the textbook.
4. As a review activity, label figure 19.2 and study figure 19.3.
5. Examine the human torso model and locate examples of fascia, tendons, and aponeuroses. Locate examples of tendons in your body.
6. Complete Part A of Laboratory Report 19.

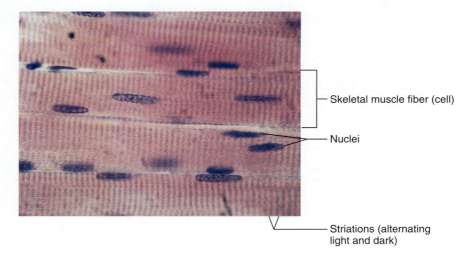

Skeletal muscle fiber (cell)

Nuclei

Striations (alternating light and dark)

Figure 19.1   Structures found in skeletal muscle fibers (cells) (250× micrograph enlarged to 700×).

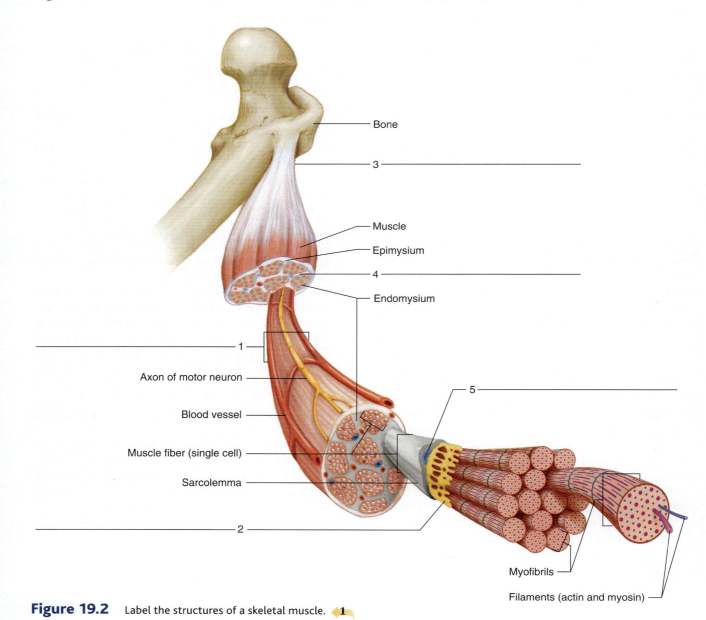

Bone

3

Muscle

Epimysium

4

Endomysium

Axon of motor neuron

1

Blood vessel

5

Muscle fiber (single cell)

Sarcolemma

2

Myofibrils

Filaments (actin and myosin)

Figure 19.2   Label the structures of a skeletal muscle.   1

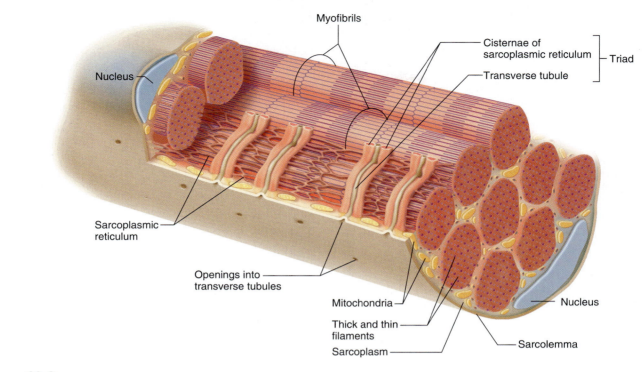

Myofibrils

Cisternae of
sarcoplasmic reticulum — Triad

Nucleus

Transverse tubule

Sarcoplasmic
reticulum

Openings into
transverse tubules

Mitochondria

Nucleus

Thick and thin
filaments

Sarcolemma

Sarcoplasm

Figure 19.3 Structures of a skeletal muscle fiber.

7. Examine the model of the skeletal muscle fiber and locate the following:

sarcolemma

sarcoplasm

myofibril

    thick (myosin) filament

    thin (actin) filament

sarcomere

    A band (dark)

    I band (light)

    Z line (disc)

sarcoplasmic reticulum

    cisternae

transverse tubules

8. Complete Part B of the laboratory report.
9. Review the section entitled "Skeletal Muscle Actions" in chapter 8 of the textbook.
10. Locate the biceps brachii muscle and its origin and insertion in the human torso model and in your body.
11. Make various movements with your upper limb at the shoulder and elbow. For each movement, determine the location of the muscles functioning as prime movers (agonists) and as antagonists. **(Remember, when a prime mover contracts (shortens) and a joint moves, its antagonist relaxes (lengthens).)** Synergist muscles often assist the contraction force of a prime mover, or by acting as fixators, might also stabilize nearby joints. Antagonistic muscle pairs pull from opposite sides of the same body region. The role of a muscle as a prime mover, antagonist, or a synergist depends upon the movement under consideration, as their roles can be changed.
12. Complete Part C of the laboratory report.

Notes

Name _____

Date _____

Section _____

The ⬅ corresponds to the indicated outcome(s) found at the beginning of the laboratory exercise.

Skeletal Muscle Structure

Part A Assessments

Match the terms in column A with the definitions in column B. Place the letter of your choice in the space provided. ⬅**1** ⬅**2**

Column A	Column B
a. Endomysium	_____ **1.** Membranous channel extending inward from muscle fiber membrane
b. Epimysium	
c. Fascia	_____ **2.** Cytoplasm of a muscle fiber
d. Fascicle	_____ **3.** Network of connective tissue that extends throughout the muscular system
e. Myosin	
f. Perimysium	_____ **4.** Layer of connective tissue that separates a muscle into small bundles called fascicles
g. Sarcolemma	
h. Sarcomere	_____ **5.** Cell membrane of a muscle fiber
i. Sarcoplasm	
j. Sarcoplasmic reticulum	_____ **6.** Layer of connective tissue that surrounds a skeletal muscle
k. Tendon	
l. Transverse (T) tubule	_____ **7.** Unit of alternating light and dark striations between Z lines

_____ **8.** Layer of connective tissue that surrounds an individual muscle fiber

_____ **9.** Cellular organelle in muscle fiber corresponding to the endoplasmic reticulum

_____ **10.** Cordlike part that attaches a muscle to a bone

_____ **11.** Protein found within thick myofibril

_____ **12.** A small bundle of muscle fibers

Part B Assessments

Provide the labels for the electron micrograph in figure 19.4. ⬅**1**

1. _____

2. _____

3. _____

4. _____

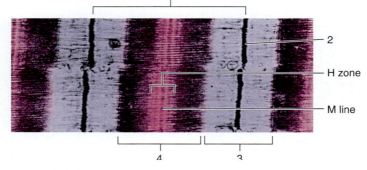

Terms:
A band (dark)
I band (light)
Sarcomere
Z line

1

2

H zone

M line

4   3

Figure 19.4 Using the terms provided, identify the bands and lines of the striations in this transmission electron micrograph of relaxed sarcomeres (16,000×).

Part C Assessments

Complete the following statements:

1. The _____ of a muscle is usually attached to a fixed location. **◀3**

2. The _____ of a muscle is usually attached to a movable location. **◀3**

3. The name biceps means _____ .

4. The forearm is flexed at the elbow when the _____ muscle contracts. **◀4**

5. A muscle responsible for most of a movement is called a(n) _____ . **◀4**

6. Assisting muscles are called _____ . **◀4**

7. Antagonists are muscles that resist the actions of _____ and cause movement in the opposite direction. **◀4**

8. When the forearm is extended at the elbow joint, the _____ muscle acts as the prime mover (agonist). **◀4**

Muscles of the Face, Head, and Neck

Materials Needed

Textbook
Human torso model with musculature
Human skull
Human skeleton, articulated

For Learning Extension:
Long rubber bands

The skeletal muscles of the face and head include the muscles of facial expression, which lie just beneath the skin; the muscles of mastication, attached to the mandible; and the muscles that move the head, located in the neck.

Purpose of the Exercise

To review the locations, actions, origins, and insertions of the muscles of the face, head, and neck.

Learning Outcomes

After completing this exercise, you should be able to

1. Locate and identify the muscles of facial expression, the muscles of mastication, and the muscles that move the head.
2. Describe and demonstrate the action of each of these muscles.
3. Locate the origin and insertion of each of these muscles in a human skeleton and the musculature of the human torso model.

Procedure—Muscles of the Face, Head, and Neck

1. Review the sections entitled "Muscles of Facial Expression," "Muscles of Mastication," and "Muscles That Move the Head" in chapter 8 of the textbook.
2. As a review activity, label figures 20.1 and 20.2.
3. Locate the following muscles in the human torso model and in your body whenever possible:

epicranius (frontalis and occipitalis)
orbicularis oculi
orbicularis oris
buccinator
zygomaticus major
zygomaticus minor
platysma
masseter
temporalis
sternocleidomastoid
splenius capitis
semispinalis capitis

4. Demonstrate the actions of these muscles in your body.
5. Locate the origins and insertions of these muscles in the human skull and skeleton.
6. Complete Parts A, B, and C of Laboratory Report 20.

Learning Extension

A long rubber band can be used to simulate muscle locations, actions, origins, and insertions on the human torso model, the skeleton, or a laboratory partner. Hold one end of the rubber band firmly on the origin location of a muscle, then slightly stretch the rubber band and hold the other end on the insertion site. Allow the insertion end to slowly move toward the origin end to simulate the contraction and action of the muscle.

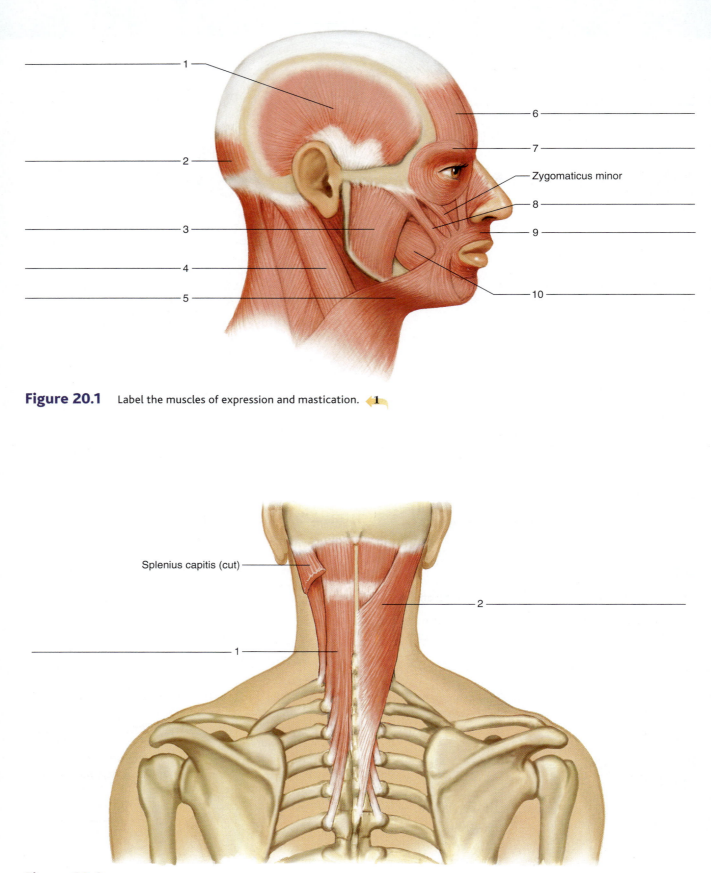

Figure 20.1   Label the muscles of expression and mastication. 1

Zygomaticus minor

Figure 20.2   Label these deep muscles of the posterior neck (trapezius removed). 1

Splenius capitis (cut)

The ◄ corresponds to the indicated outcome(s) found at the beginning of the laboratory exercise.

Muscles of the Face, Head, and Neck

Part A Assessments

Complete the following statements:

1. When the _____ contracts, the corner of the mouth is drawn upward. ◄2

2. The _____ acts to compress the wall of the cheeks when air is blown out of the mouth. ◄2

3. The _____ causes the lips to protrude or close. ◄2

4. The temporalis acts to _____ . ◄2

5. The _____ can close the eye, as in blinking. ◄2

6. The _____ can pull the head toward the chest. ◄2

7. The _____ can pull the corners of the mouth downward, as when pouting. ◄2

Part B Assessments

Name the muscle indicated by the following combinations of origin and insertion. ◄3

Origin	Insertion	Muscle
1. Occipital bone	Skin and muscle around eye	_____
2. Zygomatic bone	Orbicularis oris	_____
3. Zygomatic arch	Lateral surface of mandible	_____
4. Anterior surface of sternum and upper clavicle	Mastoid process of temporal bone	_____
5. Outer surfaces of mandible and maxilla	Orbicularis oris	_____
6. Fascia in upper chest	Lower border of mandible and skin around corner of mouth	_____
7. Temporal bone	Coronoid process and lateral surface of mandible	_____
8. Spinous processes of cervical and thoracic vertebrae	Mastoid process of temporal bone	_____
9. Processes of cervical and thoracic vertebrae	Occipital bone	_____

Part C Assessments

Critical Thinking Application

Identify the muscles of various facial expressions in the photographs of figure 20.3. ◀**1**

1. _____

2. _____

3. _____

4. _____

5. _____

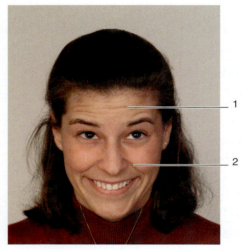

(a)

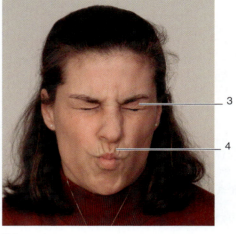

(b)

Terms:
Epicranius/frontalis
Orbicularis oculi
Orbicularis oris
Platysma
Zygomaticus

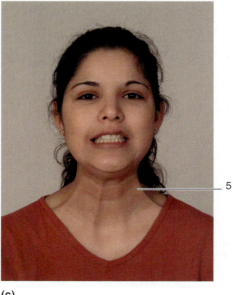

(c)

Figure 20.3   Using the terms provided, identify the muscles of facial expression being contracted in each of these photographs (a–c).

Muscles of the Chest, Shoulder, and Upper Limb

Materials Needed

Textbook
Human torso model
Human skeleton, articulated
Muscular models of the upper limb

For Learning Extension:
Long rubber bands

The muscles of the chest and shoulder are responsible for moving the scapula and arm, whereas those within the arm and forearm move joints in the elbow and hand.

Purpose of the Exercise

To review the locations, actions, origins, and insertions of the muscles in the chest, shoulder, and upper limb.

Learning Outcomes

After completing this exercise, you should be able to

1. Locate and identify the muscles of the chest, shoulder, and upper limb.

2. Describe and demonstrate the action of each of these muscles.

3. Locate the origin and insertion of each of these muscles in a human skeleton and on the muscular models.

Procedure—Muscles of the Chest, Shoulder, and Upper Limb

1. Review the sections entitled "Muscles That Move the Pectoral Girdle," "Muscles That Move the Arm," "Muscles That Move the Forearm," and "Muscles That Move the Hand" in chapter 8 of the textbook.

2. As a review activity, label figures 21.1, 21.2, 21.3, and 21.4.

3. Locate the following muscles in the human torso model and models of the upper limb. Also locate in your body as many of the muscles as you can.

muscles that move the pectoral girdle
    anterior muscles
        pectoralis minor
        serratus anterior
    posterior muscles
        trapezius
        rhomboid major
        rhomboid minor
        levator scapulae
muscles that move the arm
    origins on axial skeleton
        pectoralis major
        latissimus dorsi
    origins on scapula
        deltoid
        teres major
        coracobrachialis
    rotator cuff (SITS) muscles (origins also on scapula)
        supraspinatus
        infraspinatus
        teres minor
        subscapularis
muscles that move the forearm
    muscle bellies in arm
        biceps brachii
        brachialis
        triceps brachii
    muscle bellies in forearm
        brachioradialis
        supinator
        pronator teres
        pronator quadratus

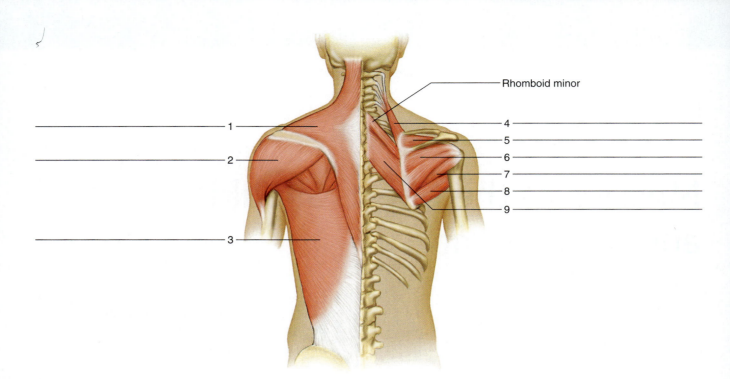

Figure 21.1 Label the muscles of the posterior shoulder. Superficial muscles are illustrated on the left side and deep muscles on the right side. ◄1

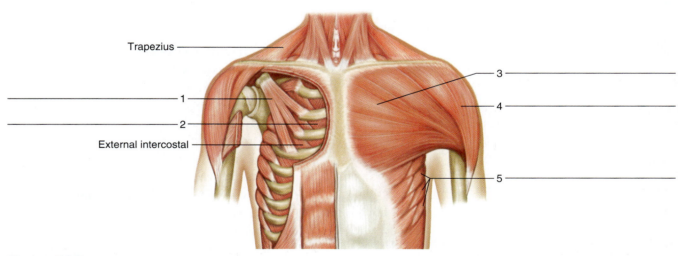

Figure 21.2 Label the muscles of the anterior chest. Superficial muscles are illustrated on the left side and deep muscles on the right side. ◄1

muscles that move the hand

    anterior flexor muscles

        flexor carpi radialis

        flexor carpi ulnaris

        palmaris longus

        flexor digitorum profundus

posterior extensor muscles

    extensor carpi radialis longus

    extensor carpi radialis brevis

    extensor carpi ulnaris

    extensor digitorum

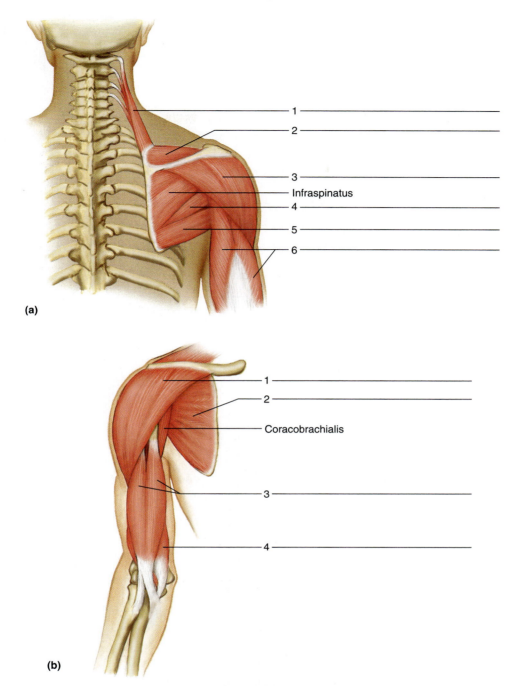

(a)

1
2
3
Infraspinatus
4
5
6

1
2
Coracobrachialis
3
4

(b)

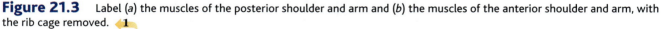

Figure 21.3 Label (a) the muscles of the posterior shoulder and arm and (b) the muscles of the anterior shoulder and arm, with the rib cage removed. ◀**1**

4. Demonstrate the actions of these muscles in your body.
5. Locate the origins and insertions of these muscles in the human skeleton.
6. Complete Parts A, B, and C of Laboratory Report 21.

Learning Extension

A long rubber band can be used to simulate muscle locations, actions, origins, and insertions on muscular models, the skeleton, or laboratory partner. Hold one end of the rubber band firmly on the origin location of a muscle, then slightly stretch the rubber band and hold the other end on the insertion site. Allow the insertion end to slowly move toward the origin end to simulate the contraction and action of the muscle. ◀**2**

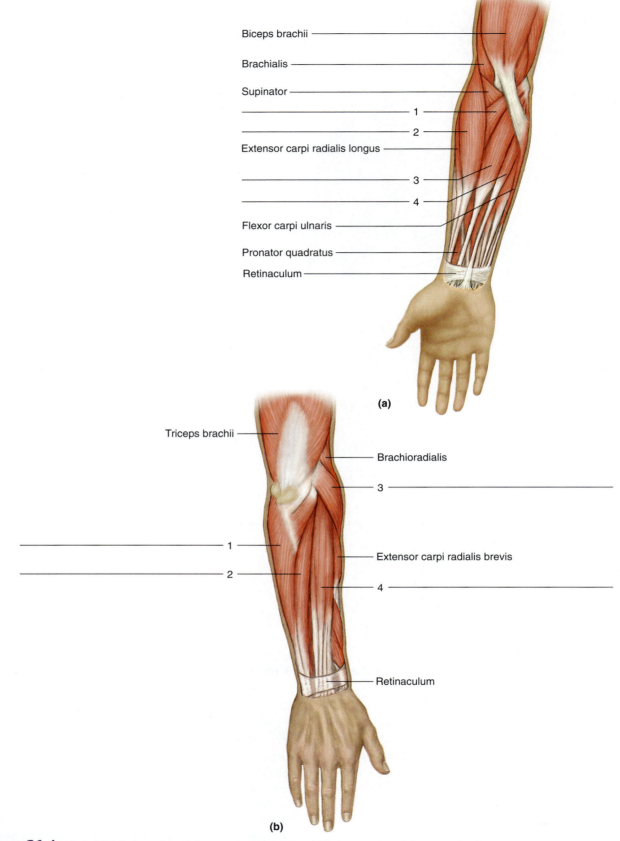

Biceps brachii

Brachialis

Supinator

_____ 1

_____ 2

Extensor carpi radialis longus

_____ 3

_____ 4

Flexor carpi ulnaris

Pronator quadratus

Retinaculum

(a)

Triceps brachii

Brachioradialis

3 _____

1 _____

2 _____

Extensor carpi radialis brevis

4 _____

Retinaculum

(b)

Figure 21.4   Label (*a*) the muscles of the anterior forearm and (*b*) the muscles of the posterior forearm.   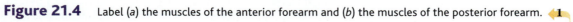 **1**

Name _____

Date _____

Section _____

The ◀ corresponds to the indicated outcome(s) found at the beginning of the laboratory exercise.

Muscles of the Chest, Shoulder, and Upper Limb

Part A Assessments

Match the muscles in column A with the actions in column B. Place the letter of your choice in the space provided. ◀2

Column A	Column B
a. Brachialis	_____ **1.** Abducts arm
b. Coracobrachialis	_____ **2.** Pulls arm forward and across chest
c. Deltoid	_____ **3.** Flexes and adducts hand at the wrist
d. Extensor carpi ulnaris	_____ **4.** Raises and adducts scapula
e. Flexor carpi ulnaris	_____ **5.** Raises ribs in forceful inhalation or pulls scapula forward and
f. Infraspinatus	downward
g. Pectoralis major	_____ **6.** Used to thrust shoulder anteriorly, as when pushing something
h. Pectoralis minor	_____ **7.** Flexes the forearm at the elbow
i. Rhomboid major	_____ **8.** Flexes and adducts arm at the shoulder along with pectoralis major
j. Serratus anterior	_____ **9.** Extends the forearm at the elbow
k. Teres major	_____ **10.** Extends, adducts, and rotates arm medially
l. Triceps brachii	_____ **11.** Extends and adducts hand at the wrist
	_____ **12.** Rotates arm laterally

Part B Assessments

Name the muscle indicated by the following combinations of origin and insertion. ◀3

Origin	Insertion	Muscle
1. Spines of upper thoracic vertebrae	Medial border of scapula	_____
2. Outer surfaces of upper ribs	Ventral surfaces of scapula	_____
3. Sternal ends of upper ribs	Coracoid process of scapula	_____
4. Coracoid process of scapula	Shaft of humerus	_____
5. Lateral border of scapula	Intertubercular groove of humerus	_____
6. Anterior surface of scapula	Lesser tubercle of humerus	_____
7. Lateral border of scapula	Greater tubercle of humerus	_____
8. Anterior shaft of humerus	Coronoid process of ulna	_____
9. Medial epicondyle of humerus and coronoid process of ulna	Lateral surface of radius	_____
10. Distal lateral end of humerus	Lateral surface of radius above styloid process	_____
11. Medial epicondyle of humerus	Base of second and third metacarpals	_____
12. Medial epicondyle of humerus	Fascia of palm	_____

Part C Assessments

Critical Thinking Application

Identify the muscles indicated in figure 21.5. ◄1

1. _____

2. _____

3. _____

4. _____

5. _____

6. _____

7. _____

8. _____

9. _____

10. _____

11. _____

12. _____

13. _____

14. _____

15. _____

16. _____

17. _____

18. _____

19. _____

20. _____

21. _____

Terms:
Biceps brachii
Deltoid
External oblique
Pectoralis major
Rectus abdominis
Serratus anterior
Sternocleidomastoid
Trapezius

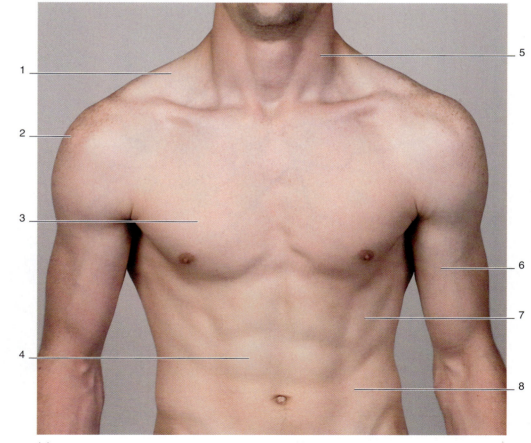

(a)

Figure 21.5   Using the terms provided, identify the muscles that appear as body surface features in these photographs (a, b, and c).

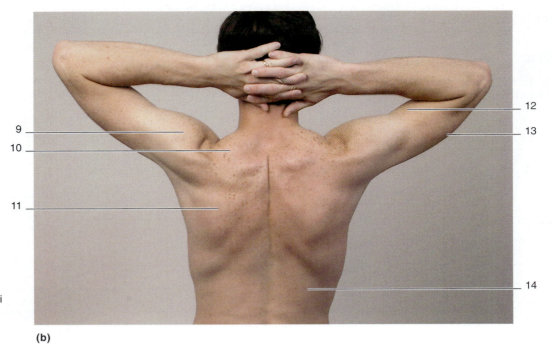

(b)

Terms:
Biceps brachii
Deltoid
Infraspinatus
Latissimus dorsi
Trapezius
Triceps brachii

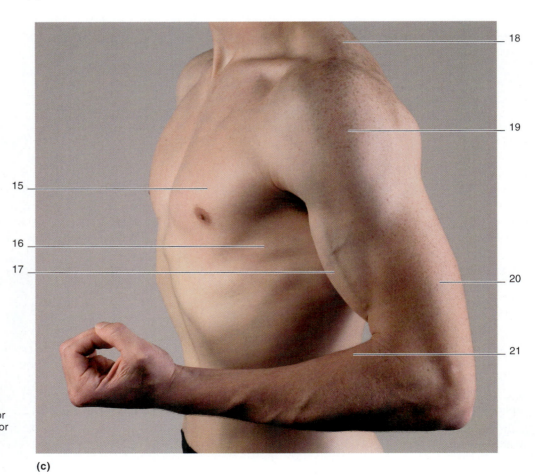

(c)

Terms:
Biceps brachii
Brachioradialis
Deltoid
Pectoralis major
Serratus anterior
Trapezius
Triceps brachii

Figure 21.5 *Continued.*

Laboratory Exercise 22

Muscles of the Abdominal Wall and Pelvic Outlet

Materials Needed

Textbook
Human torso model with musculature
Human skeleton, articulated
Muscular models of male and female pelves

The anterior and lateral walls of the abdomen contain broad, flattened muscles arranged in layers. These muscles connect the rib cage and vertebral column to the pelvic girdle. The abdominal wall muscles compress the abdominal visceral organs, help maintain posture, assist in forceful exhalation, and contribute in trunk flexion and waist rotation.

The muscles of the pelvic outlet are arranged in two muscular sheets: (1) a deeper pelvic diaphragm that forms the floor of the pelvic cavity and (2) a urogenital diaphragm that fills the space within the pubic arch. The pelvic floor is penetrated by the urethra, vagina, and anus in a female, thus are important in obstetrics.

Purpose of the Exercise

To review the actions, origins, and insertions of the muscles of the abdominal wall and pelvic outlet.

Learning Outcomes

After completing this exercise, you should be able to

1. Locate and identify the muscles of the abdominal wall and pelvic outlet.
2. Describe the action of each of these muscles.
3. Locate the origin and insertion of each of these muscles in a human skeleton or on muscular models.

Procedure—Muscles of the Abdominal Wall and Pelvic Outlet

1. Review the sections entitled "Muscles of the Abdominal Wall" and "Muscles of the Pelvic Outlet" in chapter 8 of the textbook.
2. As a review activity, label figures 22.1, 22.2, and 22.3.
3. Locate the following muscles in the human torso model:

   external oblique
   internal oblique
   transversus abdominis
   rectus abdominis

4. Demonstrate the actions of these muscles in your body.
5. Locate the origin and insertion of each of these muscles in the human skeleton. 3
6. Complete Part A of Laboratory Report 22.
7. Locate the following muscles in the models of the male and female pelves:

   levator ani
   superficial transversus perinei
   bulbospongiosus
   ischiocavernosus

8. Locate the origin and insertion of each of these muscles in the human skeleton. 3
9. Complete Part B of the laboratory report.

Critical Thinking Application

List the muscles from superficial to deep for an appendectomy incision. 1

_____

_____

_____

_____

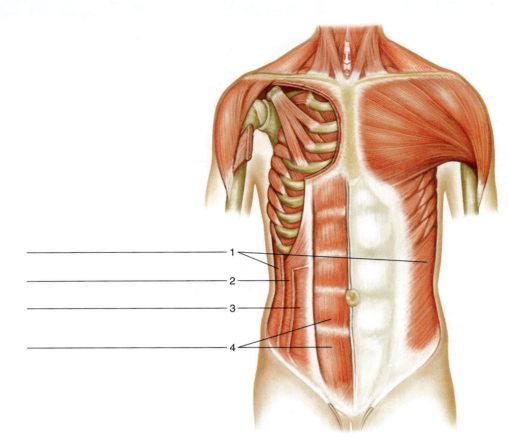

1 _____

2 _____

3 _____

4 _____

Figure 22.1   Label the muscles of the abdominal wall.

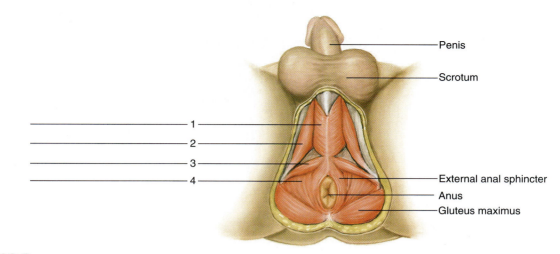

Penis

Scrotum

1

2

3

4

External anal sphincter

Anus

Gluteus maximus

Figure 22.2   Label the muscles of the male pelvic outlet. ◀1

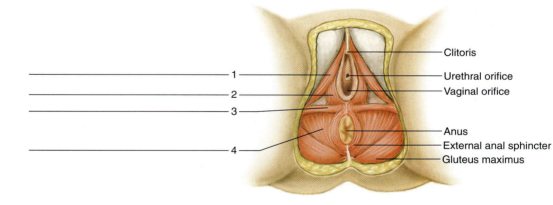

Clitoris

Urethral orifice

Vaginal orifice

1

2

3

Anus

External anal sphincter

Gluteus maximus

4

Figure 22.3   Label the muscles of the female pelvic outlet. ◀1

Name _____

Date _____

Section _____

The ⬅ corresponds to the indicated outcome(s) found at the beginning of the laboratory exercise.

Muscles of the Abdominal Wall and Pelvic Outlet

22

Part A Assessments

Complete the following statements:

1. A band of tough connective tissue in the midline of the anterior abdominal wall called the _____ serves as a muscle attachment. ⬅**1**

2. The _____ muscle spans from the ribs and sternum to the pubic bones. ⬅**3**

3. The _____ forms the third layer (deepest layer) of the abdominal wall muscles. ⬅**1**

4. The action of the external oblique muscle is to _____ . ⬅**2**

5. The action of the rectus abdominis is to _____ . ⬅**2**

Part B Assessments

Complete the following statements:

1. The levator ani forms the _____ diaphragm. ⬅**1**

2. The levator ani provides a sphincterlike action in the _____ . ⬅**2**

3. The action of the superficial transversus perinei is to _____ . ⬅**2**

4. The _____ assists in emptying the male urethra. ⬅**2**

5. In females, the bulbospongiosus acts to _____ . ⬅**2**

6. The ischiocavernosus extends from the margin of the pubic arch to the _____ . ⬅**3**

Notes

Laboratory Exercise 23

Muscles of the Hip and Lower Limb

Materials Needed

Textbook
Human torso model with musculature
Human skeleton, articulated
Muscular models of the lower limb

For Learning Extension:
Long rubber bands

The muscles that move the thigh are attached to the femur and to some part of the pelvic girdle. Those attached anteriorly primarily act to flex the thigh at the hip, whereas those attached posteriorly act to extend, abduct, or rotate the thigh.

The muscles that move the leg connect the tibia or fibula to the femur or to the pelvic girdle. They function to flex or extend the leg at the knee. Other muscles, located in the leg, act to move the foot.

Purpose of the Exercise

To review the actions, origins, and insertions of the muscles that move the thigh, leg, and foot.

Learning Outcomes

After completing this exercise, you should be able to

1. Locate and identify the muscles that move the thigh, leg, and foot.
2. Describe and demonstrate the actions of each of these muscles.
3. Locate the origin and insertion of each of these muscles in a human skeleton and on muscular models.

Procedure—Muscles of the Hip and Lower Limb

1. Review the sections entitled "Muscles That Move the Thigh," "Muscles That Move the Leg," and "Muscles That Move the Foot" in chapter 8 of the textbook.
2. As a review activity, label figures 23.1, 23.2, 23.3, 23.4, 23.5, and 23.6.

3. Locate the following muscles in the human torso model and in the lower limb models. Also locate as many of them as possible in your body.

muscles that move the thigh
 anterior hip muscles
  iliopsoas group
   psoas major
   iliacus
 posterior hip muscle
  gluteus maximus
 lateral hip muscles
  gluteus medius
  gluteus minimus
  tensor fasciae latae
 medial aductor muscles
  adductor longus
  adductor magnus
  gracilis
muscles that move the leg
 anterior thigh muscles
  sartorius
  quadriceps femoris group
   rectus femoris
   vastus lateralis
   vastus medialis
   vastus intermedius
 posterior thigh muscles
  hamstring group
   biceps femoris
   semitendinosus
   semimembranosus
muscles that move the foot
 anterior leg muscles
  tibialis anterior
  fibularis (peroneus) tertius
  extensor digitorum longus
 posterior leg muscles
  gastrocnemius
  soleus
  flexor digitorum longus
 lateral leg muscle
  fibularis (peroneus) longus

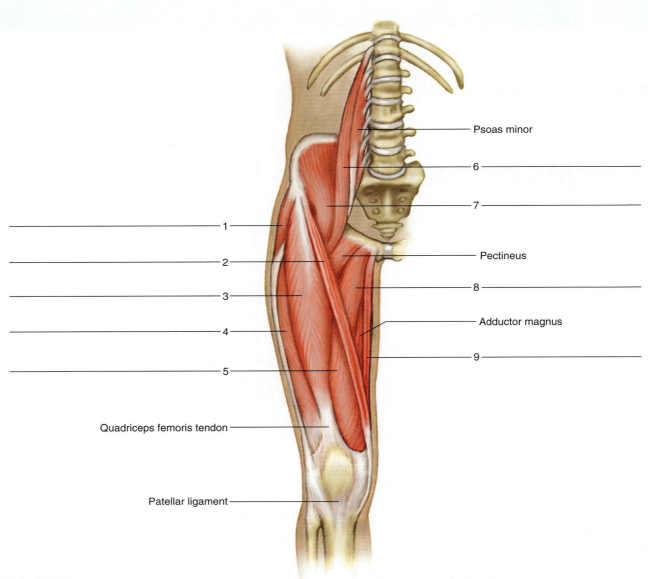

Psoas minor

6

7

Pectineus

8

Adductor magnus

9

1

2

3

4

5

Quadriceps femoris tendon

Patellar ligament

Figure 23.1 Label the muscles of the anterior right hip and thigh. ◀1

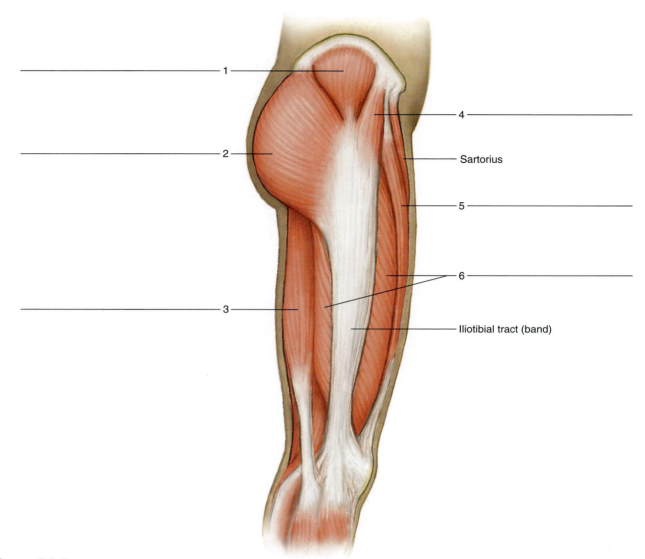

Sartorius

Iliotibial tract (band)

Figure 23.2 Label the muscles of the lateral right hip and thigh. ◀ **1**

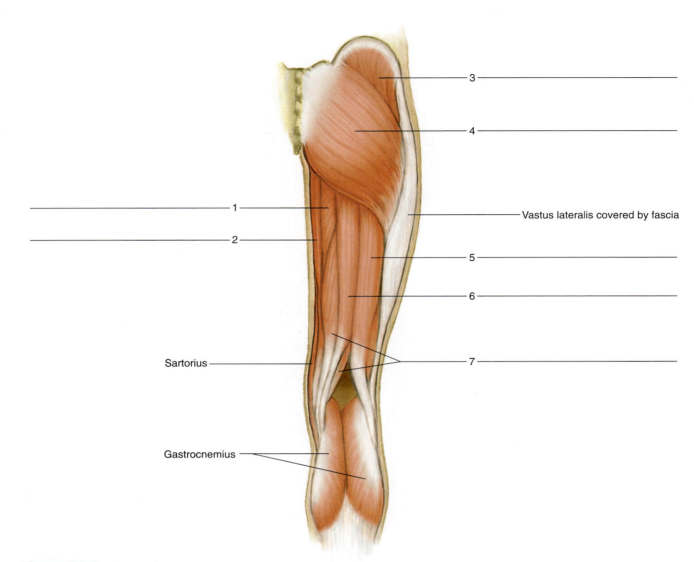

Figure 23.3 Label the muscles of the posterior right hip and thigh. ◀ **1**

Vastus lateralis covered by fascia

Sartorius

Gastrocnemius

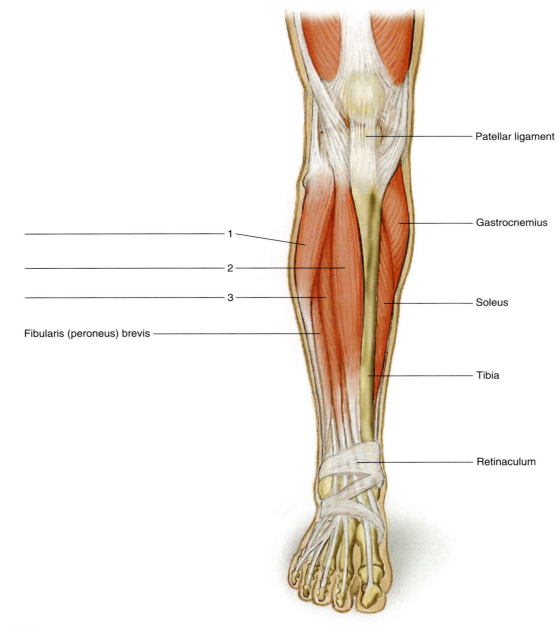

Patellar ligament

Gastrocnemius

Soleus

Tibia

Retinaculum

1

2

3

Fibularis (peroneus) brevis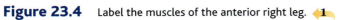

Figure 23.4 Label the muscles of the anterior right leg. 1

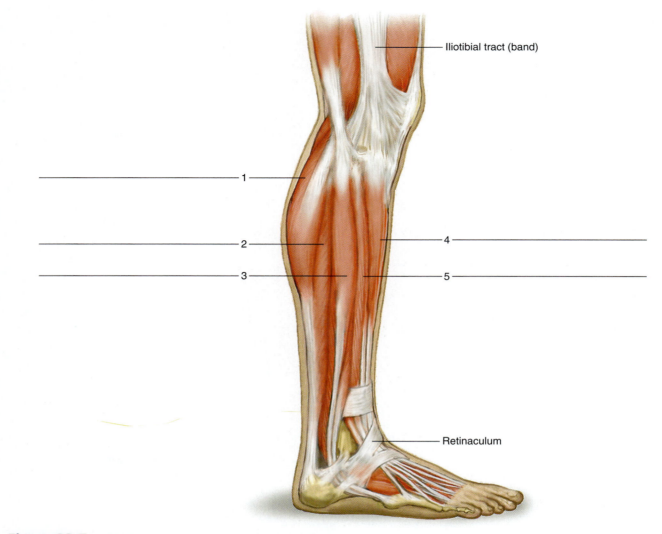

Iliotibial tract (band)

1

2

3

4

5

Retinaculum

Figure 23.5   Label the muscles of the lateral right leg.

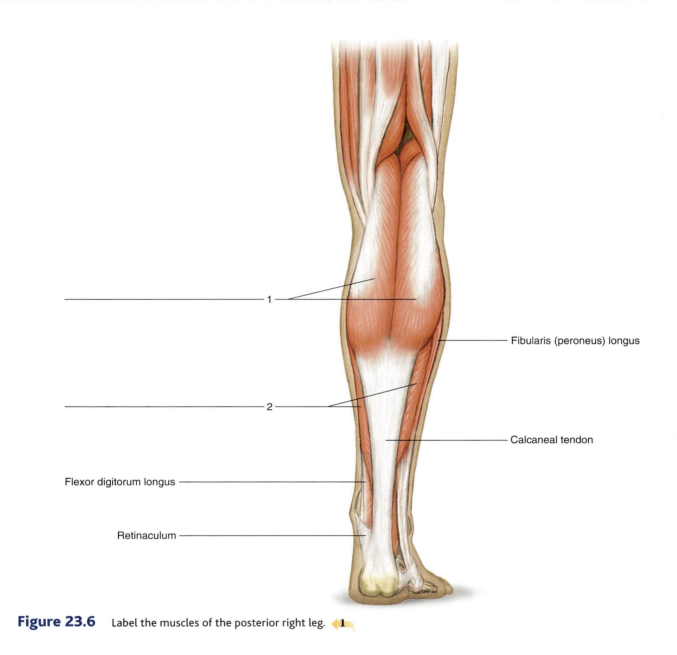

Figure 23.6 Label the muscles of the posterior right leg.

Labels on figure:
- 1
- 2
- Fibularis (peroneus) longus
- Calcaneal tendon
- Flexor digitorum longus
- Retinaculum

4. Demonstrate the action of each of these muscles in your body.
5. Locate the origin and insertion of each of these muscles in the human skeleton.
6. Complete Parts A, B, and C of Laboratory Report 23.

Learning Extension

A long rubber band can be used to simulate muscle locations, origins, insertions, and actions on muscular models, the skeleton, or on a laboratory partner. Hold one end of the rubber band firmly on the origin location of a muscle, then slightly stretch the rubber band and hold the other end on the insertion site. Allow the insertion end to slowly move toward the origin end to simulate the contraction and action of the muscle. ◀2

Notes

The ← corresponds to the indicated outcome(s) found at the beginning of the laboratory exercise.

Muscles of the Hip and Lower Limb

Part A Assessments

Match the muscles in column A with the actions in column B. Place the letter of your choice in the space provided. ←2

Column A

a. Biceps femoris
b. Fibularis (peroneus) longus
c. Gluteus medius
d. Gracilis
e. Psoas major and iliacus
f. Quadriceps femoris group
g. Sartorius
h. Tibialis anterior

Column B

_____ 1. Adducts thigh

_____ 2. Plantar flexion and eversion of foot

_____ 3. Flexes thigh at the hip

_____ 4. Abducts thigh and rotates it laterally

_____ 5. Abducts thigh and rotates it medially

_____ 6. Flexes the leg at the knee

_____ 7. Extends the leg at the knee

_____ 8. Dorsiflexion and inversion of foot

Part B Assessments

Name the muscle indicated by the following combinations of origin and insertion. ←3

Origin	Insertion	Muscle
1. Lateral surface of ilium	Greater trochanter of femur	_____
2. Ischial tuberosity	Posterior surface of femur	_____
3. Anterior superior iliac spine	Medial surface of proximal tibia	_____
4. Lateral and medial condyles of femur	Posterior surface of calcaneus	_____
5. Anterior iliac crest	Fascia (iliotibial tract) of the thigh	_____
6. Greater trochanter and posterior surface of femur	Patella to tibial tuberosity	_____
7. Ischial tuberosity	Medial surface of tibia	_____
8. Medial surface of femur	Patella to tibial tuberosity	_____
9. Posterior surface of tibia	Distal phalanges of four lateral toes	_____
10. Lateral condyle and lateral surface of tibia	Medial cuneiform and first metatarsal	_____

Part C Assessments

Critical Thinking Application

Identify the muscles indicated in figure 23.7. ◀**1**

1. _____

2. _____

3. _____

4. _____

5. _____

6. _____

7. _____

8. _____

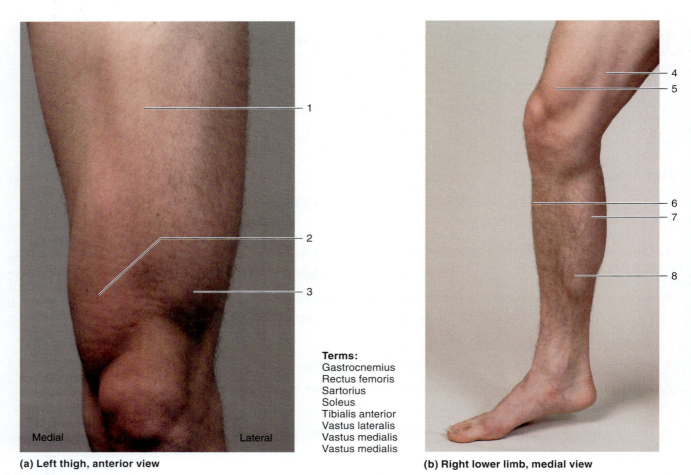

Terms:
Gastrocnemius
Rectus femoris
Sartorius
Soleus
Tibialis anterior
Vastus lateralis
Vastus medialis
Vastus medialis

Medial   Lateral

(a) Left thigh, anterior view   **(b) Right lower limb, medial view**

Figure 23.7   Using the terms provided, identify the muscles that appear as lower limb surface features in these photographs (*a* and *b*).

Laboratory Exercise

Surface Anatomy

Materials needed
Textbook
Small round stickers
Colored pencils (black and red)

External landmarks, called surface anatomy, located on the human body provide an opportunity to examine surface features and help locate other internal structures. This exercise will focus on bony landmarks and superficial soft tissue structures, primarily related to the skeletal and muscular systems. The technique used for the examination of surface features is called *palpation,* accomplished by touching with the hands or fingers. This enables the determination of the location, size, and texture of the structure. Some of the surface features can be visible without the need of palpation, especially on body builders and very thin individuals. If the individual involved has considerable subcutaneous adipose tissue, additional pressure might be necessary to palpate some of the structures.

These surface features will be of assistance to locate additional surface features and deeper structures during the study of the additional systems. Some of the respiratory structures in the anterior neck can be palpated. Surface features will be of help to determine pulse locations of superficial arteries and for proper assessments of injection sites; descriptions of pain and injury sites; insertion of examination tubes; and listening to heart, lung, and bowel sounds with a stethoscope. For those selecting careers in emergency medical services, nursing, physicians, physical education, physical therapists, chiropractors, and massage therapists, surface anatomy is especially valuable.

Purpose of the Exercise

To examine the surface features of the human body and the terms used to describe them.

Learning Outcomes

After completing this exercise you should be able to

1. Locate and identify major surface features of the human body.
2. Identify surface features by body region.
3. Distinguish between surface features that are bony landmarks and soft tissue.

Procedure—Surface Anatomy

1. Review textbook figures on body regions, human organism reference plates, skeletal system, and muscular system. These figures are located in chapters 1, 7, and 8.
2. Use the textbook figures and the other figures provided to complete figures 24.1 through 24.8. Many features have already been labeled. Features in boldface are bony features. Other labeled features (not boldface) are soft tissue features. Bony features are a part of a bone and the skeletal system. Soft tissue features are comprised of tissue other than bone such as muscle, connective tissue, or skin. The missing features are listed by each figure.
3. Complete Parts A and B of Laboratory Report 24.
4. Use figures 24.1 through 24.8 to determine which items listed in table 24.1 are bony features and which items are soft tissue features. Circle bony features with a black colored pencil. Circle soft tissue features with a red colored pencil. ◀3
5. Complete Part C of the laboratory report.

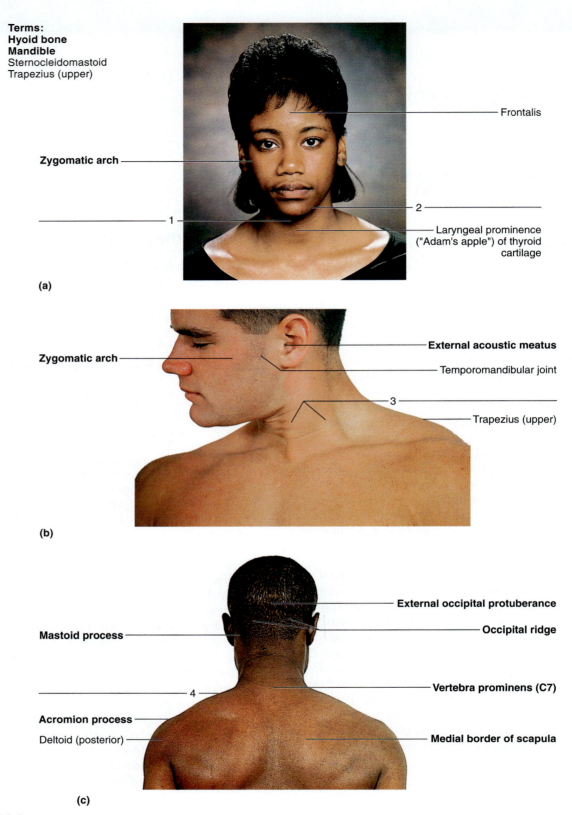

Frontalis

Zygomatic arch

1

2

Laryngeal prominence ("Adam's apple") of thyroid cartilage

(a)

Zygomatic arch

External acoustic meatus

Temporomandibular joint

3

Trapezius (upper)

(b)

External occipital protuberance

Occipital ridge

Mastoid process

Vertebra prominens (C7)

4

Acromion process

Deltoid (posterior)

Medial border of scapula

(c)

Figure 24.1   Using the terms provided, label the surface features of (*a*) the anterior head and neck, (*b*) the lateral head and neck, and (*c*) the posterior head and neck. (**Boldface** indicates bony features; not boldface indicates soft tissue features.) 3

Table 24.1   Representative bony and soft tissue surface features

Biceps brachii	Iliotibial tract (band)	Sacrum
Calcaneal tendon	Ischial tuberosity	Serratus anterior
Deltoid	Jugular (suprasternal) notch	Sternocleidomastoid
Distal interphalangeal joint	Lateral malleolus	Tibialis anterior
Gastrocnemius	Mandible	Vastus lateralis
Greater trochanter	Mastoid process	Zygomatic arch
Head of fibula	Medial border of scapula	
Iliac crest	Rectus abdominis	

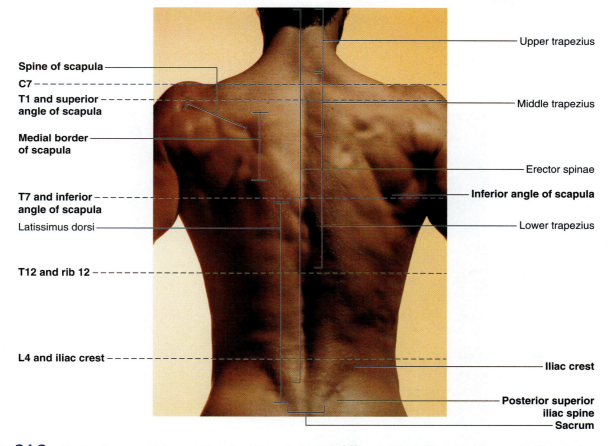

Figure 24.2   Surface features of the posterior shoulder and torso. (**Boldface** indicates bony features; not boldface indicates soft tissue features.)

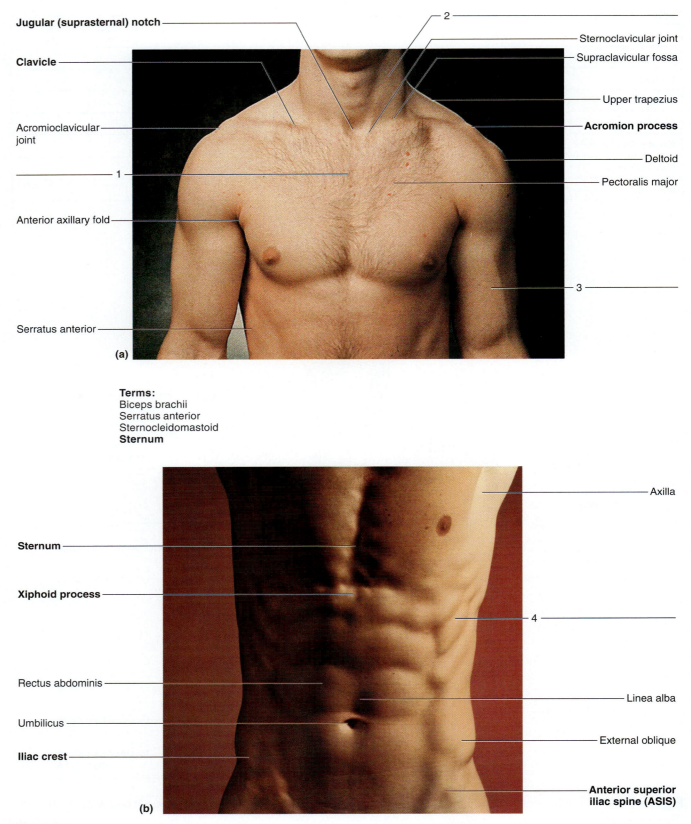

Jugular (suprasternal) notch

Clavicle

Acromioclavicular joint

Anterior axillary fold

Serratus anterior

(a)

2

Sternoclavicular joint

Supraclavicular fossa

Upper trapezius

Acromion process

Deltoid

Pectoralis major

1

3

Terms:
Biceps brachii
Serratus anterior
Sternocleidomastoid
Sternum

Sternum

Xiphoid process

Rectus abdominis

Umbilicus

Iliac crest

(b)

Axilla

4

Linea alba

External oblique

**Anterior superior
iliac spine (ASIS)**

Figure 24.3   Using the terms provided, label the anterior surface features of (*a*) upper torso and (*b*) lower torso. (**Boldface** indicates bony features; not boldface indicates soft tissue features.) ◀ **3**

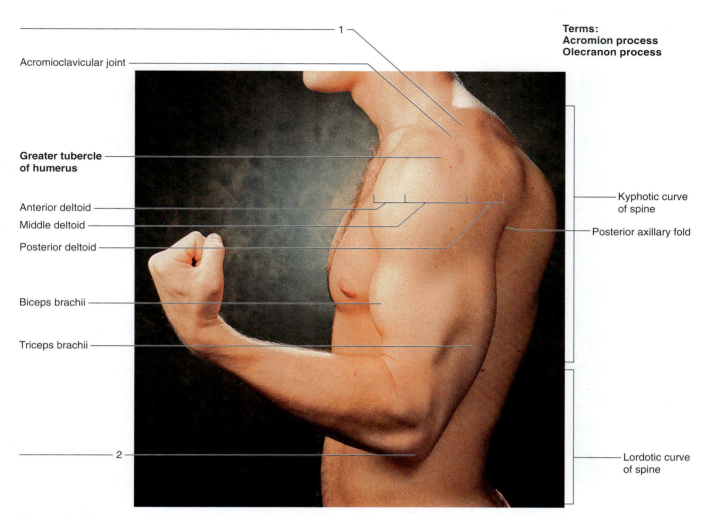

1

Acromioclavicular joint

Greater tubercle of humerus

Anterior deltoid

Middle deltoid

Posterior deltoid

Biceps brachii

Triceps brachii

2

Kyphotic curve of spine

Posterior axillary fold

Lordotic curve of spine

Figure 24.4 Using the terms provided, label the surface features of the lateral shoulder and upper limb. (**Boldface** indicates bony features; not boldface indicates soft tissue features.) 1

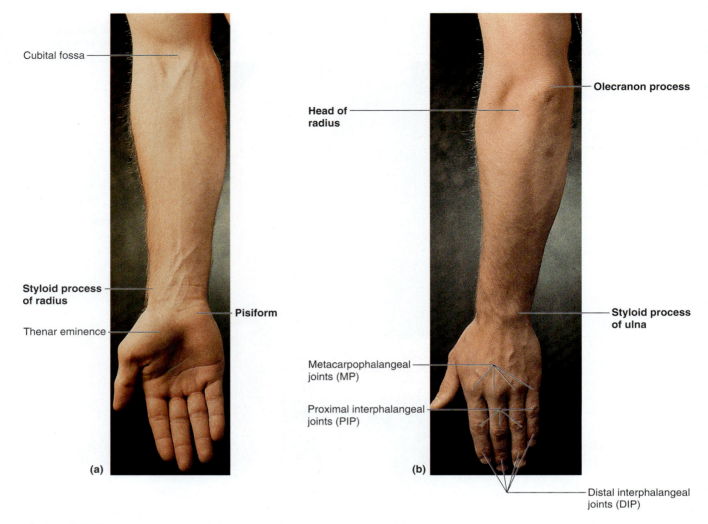

Cubital fossa

Head of radius

Olecranon process

Styloid process of radius

Thenar eminence

Pisiform

Styloid process of ulna

Metacarpophalangeal joints (MP)

Proximal interphalangeal joints (PIP)

(a)

(b)

Distal interphalangeal joints (DIP)

Figure 24.5 Surface features of the upper limb (*a*) anterior view and (*b*) posterior view. (**Boldface** indicates bony features; not boldface indicates soft tissue features.)

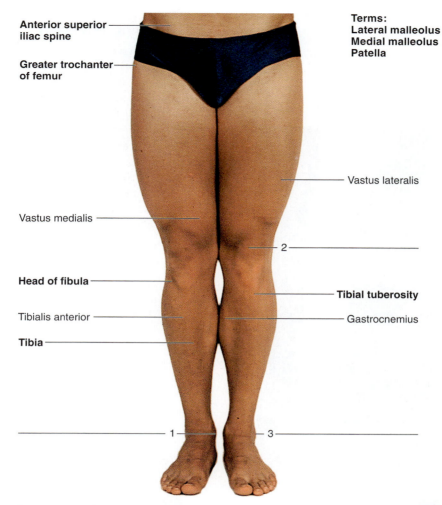

Terms:
Lateral malleolus
Medial malleolus
Patella

Anterior superior iliac spine

Greater trochanter of femur

Vastus lateralis

Vastus medialis

2

Head of fibula

Tibial tuberosity

Tibialis anterior

Gastrocnemius

Tibia

1

3

Figure 24.6 Using the terms provided, label the surface features of the anterior view of the body. (**Boldface** indicates bony features; not boldface indicates soft tissue features.) ◄ 1

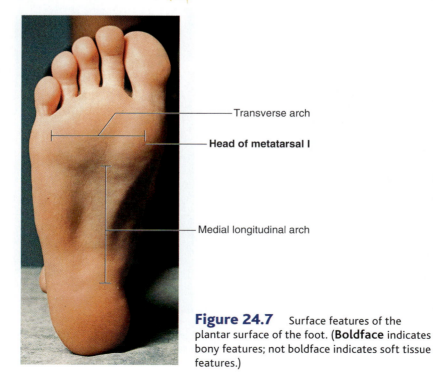

Transverse arch

Head of metatarsal I

Medial longitudinal arch

Figure 24.7 Surface features of the plantar surface of the foot. (**Boldface** indicates bony features; not boldface indicates soft tissue features.)

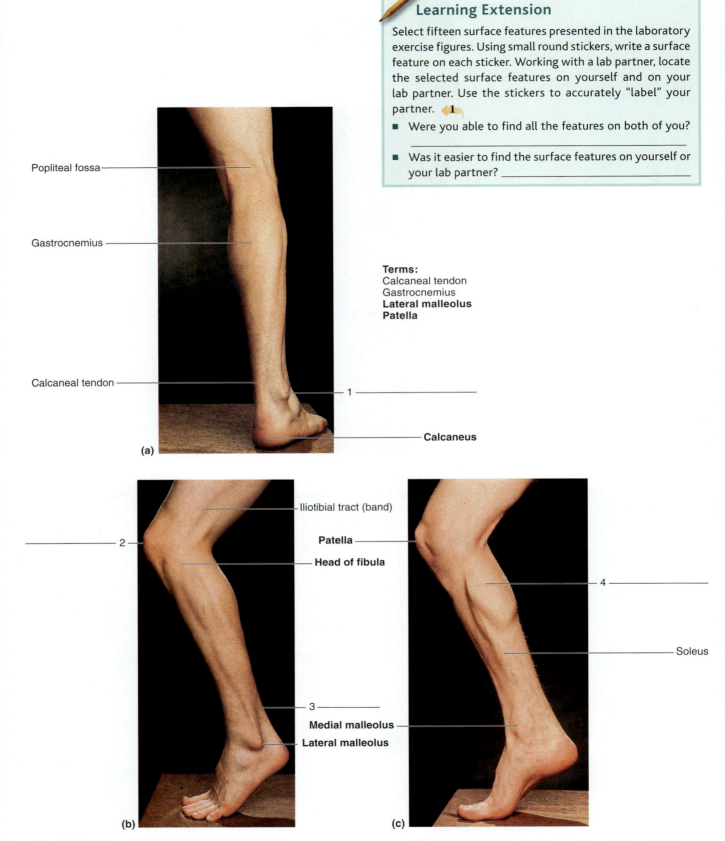

Learning Extension

Select fifteen surface features presented in the laboratory exercise figures. Using small round stickers, write a surface feature on each sticker. Working with a lab partner, locate the selected surface features on yourself and on your lab partner. Use the stickers to accurately "label" your partner. ◀**1**

■ Were you able to find all the features on both of you?

_____

■ Was it easier to find the surface features on yourself or your lab partner? _____

Popliteal fossa

Gastrocnemius

Terms:
Calcaneal tendon
Gastrocnemius
Lateral malleolus
Patella

Calcaneal tendon

1

Calcaneus

(a)

Iliotibial tract (band)

2

Patella

Head of fibula

4

Soleus

3

Medial malleolus

Lateral malleolus

(b)

(c)

Figure 24.8   Using the terms provided, label the surface features of the (a) posterior lower limb, (b) lateral lower limb, and (c) medial lower limb. (**Boldface** indicates bony features; not boldface indicates soft tissue features.) ◀**3**

The ⬅ corresponds to the indicated outcome(s) found at the beginning of the laboratory exercise.

Surface Anatomy

Part A Assessments

Match the body region in column A with the surface feature in column B. Place the letter of your choice in the space provided. ⬅2

Column A	Column B
a. Head	_____ **1.** Umbilicus
b. Upper limb	_____ **2.** Medial malleolus
c. Lower limb	_____ **3.** Iliac crest
d. Trunk	_____ **4.** Transverse arch
	_____ **5.** Spine of scapula
	_____ **6.** External occipital protuberance
	_____ **7.** Metacarpophalangeal joints
	_____ **8.** Sternum
	_____ **9.** Olecranon process
	_____ **10.** Zygomatic arch

Part B Assessments

Label figure 24.9 with the surface features provided.

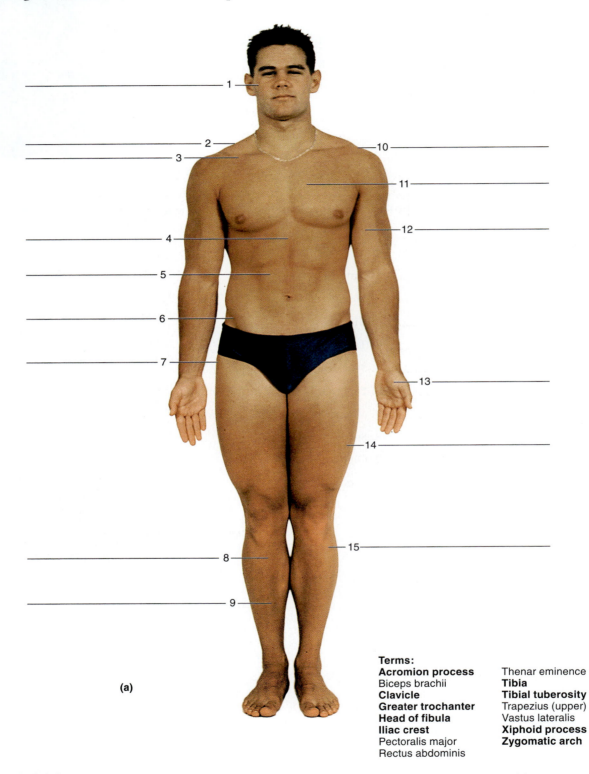

(a)

Terms:

Acromion process	Thenar eminence
Biceps brachii	**Tibia**
Clavicle	**Tibial tuberosity**
Greater trochanter	Trapezius (upper)
Head of fibula	Vastus lateralis
Iliac crest	**Xiphoid process**
Pectoralis major	**Zygomatic arch**
Rectus abdominis	

Figure 24.9 Using the terms provided, label the surface features of the (a) anterior view of body and (b) posterior view of the body. (**Boldface** indicates bony features; not boldface indicates soft tissue features.) ◀ **1**

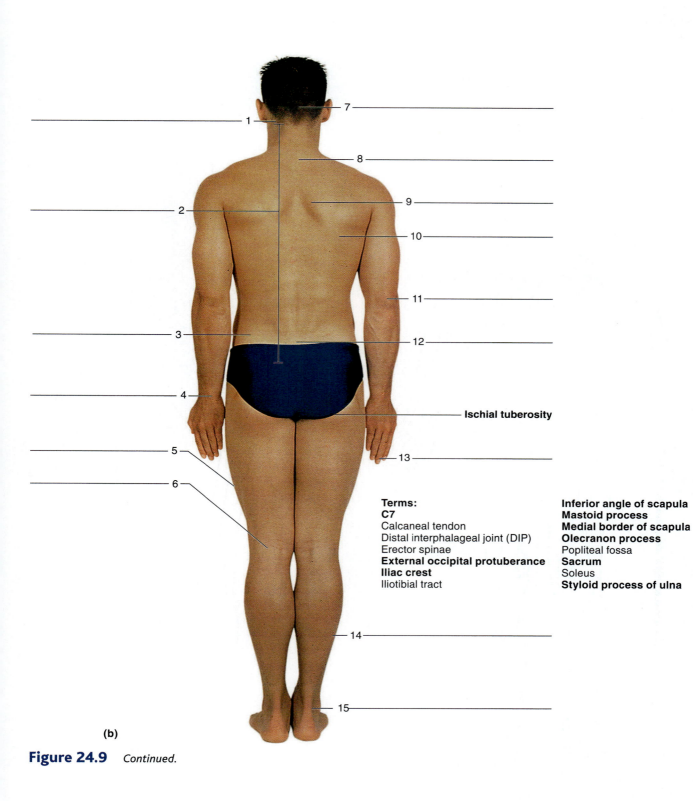

(b)

Figure 24.9 *Continued.*

Ischial tuberosity

Terms:
C7
Calcaneal tendon
Distal interphalageal joint (DIP)
Erector spinae
External occipital protuberance
Iliac crest
Iliotibial tract

Inferior angle of scapula
Mastoid process
Medial border of scapula
Olecranon process
Popliteal fossa
Sacrum
Soleus
Styloid process of ulna

Part C Assessments

Using table 24.1 from Laboratory Exercise 24, indicate the locations of the surface features with an "X" on figure 24.10. Use a black X for the bony features and a red X for the soft tissue features.

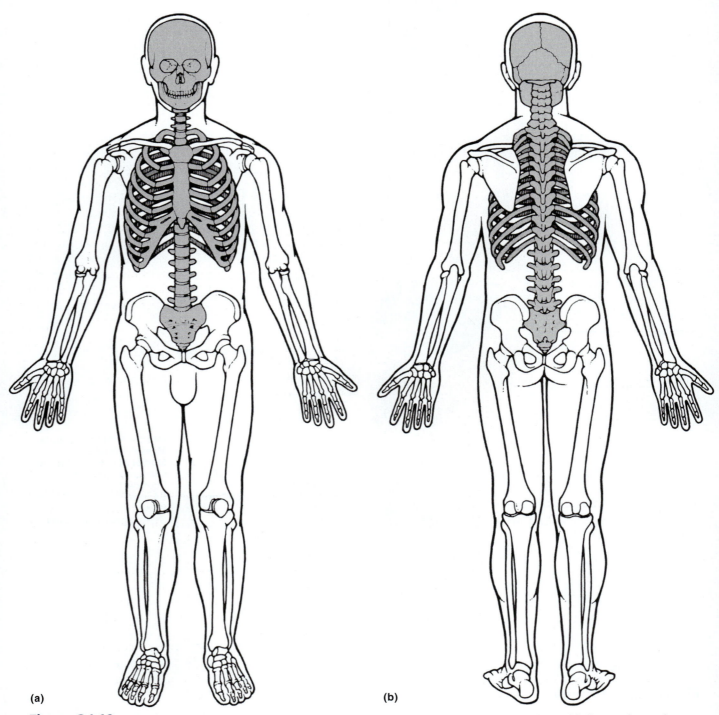

(a)   (b)

Figure 24.10   Indicate bony surface features with a black "X" and soft tissue surface features with a red "X" using the results from table 24.1, on the (*a*) anterior and (*b*) posterior diagrams of the body.   ◀3

Laboratory Exercise 25

Nervous Tissue and Nerves

Nervous tissue, which occurs in the brain, spinal cord, and nerves, contains neurons and neuroglial cells (neuroglia). The neurons are the basic structural and functional units of the nervous system involved in decision-making processes, detecting stimuli, and conducting messages. The neuroglial cells perform various supportive and protective functions for neurons.

Purpose of the Exercise

To review the characteristics of nervous tissue and to observe neurons, neuroglial cells, and various features of the nerves.

Learning Outcomes

After completing this exercise, you should be able to

1. Describe and locate the general characteristics of nervous tissue.
2. Distinguish structural and functional characteristics between neurons and neuroglial cells.
3. Identify and sketch the major structures of a neuron and a nerve.

Procedure—Nervous Tissue and Nerves

1. Review the sections entitled "Neurons" and "Neuroglial Cells" in chapter 9 of the textbook.
2. As a review activity, label figures 25.1 and 25.2.
3. Complete Part A of Laboratory Report 25.
4. Obtain a prepared microscope slide of a spinal cord smear. Using low-power magnification, search the slide and locate the relatively large, deeply stained cell bodies of motor neurons (multipolar neurons).
5. Observe a single motor neuron using high-power magnification, and note the following features:

   cell body
       nucleus
       nucleolus
       neurofibrils (threadlike structures extending into the nerve fibers)
   dendrites
   axon (nerve fiber)

   Compare the slide to the neuron model and to figure 25.3. You also may note small, darkly stained nuclei of neuroglial cells around the motor neuron.
6. Sketch and label a single motor (efferent) neuron in the space provided in Part B of the laboratory report.
7. Obtain a prepared microscope slide of a dorsal root ganglion. Search the slide and locate a cluster of sensory neuron cell bodies. You also may note bundles of nerve fibers passing among groups of neuron cell bodies (fig. 25.4).
8. Sketch and label a single sensory (afferent) neuron cell body in the space provided in Part B of the laboratory report.
9. Obtain a prepared microscope slide of neuroglial cells. Search the slide and locate some darkly stained astrocytes with numerous long, slender processes (fig. 25.5).

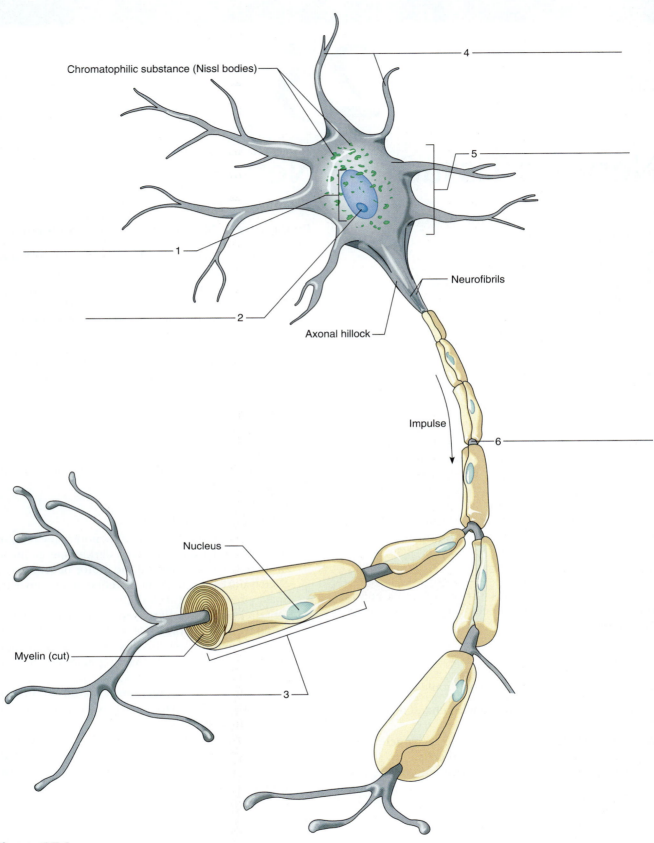

Chromatophilic substance (Nissl bodies)

4

5

1

2

Neurofibrils

Axonal hillock

Impulse

6

Nucleus

Myelin (cut)

3

Figure 25.1 Label this diagram of a motor neuron.

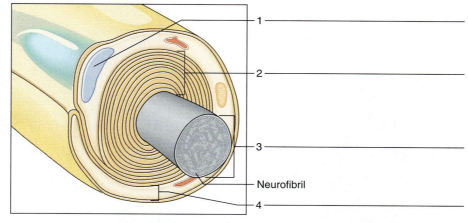

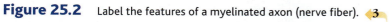

1 _____

2 _____

3 _____

Neurofibril _____

4 _____

Figure 25.2 Label the features of a myelinated axon (nerve fiber). 3

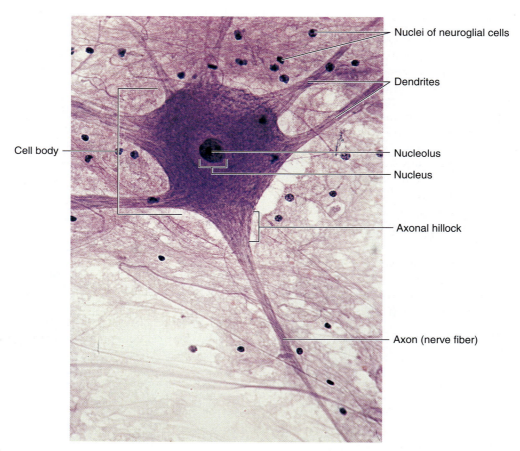

Nuclei of neuroglial cells

Dendrites

Cell body

Nucleolus

Nucleus

Axonal hillock

Axon (nerve fiber)

Figure 25.3 Micrograph of a multipolar neuron and neuroglial cells from a spinal cord smear (100× micrograph enlarged to 600×).

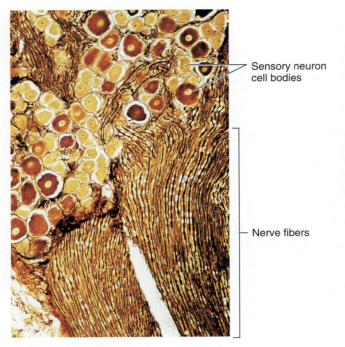

Figure 25.4 Micrograph of a dorsal root ganglion (50× micrograph enlarged to 100×).

Sensory neuron cell bodies

Nerve fibers

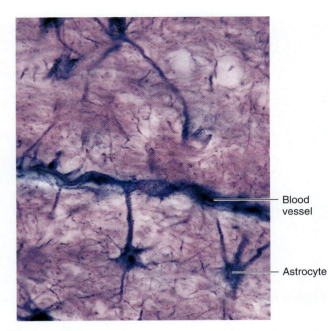

Figure 25.5 Micrograph of astrocytes (250× micrograph enlarged to 1,000×).

Blood vessel

Astrocyte

10. Sketch a single neuroglial cell in the space provided in Part B of the laboratory report.

11. Obtain a prepared microscope slide of a nerve. Locate the cross section of the nerve and note the many round nerve fibers inside. Also note the dense layer of connective tissue (perineurium) that encircles the nerve fibers and holds them together in a bundle. The individual nerve fibers are surrounded by a layer of more delicate connective tissue (endoneurium) (fig. 25.6).

12. Using high-power magnification, observe a single nerve fiber. Note the following features:

    central axon
    myelin sheath around the axon of Schwann cell (most of the myelin may have been dissolved and lost during the slide preparation)
    neurilemma of Schwann cell

13. Sketch and label a single nerve fiber with Schwann cell (cross section) in the space provided in Part B of the laboratory report.

14. Locate the longitudinal section of the nerve on the slide (fig. 25.7). Note the following:

    central axons
    myelin sheath of Schwann cells
    neurilemma of Schwann cells
    nodes of Ranvier

15. Sketch and label a single nerve fiber with Schwann cell (longitudinal section) in the space provided in Part B of the laboratory report.

Learning Extension

Obtain a prepared microscope slide of Purkinje cells. To locate these neurons, search the slide for large, flask-shaped cell bodies. Each cell body has one or two large, thick dendrites that give rise to branching networks of fibers. These cells are found in a particular region of the brain (cerebellar cortex).

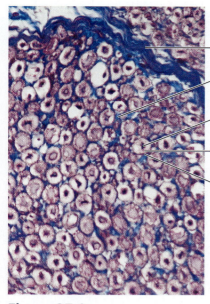

Figure 25.6   Cross section of a bundle of neurons within a nerve (400×).

Connective tissue (perineurium)

Connective tissue (endoneurium)

Axon (nerve fiber)

Myelin sheath of Schwann cell

Neurilemma of Schwann cell

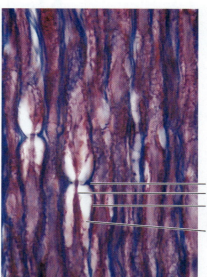

Figure 25.7   Longitudinal section of a nerve (250× micrograph enlarged to 2,000×).

Node of Ranvier
Axon (nerve fiber)
Neurilemma of Schwann cell
Myelin sheath of Schwann cell

The ⬅ corresponds to the indicated outcome(s) found at the beginning of the laboratory exercise.

Nervous Tissue and Nerves

Part A Assessments

Match the terms in column A with the descriptions in column B. Place the letter of your choice in the space provided. ◄1 ◄2

Column A	Column B
a. Astrocyte	_____ **1.** Sheath of Schwann cell containing cytoplasm and nucleus that encloses myelin
b. Axon	
c. Dendrite	_____ **2.** Network of fine threads extending into a nerve fiber
d. Interneurons	
e. Microglial cell	_____ **3.** Substance of Schwann cell composed of lipoprotein that insulates axons and increases impulse speed
f. Motor (efferent) neuron	
g. Myelin sheath	_____ **4.** Neuron process with many branches that conducts impulse toward the cell body
h. Neurilemma	
i. Neurofibrils	_____ **5.** Star-shaped neuroglial cell between neurons and blood vessels
j. Oligodendrocyte	
k. Sensory (afferent) neuron	_____ **6.** Nerve fiber arising from slight elevation of the cell body that conducts impulse away from the cell body
	_____ **7.** Transmits impulse from sensory to motor neuron
	_____ **8.** Transmits impulse out of brain or spinal cord to effectors
	_____ **9.** Transmits impulse into brain or spinal cord from receptors
	_____ **10.** Myelin-forming neuroglial cell in brain and spinal cord
	_____ **11.** Phagocytic neuroglial cell

Part B Assessments

1. Sketch a single motor neuron. Label the cell body, nucleus, nucleolus, and cell processes (dendrite and axon). ◄3

2. Sketch and label a single sensory (afferent) neuron cell body. ◄3

3. Sketch a single neuroglial cell. ◄2

4. Sketch and label a single nerve fiber with Schwann cell (cross section). ◄3

5. Sketch and label a single nerve fiber with Schwann cell (longitudinal section). ◄3

Laboratory Exercise 26

Spinal Cord and Meninges

Materials Needed

Textbook
Compound light microscope
Prepared microscope slide of a spinal cord cross section
  with spinal nerve roots
Spinal cord model with meninges

For Demonstration:
Preserved spinal cord with meninges intact

The spinal cord is a column of nerve fibers that extends down through the vertebral canal. Together with the brain, it makes up the central nervous system.

Neurons within the spinal cord provide a two-way communication system between the brain and the body parts outside the central nervous system. The cord also contains the processing centers for the spinal reflexes.

The meninges consist of layers of membranes located between the bones of the skull and vertebral column and the soft tissues of the central nervous system. They include the dura mater, the arachnoid mater, and the pia mater.

Purpose of the Exercise

To review the characteristics of the spinal cord and meninges and to observe the major features of these structures.

Learning Outcomes

After completing this exercise, you should be able to

1. Identify the major features of the spinal cord.
2. Arrange the layers of the meninges and describe the structure of each.

Procedure A—Structure of the Spinal Cord

1. Review the section entitled "Spinal Cord" in chapter 9 of the textbook.
2. As a review activity, label figures 26.1 and 26.2.
3. Complete Part A of Laboratory Report 26.
4. Obtain a prepared microscope slide of a spinal cord cross section. Use the low power of the microscope to locate the following features:

   posterior median sulcus

   anterior median fissure

   central canal

   gray matter

       gray commissure

       posterior (dorsal) horn

       lateral horn

       anterior (ventral) horn

   white matter

       posterior (dorsal) funiculus (column)

       lateral funiculus (column)

       anterior (ventral) funiculus (column)

   roots of spinal nerve

       dorsal roots

       dorsal root ganglia

       ventral roots

5. Observe the spinal cord model and locate the features listed in step 4.
6. Complete Part B of the laboratory report.

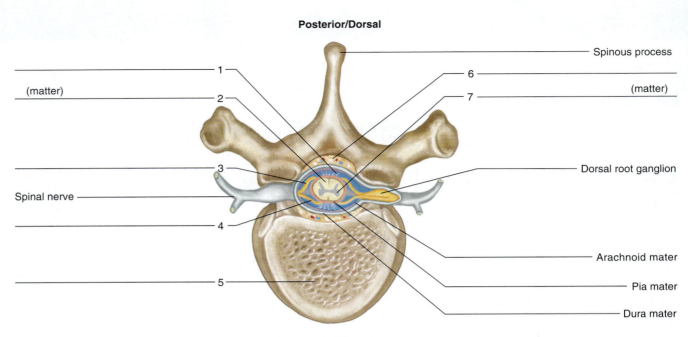

Posterior/Dorsal

Spinous process

1

6

7

(matter)

(matter)

3

Dorsal root ganglion

Spinal nerve

4

Arachnoid mater

5

Pia mater

Dura mater

Anterior/Ventral

Figure 26.1 Label the features of the spinal cord and the surrounding structures. ◀1

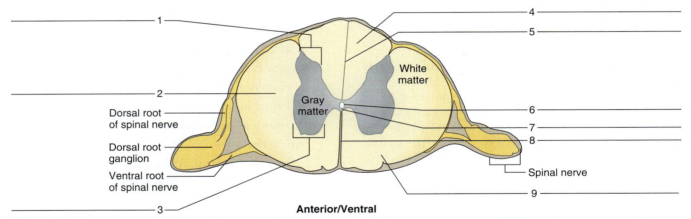

Posterior/Dorsal

1

4

5

White matter

2

Gray matter

Dorsal root of spinal nerve

6

Dorsal root ganglion

7

8

Ventral root of spinal nerve

Spinal nerve

3

9

Anterior/Ventral

Figure 26.2 Label this cross section of the spinal cord, including the features of the white and gray matter. (*Note:* A lateral horn of the gray matter is not present at the level of the spinal cord illustrated.) ◀1

Procedure B—Meninges

1. Review the section entitled "Meninges" in chapter 9 of the textbook.
2. Complete Part C of laboratory report.

Demonstration

Observe the preserved section of a spinal cord. Note the heavy covering of dura mater, firmly attached to the cord on each side by ligaments (denticulate ligaments) originating in the pia mater. The intermediate layer of meninges, the arachnoid mater, is devoid of blood vessels, but in a live human being, the space beneath this layer contains cerebrospinal fluid. The pia mater, closely attached to the surface of the spinal cord, contains many blood vessels. What are the functions of these layers?

_____

_____

Name _____

Date _____

Section _____

The ⬅ corresponds to the indicated outcome(s) found at the beginning of the laboratory exercise.

Spinal Cord and Meninges

Part A Assessments

Complete the following statements:

1. The spinal cord gives rise to 31 pairs of _____ . ⬅**1**

2. The bulge in the spinal cord that gives off nerves to the upper limbs is called the _____ . ⬅**1**

3. The bulge in the spinal cord that gives off nerves to the lower limbs is called the _____ . ⬅**1**

4. The _____ is a groove that extends the length of the spinal cord posteriorly. ⬅**1**

5. In a spinal cord cross section, the posterior _____ of the gray matter resemble the upper wings of a butterfly. ⬅**1**

6. The cell bodies of motor neurons are found in the _____ horns of the spinal cord. ⬅**1**

7. The _____ connects the gray matter on the left and right sides of the spinal cord. ⬅**1**

8. The _____ in the gray commissure of the spinal cord contains cerebrospinal fluid. ⬅**1**

9. The white matter of the spinal cord is divided into anterior, lateral, and posterior _____ . ⬅**1**

10. The longitudinal bundles of nerve fibers within the spinal cord comprise major nerve pathways called _____ . ⬅**1**

11. Collectively, the dura mater, arachnoid mater, and pia mater are called the _____ . ⬅**2**

Part B Assessments

Identify the features indicated in the spinal cord section of figure 26.3. ◀**1**

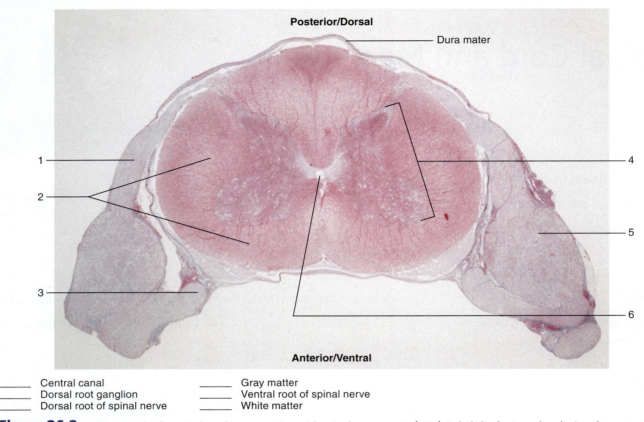

_____ Central canal   _____ Gray matter
_____ Dorsal root ganglion   _____ Ventral root of spinal nerve
_____ Dorsal root of spinal nerve   _____ White matter

Figure 26.3   Micrograph of a spinal cord cross section with spinal nerve roots (35×). Label the features by placing the correct numbers in the spaces provided.

Part C Assessments

Match the terms in column A with the descriptions in column B. Place the letter of your choice in the space provided. ◀**2**

Column A	Column B
a. Arachnoid mater	_____ **1.** Outermost layer of meninges
b. Dura mater	
c. Epidural space	_____ **2.** Follows irregular contours of spinal cord surface
d. Pia mater	_____ **3.** Contains cerebrospinal fluid
e. Subarachnoid space	
	_____ **4.** Thin, weblike membrane
	_____ **5.** Separates dura mater from bone of vertebra

Reflex Arc and Reflexes

A reflex arc represents the simplest type of nerve pathway found in the nervous system. This pathway begins with a receptor at the dendrite end of a sensory (afferent) neuron. The sensory neuron leads into the central nervous system and may communicate with one or more interneurons. Some of these interneurons, in turn, communicate with motor (efferent) neurons, whose axons (nerve fibers) lead outward to effectors.

Thus, when a sensory receptor is stimulated by a change occurring inside or outside the body, nerve impulses may pass through a reflex arc, and, as a result, effectors may respond. Such an automatic, subconscious response is called a *reflex.*

A *stretch reflex* involves a single synapse (monosynaptic) between a sensory and a motor neuron within the gray matter of the spinal cord. Examples of stretch reflexes include the patellar, calcaneal, biceps, triceps, and plantar reflexes. Other more complex *withdrawal reflexes* involve interneurons in combination with sensory and motor neurons, thus polysnaptic. Examples of withdrawal reflexes include responses to touching hot objects or stepping on sharp objects.

Reflexes demonstrated in this lab are stretch reflexes. When a muscle is stretched by a tap over its tendon, stretch receptors (proprioceptors) called *muscle spindles* are stretched within the muscle, which initiates an impulse over a reflex arc. A sensory (afferent) neuron conducts an impulse from the muscle spindle into the gray matter of the spinal cord, where it synapses with a motor (efferent) neuron, which conducts the impulse to the effector muscle. The stretched muscle responds by contracting to resist or reverse further stretching. These stretch reflexes are important to maintain proper posture, balance, and movements. Observation of many of these reflexes in clinical tests on patients may indicate damage to a level of the spinal cord or peripheral nerves of the particular reflex arc.

Purpose of the Exercise

To review the characteristics of reflex arcs and reflex behavior and to demonstrate some of the reflexes that occur in the human body.

Learning Outcomes

After completing this exercise, you should be able to

1. Describe the components of a reflex arc.
2. Demonstrate and record stretch reflexes that occur in humans.
3. Analyze the components and patterns of stretch reflexes.

Procedure—Reflex Arc and Reflexes

1. Review the section entitled "Types of Nerves" and "Nerve Pathways" in chapter 9 of the textbook.
2. As a review activity, label figure 27.1.
3. Complete Part A of Laboratory Report 27.
4. Work with a laboratory partner to demonstrate each of the reflexes listed. (See fig. 27.2 also.) *It is important that muscles involved in the reflexes be totally relaxed to observe proper responses.* If a person is trying too hard to experience the reflex or is trying to suppress the reflex, assign a multitasking activity while the stimulus with the rubber hammer occurs. For example, assign a physical task with the upper limbs along with a complex mental activity during the patellar (knee-jerk) reflex. After each demonstration, record your observations in the table provided in Part B of the laboratory report.

   a. *Patellar (knee-jerk) reflex.* Have your laboratory partner sit on a table (or sturdy chair) with legs relaxed and hanging freely over the edge without touching the floor. Gently strike your partner's patellar ligament (just below the patella) with the blunt side of a rubber percussion hammer. The normal response is a moderate extension of the leg at the knee joint.

   b. *Calcaneal (ankle-jerk) reflex.* Have your partner kneel on a chair with back toward you and with

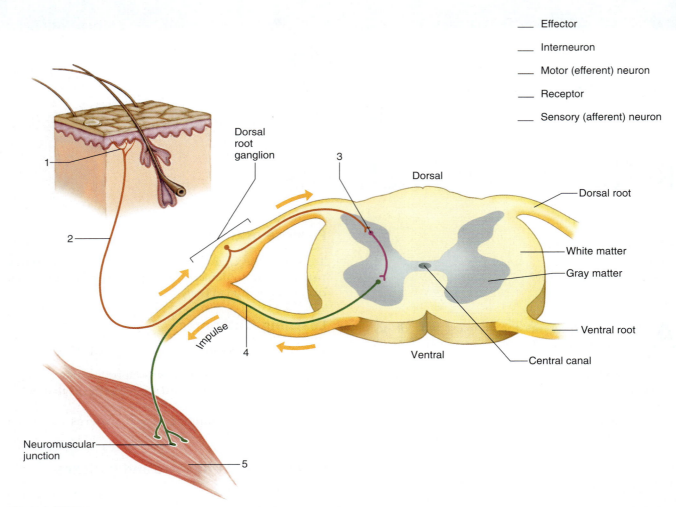

Labels (top right):

___ Effector
___ Interneuron
___ Motor (efferent) neuron
___ Receptor
___ Sensory (afferent) neuron

Dorsal root ganglion

Dorsal

Dorsal root

White matter

Gray matter

Ventral root

Central canal

Ventral

Impulse

Neuromuscular junction

Figure 27.1   Label this diagram of a withdrawal (polysynaptic) reflex arc by placing the correct numbers in the spaces provided. Reflexes demonstrated in this lab are stretch (monosynaptic) reflex arcs and lack the interneuron. ◀ **1**

feet over the edge and relaxed. Gently strike the calcaneal tendon (just above its insertion on the calcaneus) with the blunt side of the rubber hammer. The normal response is plantar flexion of the foot.

c. *Biceps (biceps-jerk) reflex.* Have your partner place a bare arm bent about 90° at the elbow on the table. Press your thumb on the inside of the elbow over the tendon of the biceps brachii, and gently strike your thumb with the rubber hammer. Watch the biceps brachii for a response. The response might be a slight twitch of the muscle or flexion of the forearm at the elbow joint.

d. *Triceps (triceps-jerk) reflex.* Have your partner lie supine with an upper limb bent about 90° across the abdomen. Gently strike the tendon of the triceps brachii near its insertion just proximal to the olecranon process at the tip of the elbow. Watch the triceps brachii for a response. The response

might be a slight twitch of the muscle or extension of the forearm at the elbow joint.

e. *Plantar reflex.* Have your partner remove a shoe and sock and lie supine with the lateral surface of the foot resting on the table. Draw the metal tip of the rubber hammer, applying firm pressure, over the sole from the heel to the base of the large toe. The normal response is flexion (curling) of the toes and plantar flexion of the foot. If the toes spread apart and dorsiflexion of the great toe occurs, the reflex is the abnormal *Babinski reflex* response (normal in infants until the nerve fibers have completed myelinization). If the Babinski reflex occurs later in life, it may indicate damage to the corticospinal tract of the CNS.

5. Complete Part B of the laboratory report.

(a) Patellar (knee-jerk) reflex

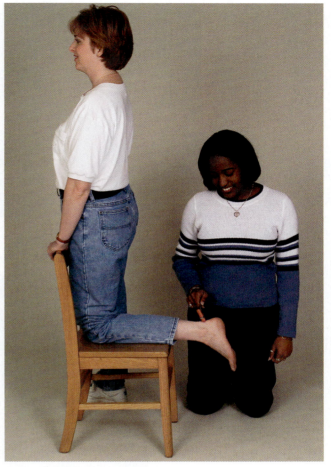

(b) Calcaneal (ankle-jerk) reflex)

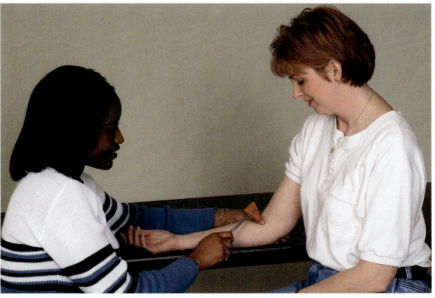

(c) Biceps reflex

Figure 27.2 Demonstrate each of the following reflexes: (*a*) patellar reflex; (*b*) calcaneal reflex; (*c*) biceps reflex; (*d*) triceps reflex; and (*e*) plantar reflex.

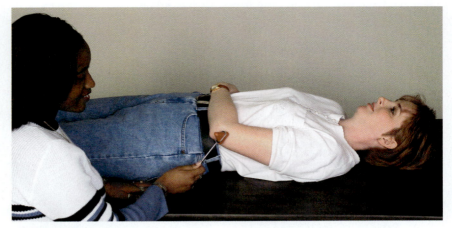

(d) Triceps reflex

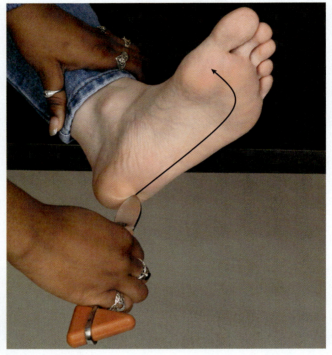

(e) Plantar reflex

Figure 27.2 *Continued.*

Name _____

Date _____

Section _____

The ◄ corresponds to the indicated outcome(s) found at the beginning of the laboratory exercise.

Reflex Arc and Reflexes

Part A Assessments

Complete the following statements:

1. _____ are routes followed by nerve impulses as they pass through the nervous system. ◄**1**

2. Interneurons in a withdrawal reflex are located in the _____ . ◄**1**

3. _____ are automatic subconscious responses to external or internal stimuli.

4. Effectors of a reflex arc are glands and _____ . ◄**1**

5. A patellar (knee-jerk) reflex employs only _____ and motor neurons. ◄**1**

6. The effector of the patellar reflex is the _____ muscle. ◄**1**

7. The sensory stretch receptors (muscle spindles) of the patellar reflex are located in the _____ . ◄**1**

8. The patellar reflex helps the body to maintain _____ . ◄**1**

9. The sensory receptors of a withdrawal reflex are located in the _____ . ◄**1**

10. _____ muscles in the limbs are the effectors of a withdrawal reflex. ◄**1**

Part B Assessments

1. Complete the following table. ◀2

Reflex Tested	Response Observed	Effector Muscle Involved
Patellar (knee-jerk)		
Calcaneal (ankle-jerk)		
Biceps		
Triceps		
Plantar		

2. List the major events that occur in the patellar (knee-jerk) reflex, from the striking of the patellar ligament to the resulting response. ◀3 _____

_____

_____

_____

_____

 **Critical Thinking Application**

What characteristics do the reflexes you demonstrated have in common? ◀3

_____

_____

_____

_____

Laboratory Exercise  28

Brain and Cranial Nerves

Materials Needed

Textbook
Dissectible model of the human brain
Preserved human brain
Anatomical charts of the human brain

The brain, the largest and most complex part of the nervous system, contains nerve centers associated with sensory functions and is responsible for sensations and perceptions. It issues motor commands to skeletal muscles and carries on higher mental activities. It also functions to coordinate muscular movements, and it contains centers and nerve pathways necessary for the regulation of internal organs.

Twelve pairs of cranial nerves arise from the ventral surface of the brain and are designated by number and name. Although most of these nerves conduct both sensory and motor impulses, some contain only sensory fibers associated with special sense organs. Others are primarily composed of motor fibers and are involved with the activities of muscles and glands.

Purpose of the Exercise

To review the structural and functional characteristics of the human brain and cranial nerves.

Learning Outcomes

After completing this exercise, you should be able to

1. Identify the major external and internal structures in the human brain.
2. Locate the major functional regions of the brain.
3. Identify each of the cranial nerves.
4. Differentiate the functions of each cranial nerve.

Procedure A—Human Brain

1. Review the section entitled "Brain" in chapter 9 of the textbook.
2. As a review activity, label figures 28.1 and 28.2.
3. Complete Part A of Laboratory Report 28.

4. Observe the anatomical charts, dissectible model, and preserved specimen of the human brain. Locate each of the following features:

cerebrum
cerebral hemispheres
corpus callosum
gyri (convolutions)
sulci
    central sulcus
    lateral sulcus
fissures
    longitudinal fissure
    transverse fissure
lobes
    frontal lobe
    parietal lobe
    temporal lobe
    occipital lobe
    insula (insular lobe)
cerebral cortex
ventricles
    lateral ventricles
    third ventricle
    fourth ventricle
    choroid plexuses
    cerebral aqueduct
diencephalon
    thalamus
    hypothalamus
    optic chiasma
    mammillary bodies
    pineal gland
cerebellum
    lateral (right and left) hemispheres
    vermis
    cerebellar cortex
brainstem
    midbrain
    pons
    medulla oblongata

5. Complete Part B of the laboratory report.

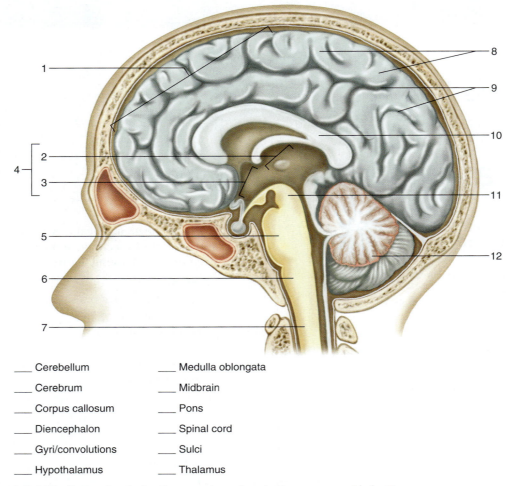

___ Cerebellum	___ Medulla oblongata
___ Cerebrum	___ Midbrain
___ Corpus callosum	___ Pons
___ Diencephalon	___ Spinal cord
___ Gyri/convolutions	___ Sulci
___ Hypothalamus	___ Thalamus

Figure 28.1 Label this diagram by placing the correct numbers in the spaces provided.

6. Label the areas in figure 28.3 that represent the following functional regions of the cerebrum:

   motor area for voluntary muscle control
   motor speech area (Broca's area)
   cutaneous sensory area
   auditory area
   visual area
   Wernicke's area

Procedure B—Cranial Nerves

1. Review the section entitled "Cranial Nerves" in chapter 9 of the textbook.
2. As a review activity, label figure 28.4.
3. Observe the model and preserved specimen of the human brain, and locate as many of the following cranial nerves as possible as you differentiate some of their associated functions:

   olfactory nerves (I)—smell
   optic nerves (II)—vision
   oculomotor nerves (III)—pupil constriction and opens eyelid
   trochlear nerves (IV)—stimulates superior oblique eye muscle

   trigeminal nerves (V)—sensory from face and teeth
   abducens nerves (VI)—lateral eye movements
   facial nerves (VII)—salivation, tear secretions, and taste
   vestibulocochlear nerves (VIII)—hearing and balance
   glossopharyngeal nerves (IX)—regulates blood pressure, salivation, and swallowing
   vagus nerves (X)—regulates many visceral organs including slows the heart rate
   accessory nerves (XI)—controls neck and shoulder muscles
   hypoglossal nerves (XII)—controls tongue movements

The following mnemonic device will help you learn the twelve pairs of cranial nerves in the proper order:

   Old **O**pie **oc**casionally **tr**ies **trig**onometry, **a**nd feels **v**ery **glo**omy, **vagu**e, and **hypo**active.[1]

4. Complete Part C of the laboratory report.

---

1. From *HAPS-EDucator*, Winter 2002. An official publication of the Human Anatomy & Physiology Society (HAPS).

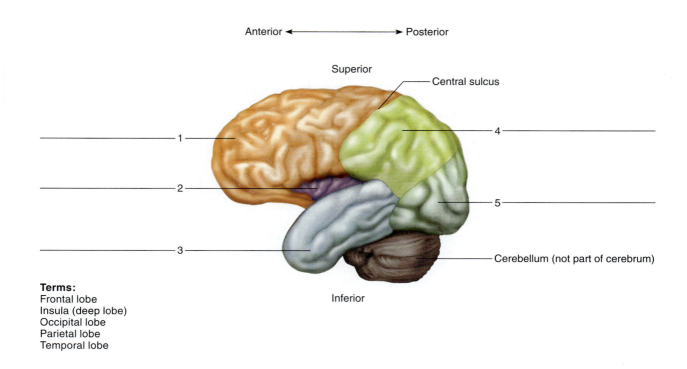

Anterior ←————→ Posterior

Superior

Central sulcus

1

2

3

4

5

Cerebellum (not part of cerebrum)

Inferior

Terms:
Frontal lobe
Insula (deep lobe)
Occipital lobe
Parietal lobe
Temporal lobe

Figure 28.2   Using the terms provided, label the five lobes of the left cerebral hemisphere. Lobe 3 is retracted to expose the deep lobe 2 in this figure. ◀**1**

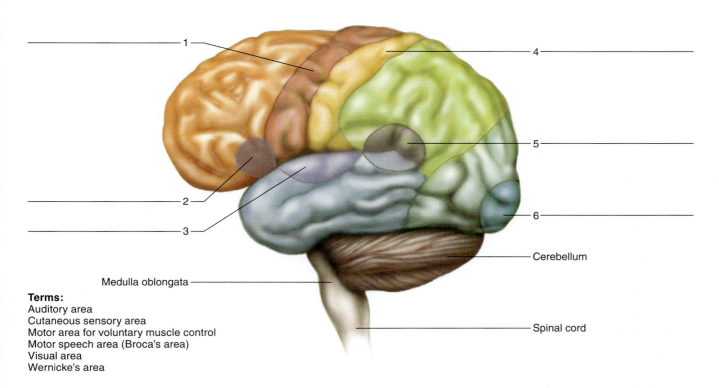

1

2

3

4

5

6

Cerebellum

Medulla oblongata

Spinal cord

Terms:
Auditory area
Cutaneous sensory area
Motor area for voluntary muscle control
Motor speech area (Broca's area)
Visual area
Wernicke's area

Figure 28.3   Using the terms provided, label the functional areas of the cerebrum. (*Note:* These areas are not visible as distinct parts of the brain.) ◀**2**

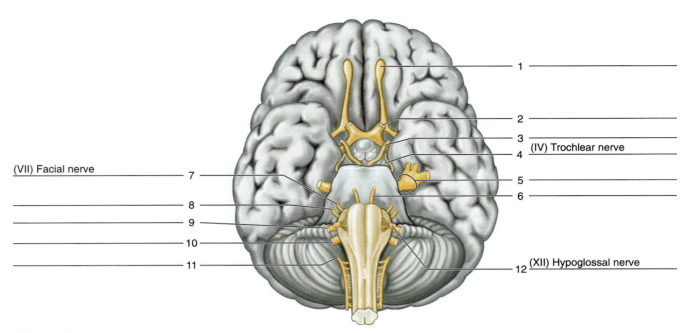

(VII) Facial nerve ————— 7

——————— 8

——————— 9

——————— 10

——————— 11

1

2

3 (IV) Trochlear nerve

4

5

6

12 (XII) Hypoglossal nerve

Figure 28.4 Provide the names of the cranial nerves in this ventral view. **3**

Name _____

Date _____

Section _____

The ⬅ corresponds to the indicated outcome(s) found at the beginning of the laboratory exercise.

Brain and Cranial Nerves

Part A Assessments

Match the terms in column A with the descriptions in column B. Place the letter of your choice in the space provided. ⬅**1**

Column A	Column B
a. Central sulcus	_____ **1.** Cone-shaped structure attached to upper posterior portion of diencephalon
b. Cerebral cortex	
c. Corpus callosum	_____ **2.** Connects cerebral hemispheres
d. Diencephalon	_____ **3.** Ridge on surface of cerebrum
e. Gyrus (convolution)	_____ **4.** Separates frontal and parietal lobes
f. Medulla oblongata	_____ **5.** Part of brainstem between diencephalon and pons
g. Midbrain	_____ **6.** Rounded bulge on underside of brainstem
h. Pineal gland	_____ **7.** Part of brainstem continuous with the spinal cord
i. Pons	_____ **8.** Internal brain chamber filled with cerebrospinal fluid
j. Ventricle	_____ **9.** Contains thalamus and hypothalamus
	_____ **10.** Thin layer of gray matter on surface of cerebrum

Part B Assessments

Identify the indicated features on the midsagittal section of the right half of the human brain in figure 28.5. ⬅**1**

1. _____
2. _____
3. _____
4. _____
5. _____
6. _____
7. _____
8. _____
9. _____
10. _____

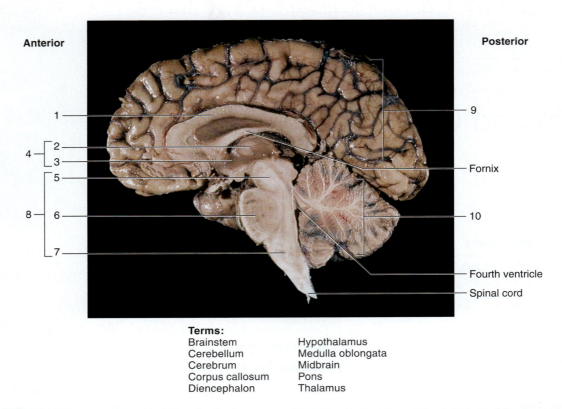

Terms:

Brainstem   Hypothalamus
Cerebellum   Medulla oblongata
Cerebrum   Midbrain
Corpus callosum Pons
Diencephalon   Thalamus

Figure 28.5 Using the terms provided, identify the features on this midsagittal section of the right half of the human brain.

Part C Assessments

Indicate by name and number which cranial nerve (nerves) is (are) most closely associated with each of the following functions:

1. Sense of hearing   _____

2. Sense of taste   _____

3. Sense of sight   _____

4. Sense of smell   _____

5. Sense of equilibrium   _____

6. Conducting sensory impulses from upper teeth _____

7. Conducting sensory impulses from lower teeth _____

8. Raising eyelids   _____

9. Focusing lenses of eyes   _____

10. Adjusting amount of light entering eye   _____

11. Moving eyes   _____

12. Stimulating salivary secretions   _____

13. Movement of trapezius and sternocleidomastoid muscles _____

14. Muscular movements associated with speech   _____

15. Muscular movements associated with swallowing _____

Dissection of the Sheep Brain

Materials Needed

Dissectible model of human brain
Preserved sheep brain
Dissecting tray
Dissection instruments
Long knife

⚠ Safety

- Wear disposable gloves when handling the sheep brains.
- Save or dispose of the brains as instructed.
- Wash your hands before leaving the laboratory.

Mammalian brains have many features in common. Human brains may not be available, so sheep brains often are dissected as an aid to understanding mammalian brain structure. However, the adaptations of the sheep differ from the adaptations of the human, so comparisons of their structural features may not be precise. The sheep is a quadruped, therefore the spinal cord is horizontal, unlike the vertical orientation in a bipedal human. Preserved sheep brains have a different appearance and are firmer than those removed from the cranial cavity, caused by the preservatives used.

Purpose of the Exercise

To observe the major features of the sheep brain and to compare these features with those of the human brain.

Learning Outcomes

After completing this exercise, you should be able to

1. Examine the major structures of the sheep brain.
2. Locate the larger cranial nerves of the sheep brain.
3. Summarize several differences and similarities between the sheep brain and the human brain.

Procedure—Dissection of the Sheep Brain

1. Obtain a preserved sheep brain and rinse it thoroughly in water to remove as much of the preserving fluid as possible.
2. Examine the surface of the brain for the presence of meninges. (The outermost layers of these membranes may have been lost during removal of the brain from the cranial cavity.) If meninges are present, locate the following:

   dura mater—the thick, opaque outer layer

   arachnoid mater—the delicate, transparent middle layer attached to the undersurface of the dura mater

   pia mater—the thin, vascular layer that adheres to the surface of the brain (should be present)

3. Remove any remaining dura mater by pulling it gently from the surface of the brain.
4. Position the brain with its ventral surface down in the dissecting tray. Study figure 29.1, and locate the following structures on the specimen:

   cerebral hemispheres
   gyri (convolutions)
   sulci
   longitudinal fissure
   frontal lobe
   parietal lobe
   temporal lobe
   occipital lobe
   cerebellum
   medulla oblongata
   spinal cord

5. Gently separate the cerebral hemispheres along the longitudinal fissure and expose the transverse band of white fibers within the fissure that connects the hemispheres. This band is the *corpus callosum*.
6. Bend the cerebellum and medulla oblongata slightly downward and away from the cerebrum (figure 29.2). This will expose the *pineal gland* in the

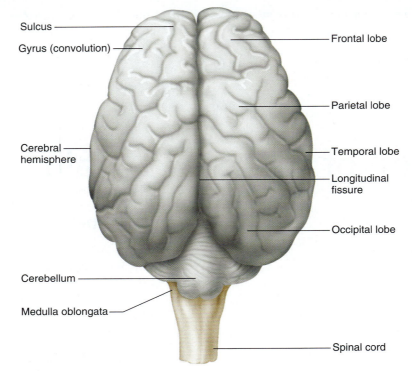

Sulcus —
Gyrus (convolution) —
Cerebral hemisphere —
Cerebellum —
Medulla oblongata —

— Frontal lobe
— Parietal lobe
— Temporal lobe
— Longitudinal fissure
— Occipital lobe
— Spinal cord

Figure 29.1 Dorsal surface of the sheep brain.

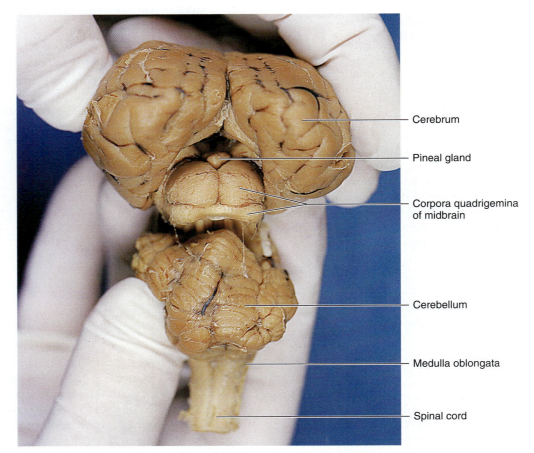

— Cerebrum
— Pineal gland
— Corpora quadrigemina of midbrain
— Cerebellum
— Medulla oblongata
— Spinal cord

Figure 29.2 Gently bend the cerebellum and medulla oblongata away from the cerebrum to expose the pineal gland and the corpora quadrigemina.

upper midline and the *corpora quadrigemina,* which consists of four rounded structures associated with the midbrain.

7. Position the brain with its ventral surface upward. Study figures 29.3 and 29.4, and locate the following structures on the specimen:

longitudinal fissure

olfactory bulbs

optic nerves

optic chiasma

optic tract

mammillary bodies

infundibulum (pituitary stalk)

midbrain

pons

8. Although some of the cranial nerves may be missing or are quite small and difficult to find, locate as many of the following as possible, using figures 29.3 and 29.4 as references:

oculomotor nerves

trochlear nerves

trigeminal nerves

abducens nerves

facial nerves

vestibulocochlear nerves

glossopharyngeal nerves

vagus nerves

accessory nerves

hypoglossal nerves

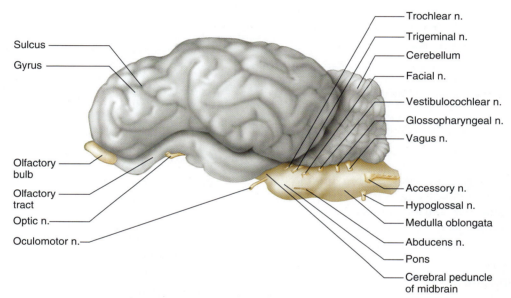

Figure 29.3   Lateral surface of the sheep brain.

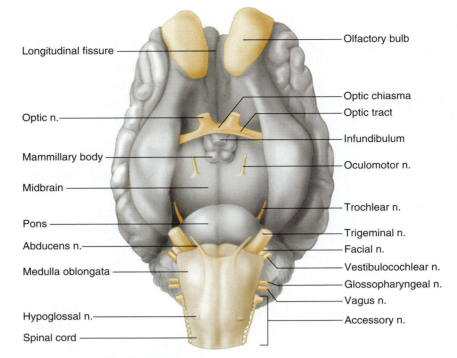

Longitudinal fissure

Optic n.

Mammillary body

Midbrain

Pons

Abducens n.

Medulla oblongata

Hypoglossal n.

Spinal cord

Olfactory bulb

Optic chiasma

Optic tract

Infundibulum

Oculomotor n.

Trochlear n.

Trigeminal n.

Facial n.

Vestibulocochlear n.

Glossopharyngeal n.

Vagus n.

Accessory n.

Figure 29.4   Ventral surface of the sheep brain.

9. Using a long, sharp knife, cut the sheep brain along the midline to produce a midsagittal section. Study figures 29.5 and 29.6, and locate the following structures on the specimen:

cerebrum

olfactory bulb

corpus callosum

cerebellum

    white matter

    gray matter

lateral ventricle (one in each cerebral hemisphere)

third ventricle (within diencephalon)

fourth ventricle (between brainstem and cerebellum)

diencephalon

    optic chiasma

    infundibulum

    pituitary gland (this structure may be missing)

    thalamus

    hypothalamus

    pineal gland

midbrain

pons

medulla oblongata

10. Dispose of the sheep brain as directed by the laboratory instructor.

11. Complete Parts A and B of Laboratory Report 29.

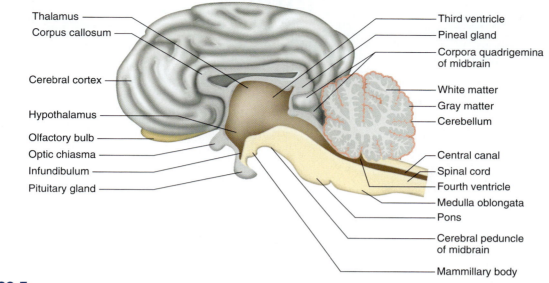

Figure 29.5 Midsagittal section of the right half of the sheep brain.

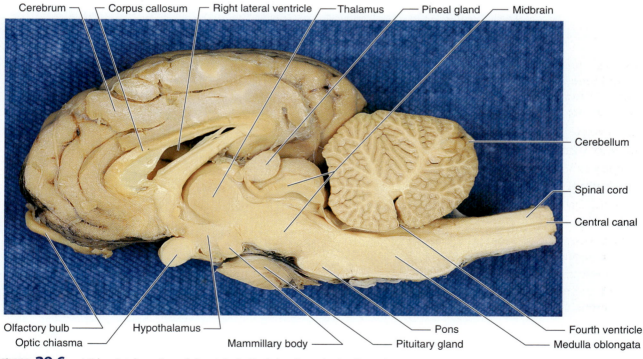

Figure 29.6 Midsagittal section of the right half of the sheep brain dissection.

Notes

The ◀ corresponds to the indicated outcome(s) found at the beginning of the laboratory exercise.

Dissection of the Sheep Brain

Part A Assessments

Answer the following questions:

1. Compare the relative sizes of the sheep and human cerebral hemispheres. ◀1 ◀3 _____
   _____
   _____

2. How do the gyri (convolutions) and sulci of the sheep cerebrum compare with the human cerebrum in numbers? ◀1 ◀3
   _____
   _____
   _____

3. What is the significance of the differences you noted in your answers for questions 1 and 2? ◀3
   _____
   _____

4. What difference did you note in the structures of the sheep cerebellum and the human cerebellum? ◀3
   _____
   _____

5. How do the sizes of the olfactory bulbs of the sheep brain compare with those of the human brain? ◀2
   _____
   _____

6. Based on their relative sizes, which of the cranial nerves seem to be most highly developed in the sheep brain? ◀2
   _____
   _____

7. What is the significance of the observations you noted in your answers for questions 5 and 6? ◀2
   _____
   _____

Critical Thinking Application

Prepare a list of at least six features to illustrate ways in which the brains of sheep and humans are similar. ◀ 1   ◀ 3

1. _____

2. _____

3. _____

4. _____

5. _____

6. _____

Interpret the significance of these similarities. ◀ 3 _____

_____

_____

_____

Ear and Hearing

The ear is composed of outer (external), middle, and inner (internal) parts. The outer structures gather sound waves and direct them inward to the eardrum (tympanic membrane). The parts of the middle ear, in turn, transmit vibrations from the eardrum to the inner ear, where the hearing receptors are located. As they are stimulated, these receptors initiate nerve impulses to pass over the vestibulocochlear nerve into the auditory cortex of the brain, where the impulses are interpreted and the sensations of hearing are created.

Purpose of the Exercise

To review the structural and functional characteristics of the ear and to conduct some ordinary hearing tests.

Learning Outcomes

After completing this exercise, you should be able to

1. Locate the major structures of the ear.
2. Describe the functions of the structures of the ear.
3. Trace the pathway of sound vibrations from the eardrum to the hearing receptors.
4. Conduct four ordinary hearing tests and summarize the results.

Procedure A—Structure and Function of the Ear

1. Review the section entitled "Sense of Hearing" in chapter 10 of the textbook.
2. As a review activity, label figures 30.1, 30.2, and 30.3.
3. Examine the dissectible model of the ear and locate the following features:

outer (external) ear
    auricle (pinna)
    external acoustic meatus (external auditory canal)
    eardrum (tympanic membrane)
middle ear
    tympanic cavity
    auditory ossicles (fig. 30.4)
        malleus
        incus
        stapes
    oval window
auditory tube (pharyngotympanic tube; eustachian tube)
inner (internal) ear
    osseous (bony) labyrinth
    membranous labyrinth
    cochlea
    semicircular canals
    vestibule
        utricle
        saccule
vestibulocochlear nerve
    vestibular nerve (balance branch)
    cochlear nerve (hearing branch)

4. Complete Parts A and B of Laboratory Report 30.

Demonstration

Observe the section of the cochlea in the microscope set up by the laboratory instructor. Locate one of the turns of the cochlea and, using figures 30.3 and 30.5 as a guide, identify the scala vestibuli, cochlear duct, scala tympani, vestibular membrane, basilar membrane, and the spiral organ (organ of Corti).

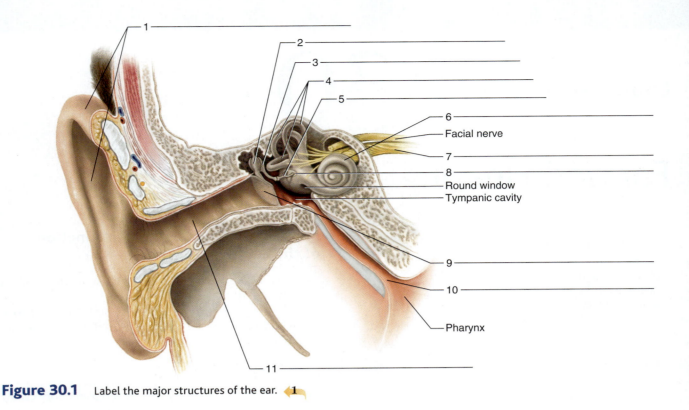

Figure 30.1 Label the major structures of the ear.

Labels on figure:
1
2
3
4
5
6 — Facial nerve
7
8 — Round window
Tympanic cavity
9
10
Pharynx
11

Procedure B—Hearing Tests

Perform the following tests in a quiet room, using your laboratory partner as the test subject.

1. *Auditory acuity test.* To conduct this test, follow these steps:
   a. Have the test subject sit with eyes closed.
   b. Pack one of the subject's ears with cotton.
   c. Hold a ticking watch close to the open ear, and slowly move it straight out and away from the ear.
   d. Have the subject indicate when the sound of the ticking can no longer be heard.
   e. Use a meterstick to measure the distance in centimeters from the ear to the position of the watch.
   f. Repeat this procedure to test the acuity of the other ear.
   g. Record the test results in Part C of the laboratory report.
2. *Sound localization test.* To conduct this test, follow these steps:
   a. Have the subject sit with eyes closed.
   b. Hold the ticking watch somewhere within the audible range of the subject's ears, and ask the subject to point to the watch.

   c. Move the watch to another position and repeat the request. In this manner, determine how accurately the subject can locate the watch when it is in each of the following positions: in front of the head, behind the head, above the head, on the right side of the head, and on the left side of the head.
   d. Record the test results in Part C of the laboratory report.
3. *Rinne test.* This test is done to assess possible conduction deafness by comparing bone and air conduction. To conduct this test, follow these steps:
   a. Obtain a tuning fork and strike it with a rubber hammer, or on the heel of your hand, causing it to vibrate.
   b. Place the end of the fork's handle against the subject's mastoid process behind one ear. Have the prongs of the fork pointed downward and away from the ear, and be sure nothing is touching them (fig. 30.6*a*). The sound sensation is that of bone conduction. If no sound is experienced, nerve deafness exists.
   c. Ask the subject to indicate when the sound is no longer heard.

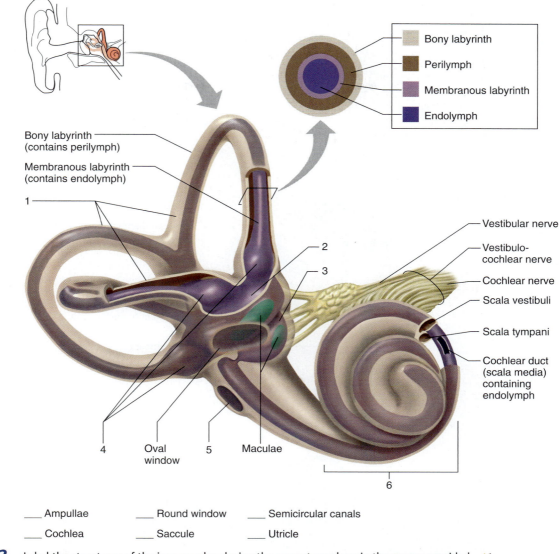

Bony labyrinth
Perilymph
Membranous labyrinth
Endolymph

Bony labyrinth
(contains perilymph)

Membranous labyrinth
(contains endolymph)

1

2

3

Vestibular nerve

Vestibulo-
cochlear nerve

Cochlear nerve

Scala vestibuli

Scala tympani

Cochlear duct
(scala media)
containing
endolymph

4   Oval   5   Maculae
         window

6

___ Ampullae   ___ Round window   ___ Semicircular canals

___ Cochlea   ___ Saccule   ___ Utricle

Figure 30.2   Label the structures of the inner ear by placing the correct numbers in the spaces provided.

d. Then quickly remove the fork from the mastoid process and position it in the air close to the opening of the nearby external acoustic meatus (fig. 30.6*b*).

If hearing is normal, the sound (from air conduction) will be heard again; if there is conductive impairment, the sound will not be heard. Conductive impairment involves outer or middle ear defects. Hearing aids can improve hearing for conductive deafness because bone conduction transmits the sound into the inner ear. Surgery could possibly correct this type of defect.

e. Record the test results in Part C of the laboratory report.

4. *Weber test.* This test is used to distinguish possible conduction or sensory deafness. To conduct this test, follow these steps:

a. Strike the tuning fork with the rubber hammer.

b. Place the handle of the fork against the subject's forehead in the midline (fig. 30.7).

c. Ask the subject to indicate if the sound is louder in one ear than in the other or if it is equally loud in both ears.

If hearing is normal, the sound will be equally loud in both ears. If there is conductive impairment, the sound will appear louder in the affected ear. If some degree of sensory (nerve) deafness exists, the sound will be louder in the normal ear.

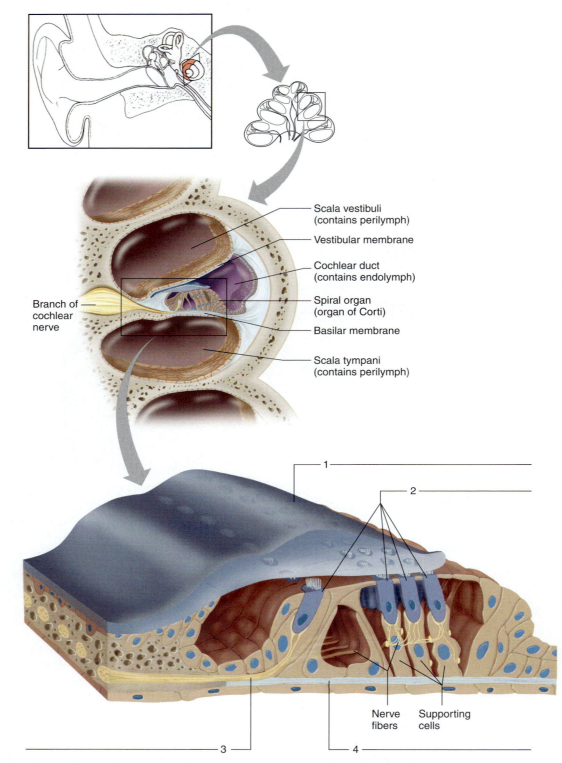

Scala vestibuli
(contains perilymph)

Vestibular membrane

Cochlear duct
(contains endolymph)

Spiral organ
(organ of Corti)

Basilar membrane

Scala tympani
(contains perilymph)

Branch of
cochlear
nerve

1

2

Nerve
fibers

Supporting
cells

3

4

Figure 30.3   Label the structures associated with the spiral organ (organ of Corti) of a cochlea.

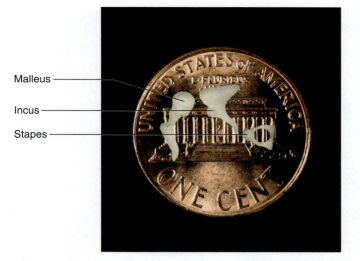

Figure 30.4 Middle ear bones (auditory ossicles) superimposed on a penny for a size relationship. The arrangement of the malleus, incus, and stapes in the enlarged photograph resembles the articulations in the middle ear.

Malleus

Incus

Stapes

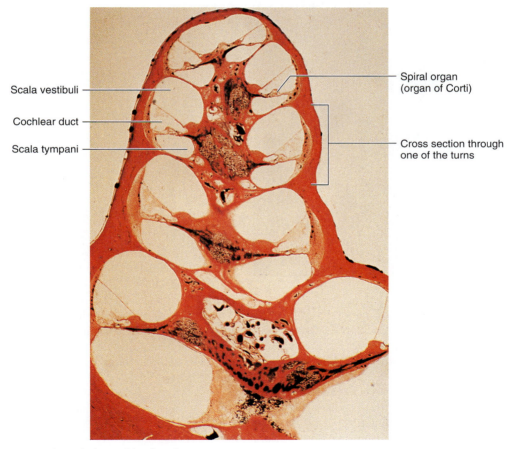

Scala vestibuli

Cochlear duct

Scala tympani

Spiral organ (organ of Corti)

Cross section through one of the turns

Figure 30.5 A section through the cochlea (22×).

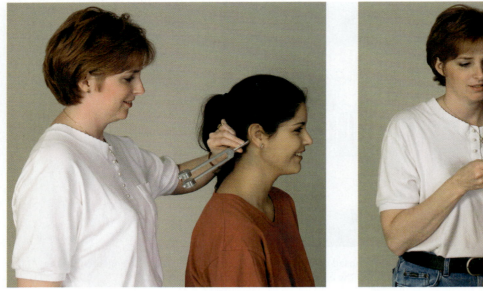

(a)

(b)

Figure 30.6   Rinne test: (a) first placement of vibrating tuning fork until sound is no longer heard; (b) second placement of tuning fork to assess air conduction.

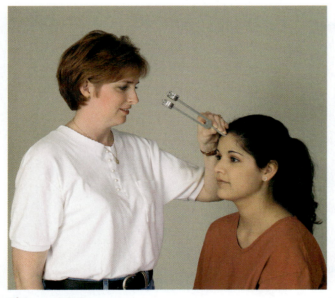

Figure 30.7   Weber test.

The impairment involves the spiral organ (organ of Corti) or the cochlear nerve. Hearing aids will not improve sensory deafness.

   d. Have the subject experience the effects of conductive impairment by packing one ear with cotton and repeating the Weber test. Usually the sound appears louder in the plugged (or impaired) ear because extraneous sounds from the room are blocked out.

   e. Record the test results in Part C of the laboratory report.

5. Complete Part C of the laboratory report.

Critical Thinking Application

Ear structures from the outer ear into the inner ear are progressively smaller. Using results obtained from the hearing tests, explain this advantage.

_____

_____

_____

_____

_____

Demonstration

Ask the laboratory instructor to demonstrate the use of the audiometer. This instrument produces sound vibrations of known frequencies transmitted to one or both ears of a test subject through earphones. The audiometer can be used to determine the threshold of hearing for different sound frequencies, and, in the case of a hearing impairment, it can be used to determine the percentage of hearing loss for each frequency.

Name _____

Date _____

Section _____

The ◀ corresponds to the indicated outcome(s) found at the beginning of the laboratory exercise.

Ear and Hearing

Part A Assessments

Match the terms in column A with the descriptions in column B. Place the letter of your choice in the space provided. ◀1 ◀2

Column A	Column B
a. Auditory tube	_____ **1.** Auditory ossicle attached to eardrum
b. Eardrum	_____ **2.** Air-filled space containing auditory ossicles
c. External acoustic meatus	_____ **3.** Contacts hairs of hearing receptors
d. Malleus	_____ **4.** Leads from oval window to apex of cochlea
e. Membranous labyrinth	
f. Osseous (bony) labyrinth	_____ **5.** S-shaped tube leading to eardrum
g. Scala tympani	_____ **6.** Cone-shaped, semitransparent membrane attached to malleus
h. Scala vestibuli	_____ **7.** Auditory ossicle attached to oval window
i. Stapes	
j. Tectorial membrane	_____ **8.** Contains endolymph
k. Tympanic cavity	_____ **9.** Bony canal of inner ear in temporal bone
	_____ **10.** Connects middle ear and pharynx
	_____ **11.** Extends from apex of cochlea to round window

Part B Assessments

Label the structures indicated in the micrograph of the spiral organ (organ of Corti) in figure 30.8. ◀1

Part C Assessments

1. Results of auditory acuity test: ◀4

Ear Tested	Audible Distance (cm)
Right	
Left	

Terms:
Basilar membrane
Hair cells
Scala media (cochlear duct)
Scala tympani
Tectorial membrane

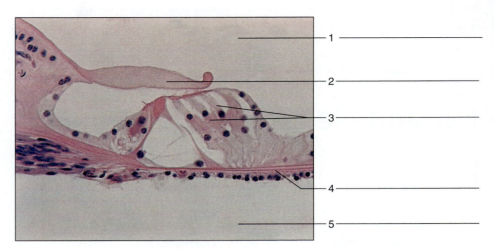

1
2
3
4
5

Figure 30.8 Uusing the terms provided, label the structures associated with this spiral organ (organ of Corti) region of a cochlea, (75× micrograph enlarged to 300×).

2. Results of sound localization test:

Actual Location	Reported Location
Front of the head	
Behind the head	
Above the head	
Right side of the head	
Left side of the head	

3. Results of experiments using tuning forks:

Test	Left Ear (normal or impaired)	Right Ear (normal or impaired)
Rinne		
Weber		

4. Summarize the results of the hearing tests you conducted on your laboratory partner.

_____

_____

_____

_____

Eye Structure

Materials Needed

Textbook
Dissectible eye model
Sheep or beef eye (fresh or preserved)
Dissecting tray
Dissecting instruments—forceps, sharp scissors,
    and dissecting needle

For Learning Extension:
Ophthalmoscope

⚠️ Safety

- Wear disposable gloves when working on the eye dissection.
- Dispose of tissue remnants and gloves as instructed.
- Wash the dissecting tray and instruments as instructed.
- Wash your laboratory table.
- Wash your hands before leaving the laboratory.

The eye contains photoreceptors, modified neurons located on its inner wall. Other parts of the eye provide protective functions or make it possible to move the eyeball. Still other structures serve to focus light entering the eye so that a sharp image is projected onto the receptor cells. Nerve impulses generated when the receptors are stimulated travel along the optic nerves to the brain, which interprets the impulses and creates the sensation of sight.

Purpose of the Exercise

To review the structure and function of the eye and to dissect a mammalian eye.

Learning Outcomes

After completing this exercise, you should be able to

1. Locate the major structures of an eye.
2. Describe the functions of the structures of an eye.
3. Trace the structures through which light passes as it travels from the cornea to the retina.
4. Dissect a mammalian eye and locate its major features.

Procedure A—Structure and Function of the Eye

1. Review the sections entitled "Visual Accessory Organs," "Structure of the Eye," and "Visual Receptors" in chapter 10 of the textbook.
2. As a review activity, label figures 31.1, 31.2, and 31.3.
3. Complete Part A of Laboratory Report 31.
4. Examine the dissectible model of the eye and locate the following features:

   eyelid
   conjunctiva
   orbicularis oculi
   levator palpebrae superioris
   lacrimal apparatus
      lacrimal gland
      canaliculi
      lacrimal sac
      nasolacrimal duct
   extrinsic muscles
      superior rectus
      inferior rectus
      medial rectus
      lateral rectus
      superior oblique
      inferior oblique

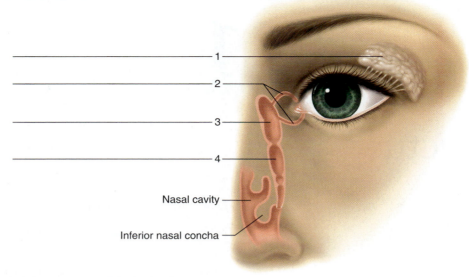

Nasal cavity

Inferior nasal concha

Figure 31.1 Label the structures of the lacrimal apparatus.

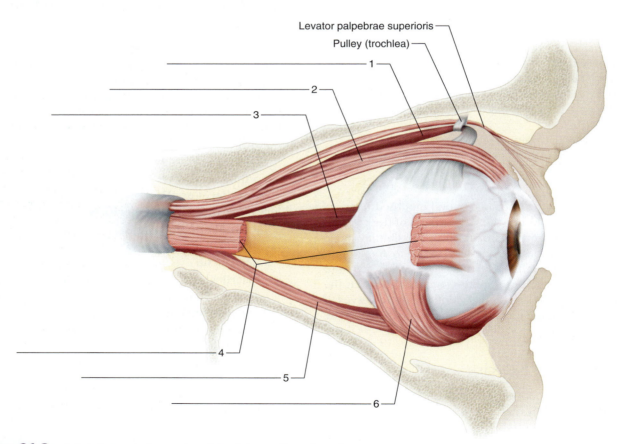

Levator palpebrae superioris

Pulley (trochlea)

Figure 31.2 Label the extrinsic muscles of the right eye (lateral view).

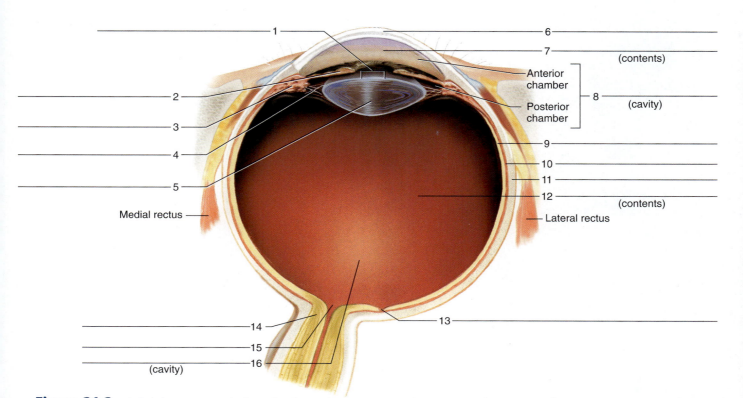

Figure 31.3 Label the structures indicated in this transverse section of the right eye (superior view). ◀1

Labels in the figure:
- 1
- 2
- 3
- 4
- 5
- 6
- 7 — Anterior chamber
- (contents)
- 8 — (cavity)
- Posterior chamber
- 9
- 10
- 11
- 12 — (contents)
- Medial rectus
- Lateral rectus
- 13
- 14
- 15
- 16
- (cavity)

outer (fibrous) layer
  sclera
  cornea
middle (vascular) layer
  choroid coat
  ciliary body
    ciliary processes
    ciliary muscles
    suspensory ligaments
  iris
  pupil (opening in center of iris)
inner (nervous) layer
  retina
    macula lutea
    fovea centralis
    optic disc
  optic nerve
lens
anterior cavity
  anterior chamber
  posterior chamber
  aqueous humor
posterior cavity
  vitreous humor

Learning Extension

Use an ophthalmoscope to examine the interior of your laboratory partner's eye. This instrument consists of a set of lenses held in a rotating disc, a light source, and some mirrors that reflect the light into the test subject's eye.

The examination should be conducted in a dimly lighted room. Have your partner seated and staring straight ahead at eye level. Move the rotating disc of the ophthalmoscope so that the *O* appears in the lens selection window. Hold the instrument in your right hand with the end of your index finger on the rotating disc (fig. 31.4). Direct the light at a slight angle from a distance of about 15 cm into the subject's right eye. The light beam should pass along the inner edge of the pupil. Look through the instrument, and you should see a reddish, circular area—the interior of the eye. Rotate the disc of lenses to higher values until sharp focus is achieved.

Move the ophthalmoscope to within about 5 cm of the eye being examined *being very careful that the instrument does not touch the eye,* and again rotate the lenses to sharpen the focus (fig. 31.5). Locate the optic disc and the blood vessels that pass through it. Also locate the yellowish macula lutea by having your partner stare directly into the light of the instrument (fig. 31.6).

Examine the subject's iris by viewing it from the side and by using a lens with a +15 or +20 value.

Figure 31.4 An ophthalmoscope is used to examine the interior of the eye.

- Eyepiece
- Rotating disc
- Lens selection window
- Handle of ophthalmoscope

(a)

(b)

Figure 31.5 (*a*) Rotate the disc of lenses until sharp focus is achieved. (*b*) Move the ophthalmoscope to within 5 cm of the eye to examine the optic disc.

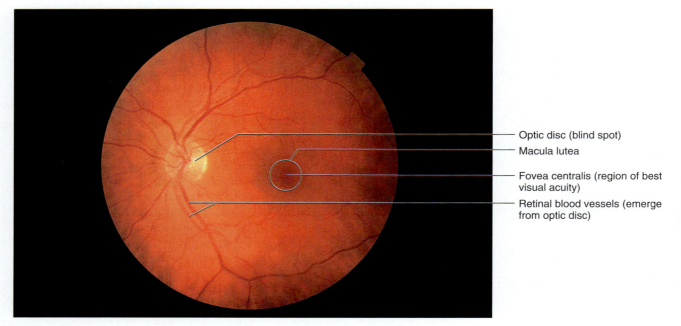

- Optic disc (blind spot)
- Macula lutea
- Fovea centralis (region of best visual acuity)
- Retinal blood vessels (emerge from optic disc)

Figure 31.6 Retina of the eye as seen through the pupil using an ophthalmoscope.

Procedure B—Eye Dissection

1. Obtain a mammalian eye, place it in a dissecting tray, and dissect it as follows:
   a. Trim away the fat and other connective tissues but leave the stubs of the *extrinsic muscles* and of the *optic nerve.* This nerve projects outward from the posterior region of the eyeball.
   b. The *conjunctiva,* which lines the inside of the eyelid, is reflected over the anterior surface of the eye, except for the cornea. Lift some of this thin membrane away from the eye with forceps and examine it.
   c. Locate and observe the *cornea, sclera,* and *iris.* Also note the *pupil* and its shape. The cornea from a fresh eye will be transparent; when preserved, it becomes opaque.
   d. Use sharp scissors to make a coronal section of the eye. To do this, cut through the wall about 1 cm from the margin of the cornea and continue all the way around the eyeball. Try not to damage the internal structures of the eye (fig. 31.7).
   e. Gently separate the eyeball into anterior and posterior portions. Usually the jellylike vitreous humor will remain in the posterior portion, and the lens may adhere to it. Place the parts in the dissecting tray with their contents facing upward.
   f. Examine the anterior portion of the eye, and locate the *ciliary body,* which appears as a dark, circular structure. Also note the *iris* and the *lens* if it remained in the anterior portion. The lens is normally attached to the ciliary body by many *suspensory ligaments,* which appear as delicate, transparent threads.
   g. Use a dissecting needle to gently remove the lens and examine it. If the lens is still transparent, hold it up and look through it at something in the distance and note that the lens inverts the image. The lens of a preserved eye is usually too opaque for this experience. If the lens of the human eye becomes opaque, the defect is called a cataract (fig. 31.8).
   h. Examine the posterior portion of the eye. Note the *vitreous humor.* This jellylike mass helps to hold the lens in place anteriorly and helps to hold the *retina* against the choroid coat.
   i. Carefully remove the vitreous humor and examine the retina. This layer will appear as a thin, nearly colorless to cream-colored membrane that detaches easily from the choroid coat. Compare the structures identified to figure 31.9.
   j. Locate the *optic disc*—the point where the retina is attached to the posterior wall of the eyeball and where the optic nerve originates. There are no receptor cells in the optic disc, so this region is also called the "blind spot."

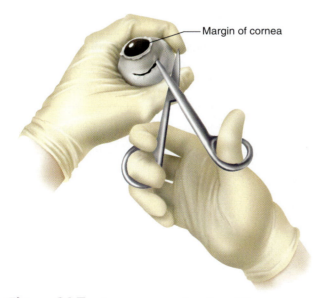

Figure 31.7 Prepare a coronal section of the eye.

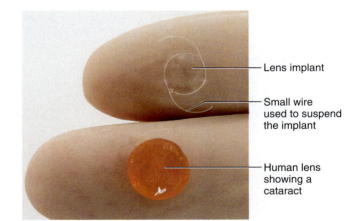

Figure 31.8 Human lens showing a cataract and one type of lens implant (intraocular lens replacement) on tips of fingers to show their relative size. A normal lens is transparent.

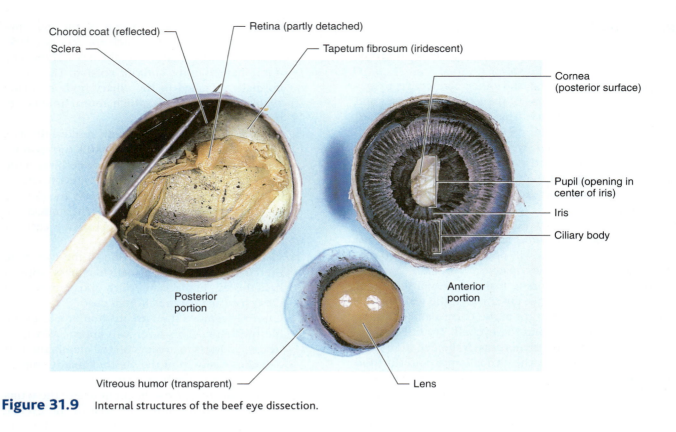

Figure 31.9 Internal structures of the beef eye dissection.

Choroid coat (reflected)

Sclera

Retina (partly detached)

Tapetum fibrosum (iridescent)

Cornea (posterior surface)

Pupil (opening in center of iris)

Iris

Ciliary body

Posterior portion

Anterior portion

Vitreous humor (transparent)

Lens

k. Note the iridescent area of the choroid coat beneath the retina. This colored surface in ungulates (mammals having hoofs) is called the *tapetum fibrosum*. It serves to reflect light back through the retina, an action thought to aid the night vision of some animals. The tapetum fibrosum is lacking in the human eye.

2. Discard the tissues of the eye as directed by the laboratory instructor.

3. Complete Part B of the laboratory report.

 **Critical Thinking Application**

A strong blow to the head might cause the retina to detach. From observations made during the eye dissection, explain why this could happen.

_____

_____

_____

_____

The ◀ corresponds to the indicated outcome(s) found at the beginning of the laboratory exercise.

Eye Structure

Part A Assessments

Match the terms in column A with the descriptions in column B. Place the letter of your choice in the space provided. ◀**1**

Column A

a. Aqueous humor
b. Choroid coat
c. Ciliary muscles
d. Conjunctiva
e. Cornea
f. Iris
g. Lacrimal gland
h. Optic disc
i. Retina
j. Sclera
k. Suspensory ligament
l. Vitreous humor

Column B

_____ **1.** Posterior five-sixths of middle (vascular) layer loosely joined to sclera

_____ **2.** White part of outer (fibrous) layer

_____ **3.** Transparent anterior portion of outer layer

_____ **4.** Inner lining of eyelid

_____ **5.** Secretes tears

_____ **6.** Fills posterior cavity of eye

_____ **7.** Area where optic nerve exits the eye

_____ **8.** Smooth muscle that controls light entering the eye

_____ **9.** Fills anterior and posterior chambers of the anterior cavity of the eye

_____ **10.** Contains visual receptors called rods and cones

_____ **11.** Connects lens to ciliary body

_____ **12.** Cause lens to change shape

Complete the following:

13. List the structures and fluids through which light passes as it travels from the cornea to the retina. ◀**3**

_____

_____

_____

14. List three ways in which rods and cones differ in structure or function. ◀**2** _____

_____

_____

_____

_____

Part B Assessments

Complete the following:

1. Which layer of the eye was the most difficult to cut? ◄4 _____

2. What kind of tissue do you think is responsible for this quality of toughness? ◄4 _____

3. How do you compare the shape of the pupil in the dissected eye with your pupil? ◄4 _____

_____

_____

4. Where do you find aqueous humor in the dissected eye? ◄4 _____

_____

_____

5. What is the function of the dark pigment in the choroid coat? ◄2 _____

_____

_____

6. Describe the lens of the dissected eye. ◄4 _____

_____

_____

_____

7. Describe the vitreous humor of the dissected eye. ◄4 _____

_____

_____

_____

Laboratory Exercise 32

Visual Tests and Demonstrations

Materials Needed

Textbook
Snellen eye chart
3″ × 5″ card (plain)
3″ × 5″ with word typed in center
Astigmatism chart
Meterstick
Metric ruler
Pen flashlight
Ichikawa's or Ishihara's color plates for colorblindness test

Normal vision (emmetropia) results when light rays from objects in the external environment are refracted by the cornea and lens of the eye and focused onto the photoreceptors of the retina. If a person has a condition of nearsightedness (myopia), the image focuses in front of the retina as the eyeball is too long, and nearby objects are clear, but distant objects are blurred. If a person has a condition of farsightedness (hyperopia), the image focuses behind the retina as the eyeball is too short, and distant objects are clear, but nearby objects are blurred. A concave lens is required to correct vision for nearsightedness; a convex lens is required to correct vision for farsightedness. If a defect occurs in the curvature of the cornea or lens, a condition called astigmatism results. Focusing in different planes cannot occur simultaneously. Corrective cylindrical lenses will allow proper refraction of light onto the retina to compensate for the unequal curvatures.

As part of a natural aging process, the lens has decreasing elasticity, which complicates nearby vision. The near point of accommodation test is used to determine the ability to accommodate. Often a person will use "reading glasses" to compensate for this condition. Hereditary defects in color vision result from lack of certain cones necessary to absorb certain wavelengths of light. Special color test plates are used to diagnose a condition of colorblindness.

As you conduct this laboratory exercise, it does not matter the order of any of the visual tests or visual demonstrations performed. Each test and demonstration does not depend upon any of the other test or demonstration

results. If you wear glasses, perform the tests with and without the corrective lenses. If you wear contact lenses, it is not necessary to run the tests under both conditions, however indicate in the laboratory report that all tests were performed with contact lenses in place.

Purpose of the Exercise

To conduct tests for visual acuity, astigmatism, accommodation, color vision, the blind spot, and certain reflexes of the eye.

Learning Outcomes

After completing this exercise, you should be able to

1. Conduct the tests used to evaluate visual acuity, astigmatism, the ability to accommodate for close vision, and color vision.
2. Describe four conditions that can lead to defective vision.
3. Demonstrate the blind spot, photopupillary reflex, accommodation pupillary reflex, and convergence reflex.

Procedure A—Visual Tests

Perform the following visual tests, using your laboratory partner as a test subject. If your partner usually wears glasses, test each eye with and without the glasses.

1. *Visual acuity test.* Visual acuity (sharpness of vision) can be measured by using a Snellen eye chart (fig. 32.1). This chart consists of several sets of letters in different sizes printed on a white card. The letters near the top of the chart are relatively large, and those in each lower set become smaller. At one end of each set of letters is an acuity value in the form of a fraction. One of the sets near the bottom of the chart, for example, is marked 20/20. The normal eye can clearly see these letters from the standard distance of 20 feet and thus is said to have 20/20 vision. The letter at the top of the chart is marked 20/200. The normal eye can read letters

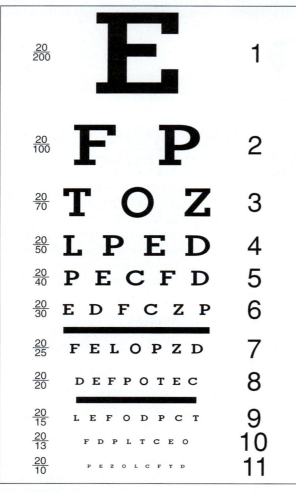

Figure 32.1   The Snellen eye chart looks similar to this but is somewhat larger.

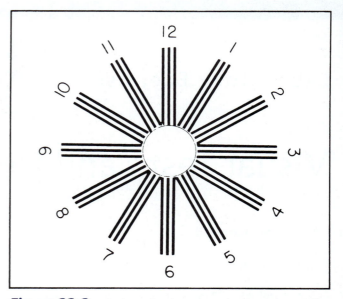

Figure 32.2   Astigmatism is evaluated using a chart such as this one.

of this size from a distance of 200 feet. Thus, an eye able to read only the top letter of the chart from a distance of 20 feet is said to have 20/200 vision. This person has less than normal vision. A line of letters near the bottom of the chart is marked 20/15. The normal eye can read letters of this size from a distance of 15 feet, but a person might be able to read it from 20 feet. This person has better than normal vision.

To conduct the visual acuity test, follow these steps:

a. Hang the Snellen eye chart on a well-illuminated wall at eye level.
b. Have your partner stand 20 feet in front of the chart, gently cover the left eye with a 3″ × 5″ card, and read the smallest set of letters possible using your right eye.
c. Record the visual acuity value for that set of letters in Part A of Laboratory Report 32.
d. Repeat the procedure, using the left eye.

2. *Astigmatism test.* Astigmatism is a condition that results from a defect in the curvature of the cornea or lens. As a consequence, some portions of the image projected on the retina are sharply focused, and other portions are blurred. Astigmatism can be evaluated by using an astigmatism chart (fig. 32.2). This chart consists of sets of black lines radiating from a central spot like the spokes of a wheel. To a normal eye, these lines appear sharply focused and equally dark; however, if the eye has an astigmatism, some sets of lines appear sharply focused and dark, while others are blurred and less dark.

To conduct the astigmatism test, follow these steps:

a. Hang the astigmatism chart on a well-illuminated wall at eye level.
b. Have your partner stand 20 feet in front of the chart, gently cover the left eye with a 3″ × 5″ card. Using your right eye, focus on the spot in the center of the radiating lines, and report which lines, if any, appear more sharply focused and darker.
c. Repeat the procedure, using the left eye.
d. Record the results in Part A of the laboratory report.

3. *Accommodation test.* Accommodation is the changing of the shape of the lens that occurs when the normal eye is focused for close vision. It involves a reflex in which muscles of the ciliary body are stimulated to contract, releasing tension on the suspensory ligaments fastened to the lens capsule. This allows the capsule to rebound elastically, causing the surface of the lens to become more convex. The

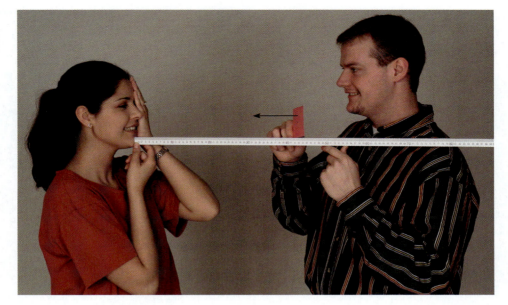

Figure 32.3   To determine the near point of accommodation, slide the 3″ × 5″ card along the meterstick toward your partner's open eye until it reaches the closest location where your partner can still see the word sharply focused.

ability to accommodate is likely to decrease with age because the tissues involved tend to lose their elasticity.

To evaluate the ability to accommodate, follow these steps:

a. Hold the end of a meterstick against your partner's chin so that the stick extends outward at a right angle to the plane of the face (fig. 32.3).

b. Have your partner close the left eye.

c. Hold a 3″ × 5″ card with a word typed in the center at the distal end of the meterstick.

d. Slide the card along the stick toward your partner's open right eye, and locate the *point closest to the eye* where your partner can still see the letters of the word sharply focused. This distance is called the *near point of accommodation,* and it tends to increase with age (table 32.1).

e. Repeat the procedure with the right eye closed.

f. Record the results in Part A of the laboratory report.

4. *Color vision test.* Some people exhibit defective color vision because they lack certain cones, usually those sensitive to the reds or greens. This trait is an X-linked (sex-linked) inheritance, so the condition is more prevalent in males (7%) than in females (0.4%). People who lack or possess decreased sensitivity to the red-sensitive cones possess protanopia colorblindness; those who lack or possess decreased sensitivity to green-sensitive cones possess deuteranopia colorblindness. The colorblindness condition is often more of a deficiency or a weakness than one of blindness. Laboratory Exercise 48 describes the genetics of colorblindness.

Table 32.1   Near Point of Accommodation

Age (years)	Average Near Point (cm)
10	7
20	10
30	13
40	20
50	45
60	90

To conduct the color vision test, follow these steps:

a. Examine the color test plates in Ichikawa's or Ishihara's book to test for any red-green color vision deficiency. Also examine figure 32.4.

b. Hold the test plates approximately 30 inches from the subject in bright light. All responses should occur within 3 seconds.

c. Compare your responses with the correct answers in Ichikawa's or Ishihara's book. Determine the percentage of males and females in your class who exhibit any color-deficient vision. If an individual exhibits color-deficient vision, determine if the condition is protanopia or deuteranopia.

d. Record the class results in Part A of the laboratory report.

5. Complete Part A of the laboratory report.

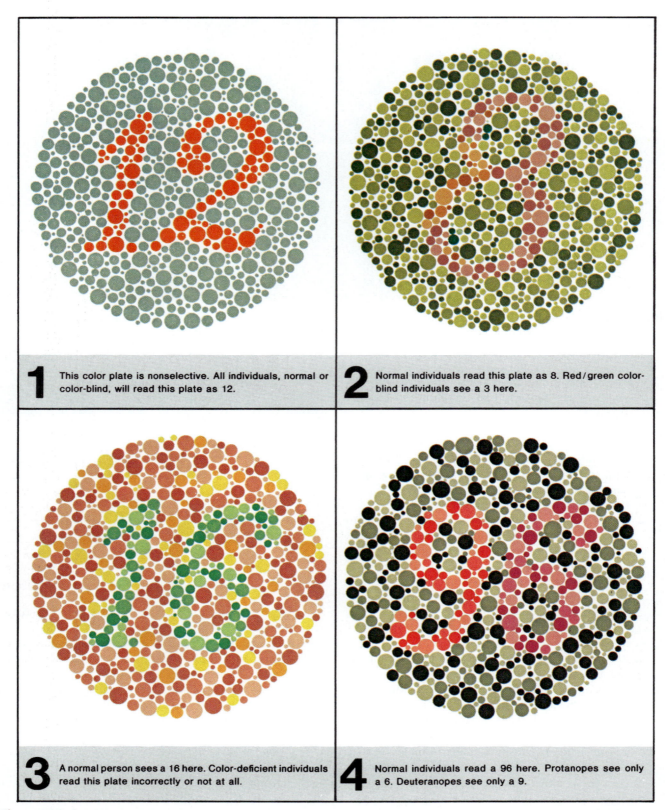

1 This color plate is nonselective. All individuals, normal or color-blind, will read this plate as 12.

2 Normal individuals read this plate as 8. Red/green color-blind individuals see a 3 here.

3 A normal person sees a 16 here. Color-deficient individuals read this plate incorrectly or not at all.

4 Normal individuals read a 96 here. Protanopes see only a 6. Deuteranopes see only a 9.

Figure 32.4 Samples of Ishihara's color plates. These plates are reproduced from *Ishihara's Tests for Colour Blindness* published by Kanehara & Co., Ltd., Tokyo, Japan, but tests for colorblindness cannot be conducted with this material. For accurate testing, the original plates should be used.

Procedure B—Visual Demonstrations

Perform the following demonstrations with the help of your laboratory partner.

1. *Blind-spot demonstration.* There are no photoreceptors in the optic disc, located where the nerve fibers of the retina leave the eye and enter the optic nerve. Consequently, this region of the retina is commonly called the *blind spot.*

   To demonstrate the blind spot, follow these steps:

   a. Close your left eye, hold figure 32.5 about 35 cm away from your face, and stare at the + sign in the figure with your right eye.

   b. Move the figure closer to your face as you continue to stare at the + until the dot on the figure suddenly disappears. This happens when the image of the dot is focused on the optic disc. Measure the right eye distance using a metric ruler or a meterstick.

   c. Repeat the procedures with your right eye closed. This time stare at the dot, and the + will disappear when the image falls on the optic disc. Measure the distance.

   d. Record the results in Part B of the laboratory report.

 ### Critical Thinking Application

Under normal visual circumstances, explain why small objects are not lost from our vision.

_____

_____

_____

_____

2. *Photopupillary reflex.* The smooth muscles of the iris control the size of the pupil. For example, when the intensity of light entering the eye increases, a photopupillary reflex is triggered, and the circular muscles of the iris are stimulated to contract. As a result, the size of the pupil decreases, and less light enters the eye.

Figure 32.5 Blind-spot demonstration.

To demonstrate this photopupillary reflex, follow these steps:

   a. Ask your partner to sit with his or her hands thoroughly covering his or her eyes for 2 minutes.

   b. Position a pen flashlight close to one eye with the light shining on the hand that covers the eye.

   c. Ask your partner to remove the hand quickly.

   d. Observe the pupil and note any change in its size.

   e. Have your partner remove the other hand, but keep that uncovered eye shielded from extra light.

   f. Observe both pupils and note any difference in their sizes.

3. *Accommodation pupillary reflex.* The pupil constricts as a normal accommodation reflex to focusing on close objects. To demonstrate the accommodation pupillary reflex, follow these steps:

   a. Have your partner stare for several seconds at some dimly illuminated object in the room that is more than 20 feet away.

   b. Observe the size of the pupil of one eye. Then hold a pencil about 25 cm in front of your partner's face, and have your partner stare at it.

   c. Note any change in the size of the pupil.

4. *Convergence reflex.* The eyes converge as a normal convergence reflex to focusing on close objects. To demonstrate the convergence reflex, follow these steps:

   a. Repeat the procedure outlined for the accommodation pupillary reflex.

   b. Note any change in the position of the eyeballs as your partner changes focus from the distant object to the pencil.

5. Complete Part B of the laboratory report.

Notes

The ⬅ corresponds to the indicated outcome(s) found at the beginning of the laboratory exercise.

Visual Tests and Demonstrations

Part A Assessments

1. Visual acuity test results: ⬅ **1**

Eye Tested	Acuity Values
Right eye	
Right eye with glasses (if applicable)	
Left eye	
Left eye with glasses (if applicable)	

2. Astigmatism test results: ⬅ **1**

Eye Tested	Darker Lines
Right eye	
Right eye with glasses (if applicable)	
Left eye	
Left eye with glasses (if applicable)	

3. Accommodation test results: ⬅ **1**

Eye Tested	Near Point (cm)
Right eye	
Right eye with glasses (if applicable)	
Left eye	
Left eye with glasses (if applicable)	

4. Color vision test results: ◀**1**

Condition	Males			Females		
	Class Number	Class Percentage	Expected Percentage	Class Number	Class Percentage	Expected Percentage
Normal color vision			93			99.6
Deficient red-green color vision			7			0.4
Protanopia (lack red-sensitive cones)			less frequent type			less frequent type
Deuteranopia (lack green-sensitive cones)			more frequent type			more frequent type

5. Complete the following: ◀**2**

a. What is meant by 20/70 vision? _____

_____

b. What is meant by 20/10 vision? _____

_____

c. What visual problem is created by astigmatism? _____

_____

d. Why does the near point of accommodation often increase with age? _____

_____

e. Describe the eye defect that causes color-deficient vision. _____

_____

Part B Assessments

1. Blind-spot results: ◀**3**
  a. Right eye distance _____
  b. Left eye distance _____

Complete the following:

2. Explain why an eye has a blind spot. ◀**3** _____

_____

3. Describe the photopupillary reflex. ◀**3** _____

_____

4. What difference did you note in the size of the pupils when one eye was exposed to bright light and the other

  eye was shielded from the light? ◀**3** _____

_____

5. Describe the accommodation pupillary reflex. ◀**3** _____

_____

6. Describe the convergence reflex. ◀**3** _____

_____

Endocrine Histology and Diabetic Physiology

Materials Needed

For Procedure A—Endocrine Histology:
Textbook
Human torso model
Compound light microscope
Prepared microscope slides of the following:
  Pituitary gland
  Thyroid gland
  Parathyroid gland
  Adrenal gland
  Pancreas

For Procedure B—Insulin Shock:
500 mL beaker
Live small fish (1″ to 1.5″ goldfish, guppy, rosy red feeder, or other)
Small fish net
Insulin (regular U-100) (HumulinR in 10 mL vials has 100 units/mL; store in refrigerator—do not freeze)
Syringes for U-100 insulin
10% glucose solution
Clock with second hand or timer

For Learning Extension:
Compound light microscope
Prepared microscope slides of the following:
  Normal human pancreas (stained for alpha and beta cells)
  Human pancreas of diabetic

⚠ Safety

- Review all the safety guidelines inside the front cover.
- Wear disposable gloves when handling the fish and syringes.
- Dispose of the used syringe and needle in the puncture-resistant container.
- Wash your hands before leaving the laboratory.

The endocrine system consists of ductless glands that act together with parts of the nervous system to help control body activities. The endocrine glands secrete hor-mones transported in body fluids and affect cells possessing appropriate receptor molecules. In this way, hormones influence the rate of metabolic reactions, the transport of substances through cell membranes, the regulation of water and electrolyte balances, and many other functions. By controlling cellular activities, endocrine glands play important roles in the maintenance of homeostasis.

The primary "fuel" for the mitochondria of cells is sugar (glucose). Insulin produced by the beta cells of the pancreatic islets has a primary regulation role in the transport of blood sugar across the cell membranes. Some individuals do not produce enough insulin; others might not have adequate transport of glucose into the body cells when cells lose insulin receptors. This rather common disease is called *diabetes mellitus*. Diabetes mellitus is a functional disease that might have its onset either during childhood or later in life.

The protein hormone insulin has several functions: it stimulates the liver and skeletal muscles to form glycogen from glucose; it inhibits the conversion of noncarbohy-drates into glucose; it promotes the facilitated diffusion of glucose across cell membranes of cells possessing insulin receptors (adipose tissue, cardiac muscle, and resting skeletal muscle); it decreases blood sugar (glucose); it increases protein synthesis; and it promotes fat storage in adipose cells. In a normal person, as blood sugar in-creases during nutrient absorption after a meal, the rising glucose levels directly stimulate beta cells of the pancreas to secrete insulin. This increase in insulin prevents blood sugar from sudden surges (hyperglycemia) by promoting glycogen production in the liver and an increase in the entry of glucose into muscle and adipose cells.

Between meals and during sleep, blood glucose levels decrease and insulin also decreases. When insu-lin concentration decreases, more glucose is available to enter cells that lack insulin receptors. Neurons, liver cells, kidney cells, and red blood cells either lack insu-lin receptors or express limited insulin receptors and obtain their glucose by facilitated diffusion. These cells depend upon blood glucose concentrations to enable cellular respiration and ATP production. When insulin is given as a medication for diabetes mellitus, blood glucose levels may drop drastically (hypoglycemia) if the individual does not eat. This can have significant effects on the nervous system, resulting in insulin shock.

Type 1 diabetes mellitus usually has a rapid onset relatively early in life, but it can occur in adults. It occurs when insulin is no longer produced; hence sometimes it is referred to as *insulin-dependent diabetes mellitus (IDDM)*. This autoimmune disorder occurs when the immune system destroys beta cells, resulting in a decrease in insulin production. This disorder affects about 10 to 15% of diabetics. The treatment for type 1 diabetes mellitus is to administer insulin, usually by injection.

Type 2 diabetes mellitus has a gradual onset, usually in those over age 40. It is sometimes called *non-insulin-dependent diabetes mellitus (NIDDM)*. This disease results as body cells become less sensitive to insulin or lose insulin receptors and cannot respond to insulin even though insulin levels may remain normal. This situation is called *insulin resistance*. Risk factors for the onset of type 2 diabetes mellitus, which afflicts about 85 to 90% of diabetics, include heredity, obesity, and lack of exercise. Obesity is a major indicator for predicting type 2 diabetes; a weight loss of a mere 10 pounds will increase the body's sensitivity to insulin as a means of prevention. The treatments for type 2 diabetes mellitus include a diet that avoids foods that stimulate insulin production, weight control, exercise, and medications. (The causes and treatments for type 2 diabetes closely resemble those for coronary artery disease!)

Purpose of the Exercise

To examine microscopically the tissues of endocrine glands and to observe behavior changes that occur during insulin shock.

Learning Outcomes

After completing this exercise, you should be able to

1. Name and locate the major endocrine glands.
2. Associate the principal hormones and functions by each of the major endocrine glands.
3. Sketch and label tissue sections from the pituitary gland, thyroid gland, parathyroid glands, adrenal glands, and pancreas.
4. Compare the causes, symptoms, and treatments for type 1 and type 2 diabetes mellitus.
5. Observe and record the behavioral changes caused by insulin shock.
6. Observe and record the recovery from insulin shock when sugar is provided to cells.

Procedure A—Endocrine Histology

1. Review the sections entitled "Pituitary Gland," "Thyroid Gland," "Parathyroid Glands," "Adrenal Glands," "Pancreas," and "Other Endocrine Glands" in chapter 11 of the textbook.
2. As a review activity, label figure 33.1.
3. Complete Part A of Laboratory Report 33.
4. Examine the human torso model and locate the following:

   hypothalamus
   pituitary stalk (infundibulum)
   pituitary gland (hypophysis)
     anterior lobe (adenohypophysis)
     posterior lobe (neurohypophysis)
   thyroid gland
   parathyroid glands
   adrenal glands (suprarenal glands)
     adrenal medulla
     adrenal cortex
   pancreas
   pineal gland
   thymus gland
   ovaries
   testes

5. Examine the microscopic tissue sections of the following glands, and identify the features described:
   Pituitary gland. To examine the pituitary tissue, follow these steps:
   a. Observe the tissues using low-power magnification (fig. 33.2).
   b. Locate the *pituitary stalk (infundibulum),* the *anterior lobe* (the largest part of the gland), and the *posterior lobe.*
   c. Observe an area of the anterior lobe with high-power magnification. Locate a cluster of relatively large cells and identify some *acidophil cells,* which contain pink-stained granules, and some *basophil cells,* which contain blue-stained granules. These acidophil and basophil cells are hormone-secreting cells. The hormones secreted include TSH, ACTH, prolactin, growth hormone, FSH, and LH.
   d. Observe an area of the posterior lobe with high-power magnification. Note the numerous unmyelinated nerve fibers present in this lobe. Also locate some *pituicytes,* a type of neuroglial cell, scattered among the nerve fibers. The posterior lobe secretes ADH and oxytocin.
   e. Prepare labeled sketches of representative portions of the anterior and posterior lobes of the pituitary gland in Part B of the laboratory report.
   Thyroid gland. To examine the thyroid tissue, follow these steps:
   a. Use low-power magnification to observe the tissue (fig. 33.3). Note the numerous *follicles,* each of which consists of a layer of cells surrounding a colloid-filled cavity.
   b. Observe the tissue using high-power magnification. The cells forming the wall of a follicle are

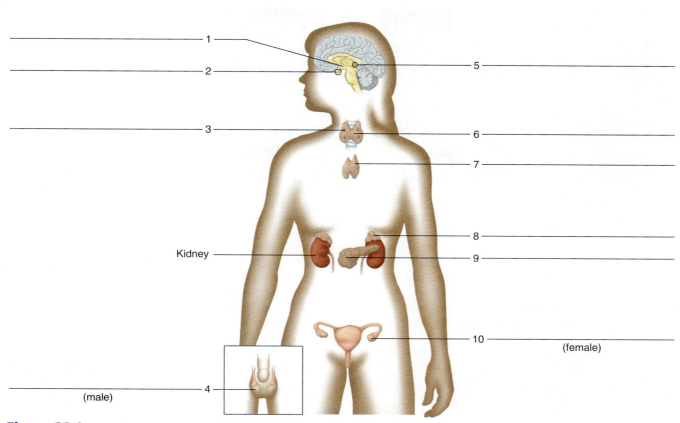

1 _____   5 _____

2 _____

Kidney _____

3 _____   6 _____

                                     7 _____

                                     8 _____

Kidney _____   9 _____

                                    10 _____
                                        (female)

(male) _____ 4

Figure 33.1   Label the major endocrine glands. 1

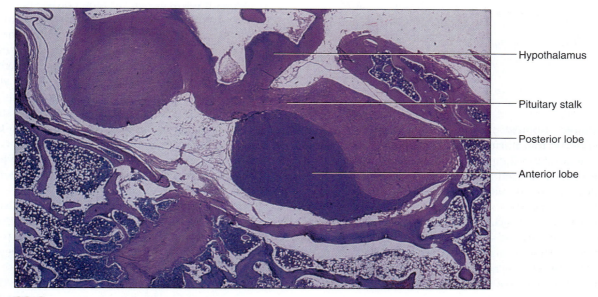

                                    Hypothalamus

                                    Pituitary stalk

                                    Posterior lobe

                                    Anterior lobe

Figure 33.2   Micrograph of the pituitary gland (6×).

239

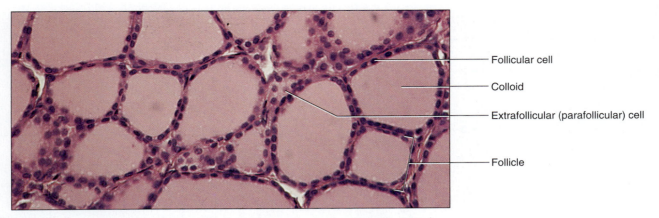

Figure 33.3 Micrograph of the thyroid gland (100× micrograph enlarged to 300×).

Follicular cell

Colloid

Extrafollicular (parafollicular) cell

Follicle

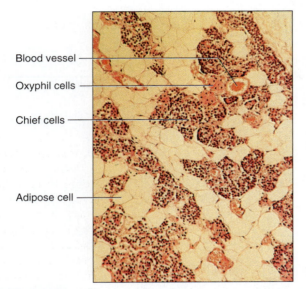

Blood vessel

Oxyphil cells

Chief cells

Adipose cell

Figure 33.4 Micrograph of the parathyroid gland (65×).

simple cuboidal epithelial cells. These cells secrete thyroid hormone (T_3 and T_4). The extrafollicular (parafollicular) cells secrete calcitonin.

c. Prepare a labeled sketch of a representative portion of the thyroid gland in Part B of the laboratory report.

Parathyroid gland. To examine the parathyroid tissue, follow these steps:

a. Use low-power magnification to observe the tissue (fig. 33.4). The gland consists of numerous tightly packed secretory cells.

b. Switch to high-power magnification and locate two types of cells—a smaller form (chief cells) arranged in cordlike patterns and a larger form (oxyphil cells) that have distinct cell boundaries

and are present in clusters. *Chief cells* secrete parathyroid hormone, whereas the function of *oxyphil cells* is not clearly understood.

c. Prepare a labeled sketch of a representative portion of the parathyroid gland in Part B of the laboratory report.

Adrenal gland. To examine the adrenal tissue, follow these steps:

a. Use low-power magnification to observe the tissue (fig. 33.5). Note the thin capsule of connective tissue that covers the gland. Just beneath the capsule, there is a relatively thick *adrenal cortex*. The central portion of the gland is the *adrenal medulla*. The cells of the cortex are in three poorly defined layers. Those of the outer layer (zona glomerulosa) are arranged irregularly; those of the middle layer (zona fasciculata) are in long cords; and those of the inner layer (zona reticularis) are arranged in an interconnected network of cords. The most noted hormones secreted from the cortex include aldosterone from the zona glomerularis, cortisol from the zona fasciculata, and estrogens and androgens from the zona reticularis. The cells of the medulla are relatively large and irregularly shaped, and they often occur in clusters. The medulla cells secrete epinephrine and norepinephrine.

b. Using high-power magnification, observe each of the layers of the cortex and the cells of the medulla.

c. Prepare labeled sketches of representative portions of the adrenal cortex and medulla in Part B of the laboratory report.

Pancreas. To examine the pancreas tissue, follow these steps:

a. Use low-power magnification to observe the tissue (fig. 33.6). The gland largely consists of deeply stained exocrine cells arranged in clusters around secretory ducts. These exocrine cells (acinar cells) secrete pancreatic juice rich in digestive enzymes.

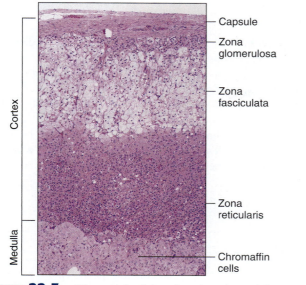

Figure 33.5 Micrograph of the adrenal cortex and the adrenal medulla (75×).

Labels: Capsule, Zona glomerulosa, Zona fasciculata, Zona reticularis, Chromaffin cells, Cortex, Medulla

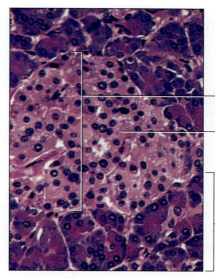

Figure 33.6 Micrograph of the pancreas (100×).

Labels: Pancreatic islet (islet of Langerhans), Endocrine cell, Exocrine (acinar) cells (digestive enzyme-secreting cells)

There are circular masses of lightly stained cells scattered throughout the gland. These clumps of cells constitute the *pancreatic islets (islets of Langerhans),* and they represent the endocrine portion of the pancreas. The beta cells secrete insulin and the alpha cells secrete glucagon.

b. Examine an islet, using high-power magnification. Unless special stains are used, it is not possible to distinguish alpha and beta cells.

c. Prepare a labeled sketch of a representative portion of the pancreas in Part B of the laboratory report.

Procedure B—Insulin Shock

1. Review the sections entitled "Pancreas" and "Diabetes Mellitus—Topic of Interest" in chapter 11 of the textbook, and the introduction to this lab.

2. Complete the table in Part C of the laboratory report.

3. A fish can be used to observe normal behavior, induced insulin shock, and recovery from insulin shock. The pancreatic islets are located in the pyloric ceca or other scattered regions in fish. The behavioral changes of fish when they experience insulin shock and recovery mimic those of humans. Observe a fish in 200 mL of aquarium water in a 500 mL beaker. Make your observations of swimming, gill cover (operculum), and mouth movements for 5 minutes. Record your observations in Part D of the laboratory report.

4. Add 200 units of room-temperature insulin slowly, using the syringe, into the water with the fish to induce insulin shock. Record the time or start the timer when the insulin is added. Watch for changes of the swimming, gill cover, and mouth movements as insulin diffuses into the blood at the gills. Signs of insulin shock might include rapid and irregular movements. As hypoglycemia occurs from the increased insulin absorption, the brain cells obtain less of the much-needed glucose, and epinephrine is secreted. Consequently, a rapid heart rate, anxiety, sweating, mental disorientation, impaired vision, dizziness, convulsions, and possible unconsciousness are complications during insulin shock. Record your observations of any behavioral changes and the time of the onset in Part D of the laboratory report.

5. After definite behavioral changes are observed, use the small fish net to move the fish into another 500 mL beaker containing 200 mL of a 10% room-temperature glucose solution. Record the time when the fish is transferred into this container. Make observations of any recovery of normal behaviors, including the amount of time involved. Record your observations in Part D of the laboratory report.

6. When the fish appears to be fully recovered, return the fish to its normal container. Dispose of the syringe according to directions from your laboratory instructor.

7. Complete Part D of the laboratory report.

Learning Extension

Examine the normal pancreas under high-power magnification (fig. 33.6). Locate the clumps of cells that constitute the *pancreatic islets (islets of Langerhans)* that represent the endocrine portions of the organ. A pancreatic islet contains four distinct cells that secrete different hormones: *alpha cells,* which secrete glucagon; *beta cells,* which secrete insulin; *delta cells,* which secrete somatostatin; and *F cells,* which secrete pancreatic polypeptide. Some of the specially stained pancreatic islets allow for distinguishing the different types of cells (figs. 33.7 and 33.8). The normal pancreatic islet has about 80% beta cells concentrated in the central region of an islet, about 15% alpha cells near the periphery of an islet, about 5% delta cells, and a small number of F cells. There are complex interactions among the four hormones affecting blood sugar, but in general, glucagon will increase blood sugar while insulin will lower blood sugar. Somatostatin acts as a paracrine secretion to inhibit alpha and beta cell secretions, and pancreatic poly-

peptide inhibits secretions from delta cells and digestive enzyme secretions from the pancreatic acini.

Examine the pancreas of a diabetic under high-power magnification (fig. 33.9). Locate a pancreatic islet and note the number of beta cells as compared to a normal number of beta cells in the normal pancreatic islet.

Describe the differences that you observed between a normal pancreatic islet and a pancreatic islet of a person with diabetes mellitus.

_____

_____

_____

_____

_____

_____

_____

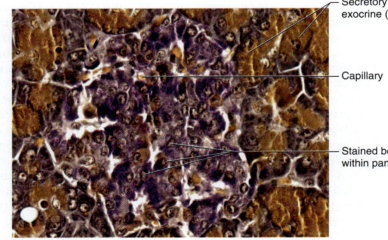

Secretory granules of exocrine (acinar) cells

Capillary

Stained beta cells within pancreatic islet

Figure 33.7   Micrograph of specially stained pancreatic islet surrounded by exocrine (acinar) cells (500×). The insulin-containing beta cells are stained dark purple.

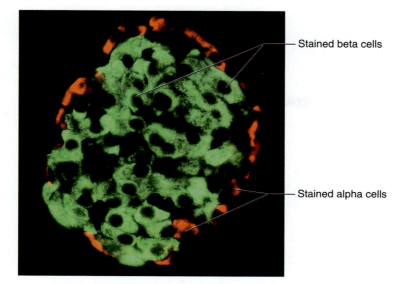

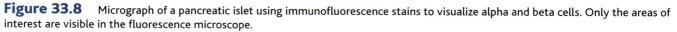

Figure 33.8   Micrograph of a pancreatic islet using immunofluorescence stains to visualize alpha and beta cells. Only the areas of interest are visible in the fluorescence microscope.

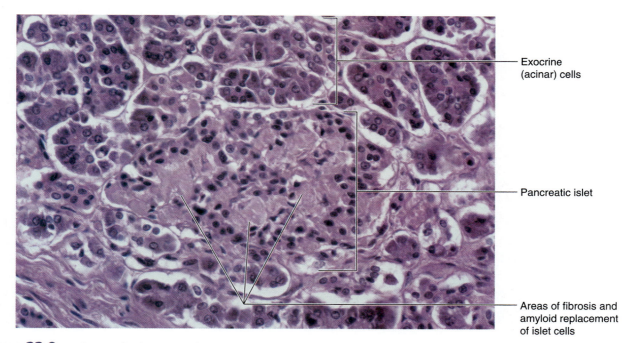

Figure 33.9   Micrograph of pancreas showing indications of changes from diabetes mellitus (100×).

Notes

Name _____

Date _____

Section _____

The ⬅ corresponds to the indicated outcome(s) found at the beginning of the laboratory exercise.

Endocrine Histology and Diabetic Physiology

Part A Assessments

Complete the following:

1. Name six hormones secreted by the anterior lobe of the pituitary gland. ⬅2 _____

_____

2. Name two hormones secreted by the posterior lobe of the pituitary gland. ⬅2 _____

3. Name the pituitary hormone responsible for the following actions: ⬅2
   a. causes kidneys to conserve water _____
   b. stimulates cells to increase in size and divide more rapidly _____
   c. stimulates secretion from thyroid gland _____
   d. causes contraction of uterine wall muscles _____
   e. stimulates secretion from adrenal cortex _____
   f. stimulates milk production _____

4. Name two thyroid hormones that affect metabolic rate. ⬅2 _____

5. Name a hormone secreted by the thyroid gland that acts to lower blood calcium. ⬅2 _____

6. Name a hormone that acts to raise blood calcium. ⬅2 _____

7. Name three target organs of parathyroid hormone. ⬅2 _____

8. Name two hormones secreted by the adrenal medulla. ⬅2 _____

9. List five different effects produced by these medullary hormones. ⬅2 _____

_____

10. Name the most important mineralocorticoid secreted by the adrenal cortex. ⬅2 _____

11. List two actions of this mineralocorticoid. ⬅2 _____

_____

12. Name the most important glucocorticoid secreted by the adrenal cortex. ⬅2 _____

13. List three actions of this glucocorticoid. ⬅2 _____

_____

14. Distinguish the two hormones secreted by the alpha and beta cells of the pancreatic islets. ⬅2 _____

_____

Critical Thinking Application

Briefly explain how the actions of pancreatic hormones complement one another. ◀ **2**

Part B Assessments

Sketch and label representative portions of the following endocrine glands: ◀ **3**

Pituitary gland (_____ ×) (anterior and posterior lobes)	Thyroid gland (_____ ×)
Parathyroid gland (_____ ×)	Adrenal gland (_____ ×) (cortex and medulla)
Pancreas (_____ ×)	

Part C Assessments

Complete the missing parts of the table of diabetes mellitus: ◀4

Characteristic	Type 1 Diabetes	Type 2 Diabetes
Onset age	Early age or adult	
Onset of symptoms		Slow
Percentage of diabetics		85–90%
Natural insulin levels	Below normal	
Beta cells of pancreatic islets		Not destroyed
Pancreatic islet cell antibodies	Present	
Risk factors of having the disease	Heredity	
Typical treatments	Insulin administration	
Untreated blood sugar levels		Hyperglycemia

Part D Assessments

1. Record your observations of normal behavior of the fish.
   a. Swimming movements: _____

   _____

   b. Gill cover movements: _____

   _____

   c. Mouth movements: _____

   _____

2. Record your observations of the fish after insulin was added to the water. ◀5
   a. Recorded time that insulin was added: _____
   b. Swimming movements: _____

   _____

   c. Gill cover movements: _____

   _____

   d. Mouth movements: _____

   _____

   e. Elapsed time until the insulin shock symptoms occurred: _____

3. Record your observations after the fish was transferred into a glucose solution. ◀6

    a. Recorded time that fish was transferred into the glucose solution: _____

    b. Describe any changes in the behavior of the fish that indicates a recovery from insulin shock. _____

    _____

    _____

    _____

    c. Elapsed time until indications of a recovery from insulin shock occurred: _____

Critical Thinking Application

Describe the importance for a type 1 diabetic of regulating insulin administration, meals, and exercise.

_____

_____

_____

_____

_____

_____

_____

_____

Blood Cells and Blood Typing

Materials Needed

Textbook
Compound light microscope
Prepared microscope slides of human blood
(Wright's stain)
Colored pencils
ABO blood-typing kit
Simulated blood-typing kits are suggested as a substitute
for collected blood

For Demonstration:

Microscope slide
Alcohol swabs
Sterile blood lancet
Toothpicks
Anti-D serum
Slide warming box (Rh blood-typing box or Rh view box)

For Learning Extension:

Prepared slides of pathological blood, such as eosinophilia,
leukocytosis, leukopenia, and lymphocytosis

Blood is a type of connective tissue whose cells are suspended in a liquid extracellular matrix. These cells are mainly formed in red bone marrow, and they include *red blood cells, white blood cells,* and some cellular fragments called *platelets.*

Red blood cells contain hemoglobin and transport gases between the body cells and the lungs, white blood cells defend the body against infections, and platelets play an important role in stoppage of bleeding (hemostasis).

Blood typing involves identifying protein substances called *antigens* present on red blood cell membranes. Although there are many different antigens associated with human red blood cells, only a few of them are of clinical importance. These include the antigens of the ABO group and those of the Rh group.

To determine which antigens are present, a blood sample is mixed with blood-typing sera that contain known types of antibodies. If a particular antibody contacts a corresponding antigen, a reaction occurs, and the red blood cells clump together (agglutination). Thus, if blood cells are mixed with serum containing antibodies that react with antigen A and the cells clump together, antigen A must be present in those cells.

Purpose of the Exercise

To review the characteristics of blood cells, to examine them microscopically, to perform a differential white blood cell count, to determine the ABO blood type of a blood sample, and to observe an Rh blood-typing test.

Learning Outcomes

After completing this exercise, you should be able to

1. Describe the structure and function of red blood cells, white blood cells, and platelets.
2. Identify and sketch red blood cells, five types of white blood cells, and platelets on a stained blood slide.
3. Perform and interpret the results of a differential white blood cell count.
4. Analyze the basis of ABO blood typing.

5. Match the ABO type of a blood sample.
6. Explain the basis of Rh blood typing.
7. Interpret how the Rh type of a blood sample is determined.

> **WARNING:** *Because of the possibility of blood infections being transmitted from one student to another if blood testing is performed in the classroom, it is suggested that commercially prepared blood slides and blood-typing kits containing virus-free human blood be used for ABO blood typing. The instructor may wish to demonstrate Rh blood typing. Observe all of the safety procedures listed for this lab.*

Procedure A—Types of Blood Cells

1. Review the sections entitled "Red Blood Cells," "White Blood Cells," and "Blood Platelets" in chapter 12 of the textbook.
2. Complete Part A of Laboratory Report 34.
3. Refer to figure 34.1 as an aid in identifying the various types of blood cells. Study the functions of the blood cells listed in table 34.1. Use the prepared slide of blood and locate each of the following:

   red blood cell (erythrocyte)
   white blood cell (leukocyte)
   >   granulocytes
   >>   neutrophil
   >>   eosinophil
   >>   basophil
   >   agranulocytes
   >>   lymphocyte
   >>   monocyte
   platelet (thrombocyte)

4. In Part B of the laboratory report, prepare sketches of single blood cells to illustrate each type. Pay particular attention to relative size, nuclear shape, and color of granules in the cytoplasm (if present). The sketches should be accomplished using either the high-power objective or the oil immersion objective of the compound light microscope.

Procedure B—Differential White Blood Cell Count

A differential white blood cell count is performed to determine the percentage of each of the various types of white blood cells present in a blood sample. The test is useful because the relative proportions of white blood cells may change in particular diseases. Neutrophils, for example, usually increase during bacterial infections, whereas eosinophils may increase during certain parasitic infections and allergic reactions.

Table 34.1 Cellular Components of Blood

Component	Function
Red blood cell (erythrocyte)	Transports oxygen and carbon dioxide
White blood cell (leukocyte)	Destroys pathogenic microorganisms and parasites and removes worn cells
Granulocytes	
1. Neutrophil	Phagocytizes small particles
2. Eosinophil	Kills parasites and helps control inflammation and allergic reactions
3. Basophil	Releases heparin and histamine
Agranulocytes	
1. Lymphocyte	Provides immunity
2. Monocyte	Phagocytizes large particles
Platelet (thrombocyte)	Helps control blood loss from broken vessels

1. To make a differential white blood cell count, follow these steps:
   a. Using high-power magnification or an oil immersion objective, focus on the cells at one end of a prepared blood slide where the cells are well distributed.
   b. Slowly move the blood slide back and forth, following a path that avoids passing over the same cells twice (fig. 34.2).
   c. Each time you encounter a white blood cell, identify its type and record it in Part C of the laboratory report.
   d. Continue searching for and identifying white blood cells until you have recorded 100 cells in the data table. *Percent* means "parts of 100," for each type of white blood cell, so the total number observed is equal to its percentage in the blood sample.
2. Complete Part C of the laboratory report.

Learning Extension

Obtain a prepared slide of pathological blood that has been stained with Wright's stain. Perform a differential white blood cell count using this slide, and compare the results with the values for normal blood listed in table 34.2. What differences do you note?

_____

_____

_____

_____

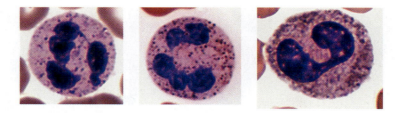

Neutrophils (3 of many variations)
•Fine light-purple granules
•Nucleus single to five lobes (highly variable)
•Younger neutrophils have a single C-shaped
  nucleus and are called bands
•Mature neutrophils have a lobed nucleus
  and are called segs
•Often called polymorphonuclear leukocytes when older

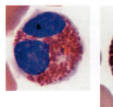
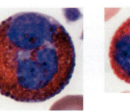

Eosinophils (3 of many variations)
•Coarse reddish granules
•Nucleus usually bilobed

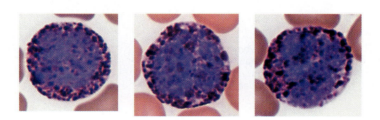

Basophils (3 of many variations)
•Coarse deep blue to almost black granules
•Nucleus often almost hidden by granules

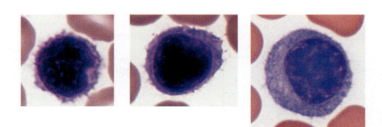

Lymphocytes (3 of many variations)
•Slightly larger than RBC's
•Thin rim of nearly clear cytoplasm
•Nearly round nucleus appears to fill most of cell
  in smaller lymphocytes
•Larger lymphocytes hard to distinguish from monocytes

Figure 34.1   Micrographs of blood cells illustrating some of the numerous variations of each type. Appearance characteristics for each cell pertain to a thin blood film using Wright's stain (1,000×).

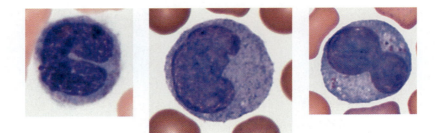

Monocytes (3 of many variations)
•Largest WBC; 2–3x larger than RBCs
•Cytoplasm nearly clear
•Nucleus round, kidney-shaped, oval, or lobed

Platelets (several variations)
•Cell fragments
•Single to small clusters

Erythrocytes (several variations)
•Lack nucleus (mature cell)
•Biconcave discs
•Thin centers appear almost hollow

Figure 34.1 *Continued.*

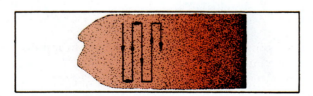

Figure 34.2 Move the blood slide back and forth to avoid passing the same cells twice.

Table 34.2 Differential White Blood Cell Count

Cell Type	Normal Value (percent)	Elevated Levels May Indicate
Neutrophil	54–62	Bacterial infections, stress
Lymphocyte	25–33	Mononucleosis, whooping cough, viral infections
Monocyte	3–9	Malaria, tuberculosis, fungal infections
Eosinophil	1–3	Allergic reactions, autoimmune diseases, parasitic worms
Basophil	<1	Cancers, chicken pox, hypothyroidism

Procedure C—ABO Blood Typing

1. Review the section entitled "ABO Blood Group" in chapter 12 of the textbook.
2. Complete Part D of the laboratory report.
3. Perform the ABO blood type test using the blood-typing kit. To do this, follow these steps:
   a. Obtain a clean microscope slide and mark across its center with a wax pencil to divide it into right and left halves. Also write "Anti-A" near the edge of the left half and "Anti-B" near the edge of the right half (fig. 34.3).
   b. Place a small drop of blood on each half of the microscope slide. Work quickly so that the blood will not have time to clot.
   c. Add a drop of anti-A serum to the blood on the left half and a drop of anti-B serum to the blood on the right half. Note the color coding of the anti-A and anti-B typing sera. To avoid contaminating the serum, avoid touching the blood with the serum while it is in the dropper; instead allow the serum to fall from the dropper onto the blood.
   d. Use separate toothpicks to stir each sample of serum and blood together, and spread each over an area about as large as a quarter. Dispose of toothpicks in an appropriate container.
   e. Examine the samples for clumping of blood cells (agglutination) after 2 minutes.

Figure 34.3 Slide prepared for ABO blood typing.

f. See table 34.3 and figure 34.4 for aid in interpreting the test results.
   g. Discard contaminated materials as directed by the laboratory instructor.
4. Complete Part E of the laboratory report.

Critical Thinking Application

Judging from the observations of the blood-typing results, predict the components in the anti-A and anti-B sera that caused clumping (agglutination).

_____

_____

_____

_____

Demonstration—Rh Blood Typing

1. Review the section entitled "Rh Blood Group" in chapter 12 of the textbook.
2. Complete Part F of the laboratory report.
3. To determine the Rh blood type of a blood sample, follow these steps:
   a. Thoroughly wash hands with soap and water, and dry them with paper towels.
   b. Cleanse the end of the middle finger with an alcohol swab, and let the finger dry in the air.
   c. Lance the tip of a finger using a sterile disposable blood lancet. Place a small drop of blood in the center of a clean microscope slide.
   d. Add a drop of anti-D serum to the blood and mix them together with a clean toothpick.
   e. Place the slide on the plate of a warming box (Rh blood-typing box or Rh view box) prewarmed to 45°C (113°F) (fig. 34.5).
   f. Slowly rock the box back and forth to keep the mixture moving, and watch for clumping (agglutination) of the blood cells. When clumping occurs in anti-D serum, the clumps usually are smaller than those that appear in anti-A or anti-B sera, so they may be less obvious. However, if clumping occurs, the blood is called Rh positive; if no clumping occurs *within 2 minutes,* the blood is called Rh negative.
   g. Discard all contaminated materials in appropriate containers.
4. Complete Part G of the laboratory report.

Table 34.3   Possible Reactions of ABO Blood-Typing Sera and Donor Compatabilities

| Reactions | | Blood Type | U.S. Frequency | Preferred Donor Type | Permissible Donor | Incompatible Donor |
Anti-A serum	Anti-B serum					
Clumping	No clumping	Type A	41%	A	A, O	B, AB
No clumping	Clumping	Type B	9%	B	B, O	A, AB
Clumping	Clumping	Type AB	3%	AB	A, B, AB, O	None
No clumping	No clumping	Type O	47%	O	O	A, B, AB

Note: The inheritance of the ABO blood groups is described in Laboratory Exercise 48.

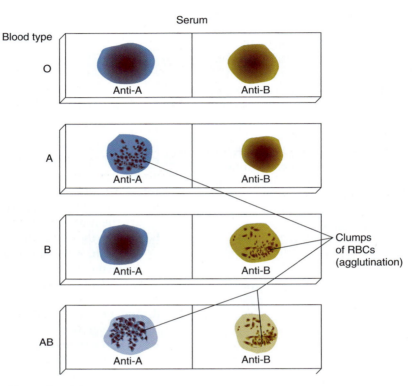

Figure 34.4   Four possible results of the ABO test.

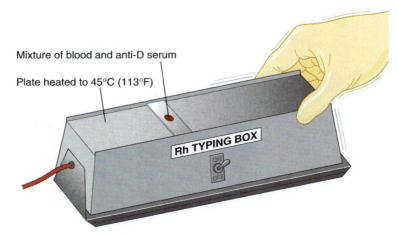

Figure 34.5   Slide warming box used for Rh blood typing.

Blood Cells and Blood Typing

Name _____

Date _____

Section _____

The ◄ corresponds to the indicated outcome(s) found at the beginning of the laboratory exercise.

Part A Assessments

Complete the following:

1. Mature red blood cells are also called _____ . ◄**1**

2. The functions of red blood cells are _____ . ◄**1**

3. _____ is the oxygen-carrying substance in a red blood cell. ◄**1**

4. White blood cells are also called _____ . ◄**1**

5. White blood cells with granular cytoplasm are called _____ . ◄**1**

6. White blood cells lacking granular cytoplasm are called _____ . ◄**1**

7. Normally, the most numerous white blood cells are _____ . ◄**1**

8. A platelet, a fragment of a cell, lacks a _____ . ◄**1**

Part B Assessments

Sketch a single blood cell of each type in the following spaces. Use colored pencils to represent the stained colors of the cells. Label any features that can be identified. ◄**2**

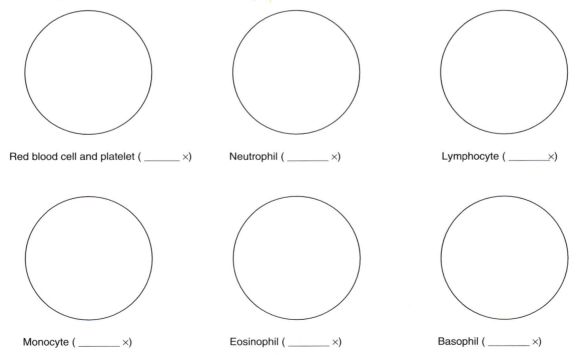

Red blood cell and platelet (_____ ×)   Neutrophil (_____ ×)   Lymphocyte (_____ ×)

Monocyte (_____ ×)   Eosinophil (_____ ×)   Basophil (_____ ×)

Part C Assessments

1. *Differential White Blood Cell Count Data Table.* As you identify white blood cells, record them on the table by using a tally system, such as ⫼ ‖. Place tally marks in the "Number Observed" column, and total each of the five WBCs when the differential count is completed. Obtain a total of all five WBCs counted to determine the percent of each WBC type. ◀3

Type of WBC	Number Observed	Total	Percent
Neutrophil			
Lymphocyte			
Monocyte			
Eosinophil			
Basophil			
		Total of column	

2. How do the results of your differential white blood cell count compare with the normal values listed in table 34.2? ◀3 _____

_____

_____

_____

Critical Thinking Application

What is the difference between a differential white blood cell count and a total white blood cell count? ◀3

_____

_____

_____

_____

Part D Assessments

Complete the following statements:

1. The antigens of the ABO blood group are located in the _____ . ◀4

2. The blood of every person contains one of (how many possible?) _____ combinations of antigens. ◀4

3. Type A blood contains antigen _____ . ◀4

4. Type B blood contains antigen _____ . ◀4

5. Type A blood contains _____ antibody in the plasma. ◀4

6. Type B blood contains _____ antibody in the plasma. ◀4

7. Persons with ABO blood type _____ are often called universal recipients. ◀4

8. Persons with ABO blood type _____ are often called universal donors. ◀4

Part E Assessments

Complete the following:

1. What was the ABO type of the blood tested? ◄5 _____

2. What ABO antigens are present in the red blood cells of this type of blood? ◄5 _____

3. What ABO antibodies are present in the plasma of this type of blood? ◄5 _____

4. If a person with this blood type needed a blood transfusion, what ABO type(s) of blood could be received safely? ◄5

   _____

5. If a person with this blood type was serving as a blood donor, what ABO blood type(s) could receive the blood safely? ◄5 _____

Part F Assessments

Complete the following statements:

1. The Rh blood group was named after the _____ . ◄6

2. Of the antigens in the Rh group, the most important is _____ . ◄6

3. If red blood cells lack Rh antigens, the blood is called _____ . ◄6

4. If an Rh-negative person sensitive to Rh-positive blood receives a transfusion of Rh-positive blood, the donor's cells are likely to _____ . ◄6

5. An Rh-negative woman, who might be carrying an _____ fetus, is given an injection of RhoGAM to prevent hemolytic disease of the fetus and newborn (erythroblastosis fetalis). ◄6

Part G Assessments

Complete the following:

1. What was the Rh type of the blood tested in the demonstration? ◄7 _____

2. What Rh antigen is present in the red blood cells of this type of blood? ◄7 _____

3. If a person with this blood type needed a blood transfusion, what type of blood could be received safely? ◄7

   _____

4. If a person with this blood type was serving as a blood donor, a person with what type of blood could receive the blood safely? ◄7 _____

Notes

Heart Structure

Materials Needed

Textbook
Dissectible human heart model
Preserved sheep or other mammalian heart
Dissecting tray
Dissecting instruments

For Learning Extension:
Colored pencils

⚠️ Safety

- Wear disposable gloves when working on the heart dissection.
- Save or dispose of the dissected heart as instructed.
- Wash the dissecting tray and instruments as instructed.
- Wash your laboratory table.
- Wash your hands before leaving the laboratory.

The heart is a muscular pump located within the mediastinum and resting upon the diaphragm. It is enclosed by the lungs, thoracic vertebrae, and sternum, and attached at its superior end (the base) are several large blood vessels. Its inferior end extends downward to the left and terminates as a bluntly pointed apex.

The heart and the proximal ends of the attached blood vessels are enclosed by a double-layered pericardium. The innermost layer of this membrane (visceral pericardium) consists of a thin covering closely applied to the surface of the heart, whereas the outer layer (parietal pericardium with fibrous pericardium) forms a tough, protective sac surrounding the heart. Between the parietal and visceral layers of the pericardium is a space, the pericardial cavity, that contains a small volume of serous (pericardial) fluid.

Purpose of the Exercise

To review the structural characteristics of the human heart and to examine the major features of a mammalian heart.

Learning Outcomes

After completing this exercise, you should be able to

1. Identify and label the major structural features of the human heart and closely associated blood vessels.
2. Match heart structures with appropriate locations and functions.
3. Compare the features of the human heart with those of another mammal.

Procedure A—The Human Heart

1. Review the section entitled "Structure of the Heart" in chapter 13 of the textbook.
2. As a review activity, label figures 35.1, 35.2, and 35.3.
3. Complete Part A of Laboratory Report 35.
4. Examine the human heart model, and locate the following features:

heart
    base (superior region where blood vessels emerge)
    apex (inferior, rounded end)
pericardium (pericardial sac)
    fibrous pericardium (outer layer)
    parietal pericardium (inner lining of fibrous pericardium)
    visceral pericardium (epicardium)
pericardial cavity (between parietal and visceral pericardial membranes)
myocardium (cardiac muscle)
endocardium (lines heart chambers)
atria
    right atrium
    left atrium
    auricles
ventricles
    right ventricle
    left ventricle
atrioventricular valves (A-V valves)
    tricuspid valve (right atrioventricular valve)
    mitral valve (bicuspid valve; left atrioventricular valve)

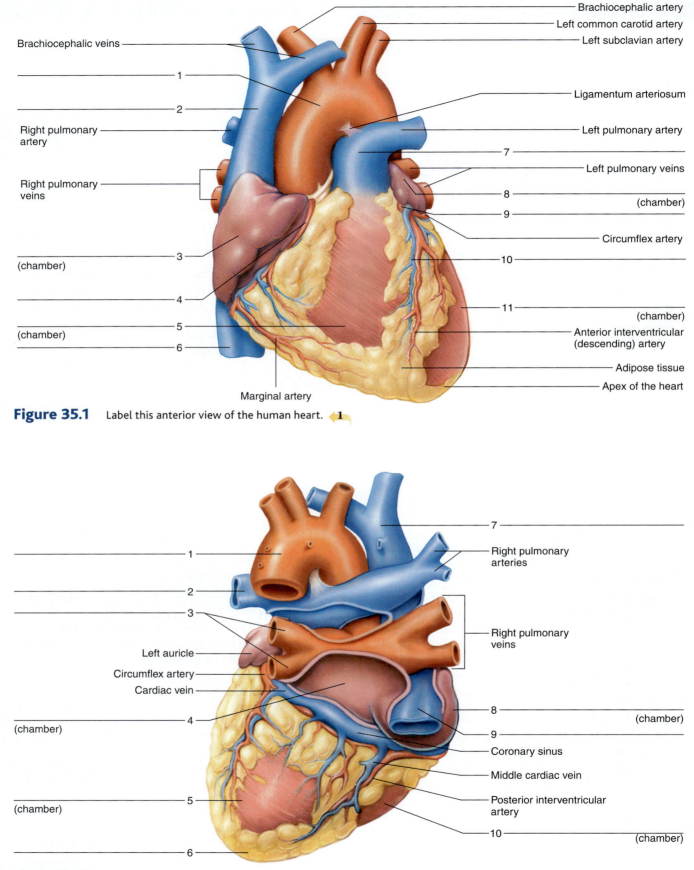

Brachiocephalic veins —————————

Brachiocephalic artery
Left common carotid artery
Left subclavian artery

——————— 1

——————— 2

Ligamentum arteriosum

Right pulmonary
artery ————

Left pulmonary artery

——————— 7

Left pulmonary veins

Right pulmonary ————
veins

——————— 8

(chamber)

——————— 9

Circumflex artery

——————— 3

(chamber)

——————— 10

——————— 4

——————— 11

(chamber)

——————— 5

(chamber)

——————— 6

Anterior interventricular
(descending) artery

Marginal artery

Adipose tissue
Apex of the heart

Figure 35.1   Label this anterior view of the human heart.  **1**

——————— 1

——————— 7

Right pulmonary
arteries

——————— 2

——————— 3

Left auricle ————

Right pulmonary
veins

Circumflex artery ————

Cardiac vein ————

——————— 4

——————— 8

(chamber)

(chamber)

——————— 9

Coronary sinus

Middle cardiac vein

Posterior interventricular
artery

——————— 5

(chamber)

——————— 10

(chamber)

——————— 6

Figure 35.2   Label this posterior view of the human heart. **1**

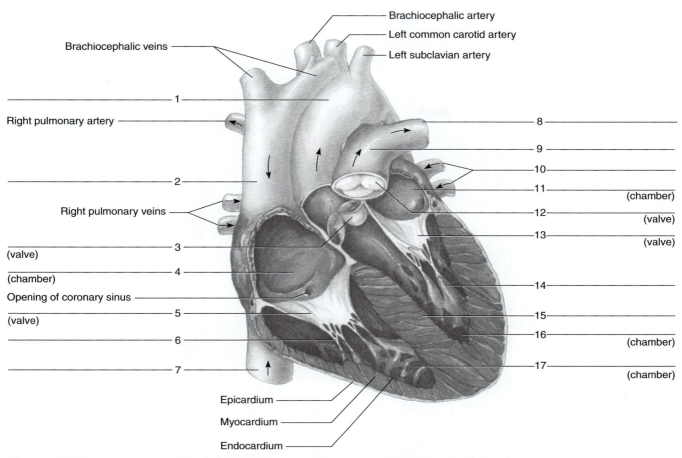

Brachiocephalic veins

Brachiocephalic artery

Left common carotid artery

Left subclavian artery

Right pulmonary artery

Right pulmonary veins

1

2

3
(valve)

4
(chamber)

Opening of coronary sinus

5
(valve)

6

7

8

9

10

11
(chamber)

12
(valve)

13
(valve)

14

15

16
(chamber)

17
(chamber)

Epicardium

Myocardium

Endocardium

Figure 35.3 Label this coronal section of the human heart. The arrows indicate the direction of blood flow.

chordae tendineae
papillary muscles
superior vena cava
inferior vena cava
pulmonary trunk
pulmonary arteries
pulmonary veins
aorta
semilunar valves
   pulmonary valve
   aortic valve
left coronary artery
right coronary artery
cardiac (coronary) veins
coronary sinus (for return of blood from cardiac
   veins into right atrium)

5. Label the human heart model in figure 35.4.

Learning Extension

Use red and blue colored pencils to color the blood vessels in figure 35.3. Use red to illustrate a blood vessel high in oxygen, and use blue to illustrate a blood vessel low in oxygen. You can check your work by referring to the corresponding figures in the textbook, presented in full color.

Procedure B—Dissection of a Sheep Heart

1. Obtain a preserved sheep heart. Rinse it in water thoroughly to remove as much of the preservative as possible. Also run water into the large blood vessels to force any blood clots out of the heart chambers.

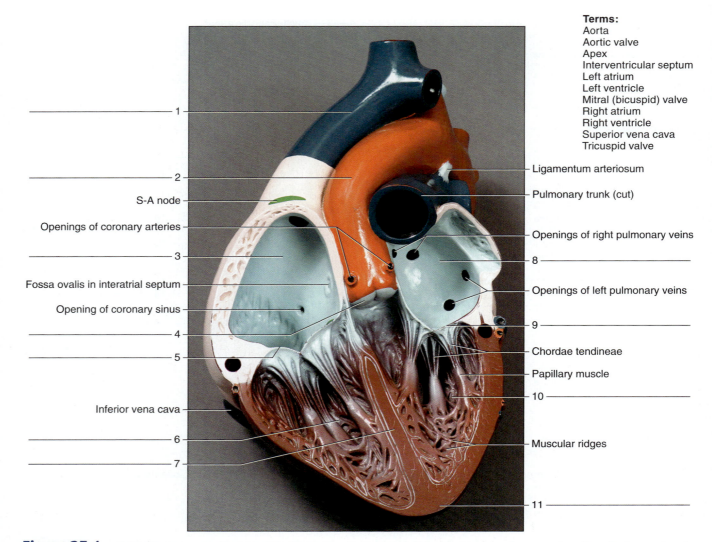

Terms:
Aorta
Aortic valve
Apex
Interventricular septum
Left atrium
Left ventricle
Mitral (bicuspid) valve
Right atrium
Right ventricle
Superior vena cava
Tricuspid valve

1

2

S-A node

Openings of coronary arteries

3

Fossa ovalis in interatrial septum

Opening of coronary sinus

4

5

Inferior vena cava

6

7

Ligamentum arteriosum

Pulmonary trunk (cut)

Openings of right pulmonary veins

8

Openings of left pulmonary veins

9

Chordae tendineae

Papillary muscle

10

Muscular ridges

11

Figure 35.4 Using the terms provided, identify the features indicated on this anterior view of a coronal section of a human heart model. (*Note: The pulmonary valve is not shown on the portion of the model photographed.*) ◄**1**

2. Place the heart in a dissecting tray with its ventral surface up (fig. 35.5), and proceed as follows:
   a. Although the relatively thick *pericardial sac* is probably missing, look for traces of this membrane around the origins of the large blood vessels.
   b. Locate the *visceral pericardium,* which appears as a thin, transparent layer on the surface of the heart. Use a scalpel to remove a portion of this layer and expose the *myocardium* beneath. Also note the abundance of fat along the paths of various blood vessels. This adipose tissue occurs in the loose connective tissue that underlies the visceral pericardium.
   c. Identify the following:

   right atrium
   right ventricle
   left atrium
   left ventricle
   coronary arteries

3. Examine the dorsal surface of the heart (fig. 35.6). Locate the stumps of two relatively thin-walled blood vessels that enter the right atrium. Demonstrate this connection by passing a slender probe through them. The upper vessel is the *superior vena cava,* and the lower one is the *inferior vena cava.*
4. Open the right atrium. To do this, follow these steps:
   a. Insert a blade of the scissors into the superior vena cava and cut downward through the atrial wall (fig. 35.6).
   b. Open the chamber, locate the *tricuspid valve (right atrioventricular valve),* and examine its cusps.
   c. Also locate the opening to the *coronary sinus* between the valve and the inferior vena cava.
   d. Run some water through the tricuspid valve to fill the chamber of the right ventricle.
   e. Gently squeeze the ventricles and watch the cusps of the valve as the water moves up against them.

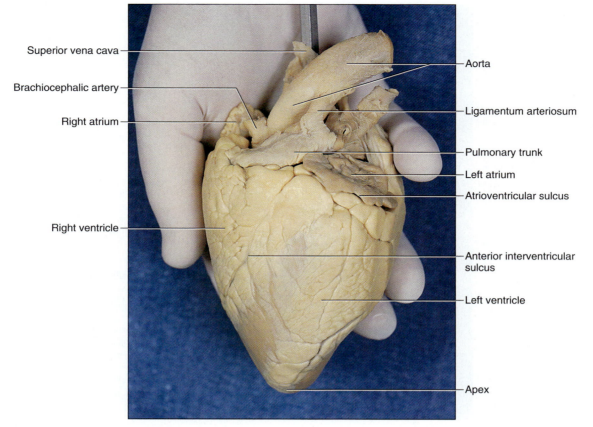

Figure 35.5 Ventral surface of sheep heart.

5. Open the right ventricle as follows:
   a. Continue cutting downward through the tricuspid valve and the right ventricular wall until you reach the apex of the heart.
   b. Locate the *chordae tendineae* and the *papillary muscles.*
   c. Find the opening to the *pulmonary trunk* and use the scissors to cut upward through the wall of the right ventricle. Follow the pulmonary trunk until you have exposed the *pulmonary valve.*
   d. Examine the valve and its cusps.
6. Open the left side of the heart. To do this, follow these steps:
   a. Insert the blade of the scissors through the wall of the left atrium and cut downward to the apex of the heart.
   b. Open the left atrium and locate the four openings of the *pulmonary veins.* Pass a slender probe through each opening and locate the stump of its vessel.

   c. Examine the *bicuspid valve (left atrioventricular valve)* and its cusps.
   d. Also examine the left ventricle and compare the thickness of its wall with that of the right ventricle.
7. Locate the aorta, which leads away from the left ventricle, and proceed as follows:
   a. Compare the thickness of the aortic wall with that of the pulmonary trunk.
   b. Use scissors to cut along the length of the aorta to expose the *aortic valve* at its base.
   c. Examine the cusps of the valve and locate the openings of the *coronary arteries* just distal to them.
8. As a review, locate and identify the stumps of each of the major blood vessels associated with the heart.
9. Discard or save the specimen as directed by the laboratory instructor.
10. Complete Part B of the laboratory report.

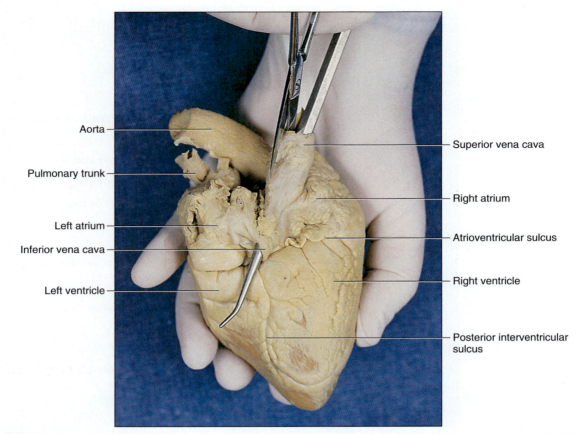

Figure 35.6 Dorsal surface of sheep heart. To open the right atrium, insert a blade of the scissors into the superior (anterior in sheep) vena cava and cut downward.

Name _____

Date _____

Section _____

The ◄ corresponds to the indicated outcome(s) found at the beginning of the laboratory exercise.

Heart Structure

Part A Assessments

Match the terms in column A with the descriptions in column B. Place the letter of your choice in the space provided. ◄2

Column A	Column B
a. Aorta	_____ **1.** Upper chamber of heart
b. Atrium	
c. Cardiac vein	_____ **2.** Structure from which chordae tendineae originate
d. Coronary artery	_____ **3.** Prevents blood movement from right ventricle to right atrium
e. Endocardium	
f. Mitral (bicuspid) valve	_____ **4.** Membranes around heart
g. Myocardium	_____ **5.** Prevents blood movement from left ventricle to left atrium
h. Papillary muscle	
i. Pericardial cavity	_____ **6.** Gives rise to left and right pulmonary arteries
j. Pericardial sac	_____ **7.** Inner lining of heart chamber
k. Pulmonary trunk	
l. Tricuspid valve	_____ **8.** Layer largely composed of cardiac muscle tissue
	_____ **9.** Space containing serous fluid
	_____ **10.** Drains blood from myocardial capillaries
	_____ **11.** Supplies blood to heart muscle
	_____ **12.** Distributes blood to body organs (systemic circuit) except lungs

Part B Assessments

Complete the following:

1. Compare the structure of the tricuspid valve with that of the pulmonary valve. ◄3 _____

_____

_____

_____

2. Describe the action of the tricuspid valve when you squeezed the water-filled right ventricle. ◄3 _____

_____

_____

_____

Notes

Blood Vessel Structure, Arteries, and Veins

The blood vessels form a closed system of tubes that carry blood to and from the heart, lungs, and body cells. These tubes include arteries and arterioles that conduct blood away from the heart; capillaries in which exchanges of substances occur between the blood and surrounding tissues; and venules and veins that return blood to the heart.

The blood vessels of the cardiovascular system can be divided into two major pathways—the pulmonary circuit and the systemic circuit. Within each circuit, arteries transport blood away from the heart. After exchanges of gases, nutrients, and wastes have occurred between the blood and the surrounding tissues, veins return the blood to the heart.

Frequent variations exist in anatomical structures among humans, especially in arteries and veins. The illustrations in this laboratory manual represent normal (normal means the most common variation) anatomy.

Purpose of the Exercise

To review the structure and functions of blood vessels, trace major circulatory pathways, and locate the major arteries and veins.

Learning Outcomes

After completing this exercise, you should be able to

1. Distinguish structures and functions of arteries, veins, and capillaries.
2. Interpret the types of blood vessels in the web of a frog's foot.
3. Distinguish and trace the pulmonary and systemic circuits.
4. Locate the major arteries in these circuits on a diagram, chart, or model.
5. Locate the major veins in these circuits on a diagram, chart, or model.

Procedure A—Blood Vessel Structure

1. Review the section entitled "Blood Vessels" in chapter 13 of the textbook.
2. As a review activity, label figure 37.1.

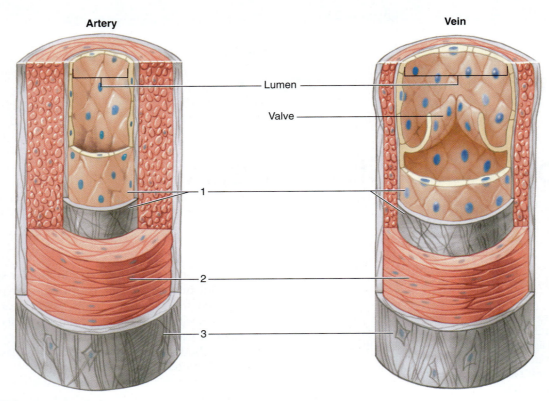

Figure 37.1 Label the tunics of the wall of this artery and vein. 🔖 **1**

Labels in figure:
- Artery
- Vein
- Lumen
- Valve
- 1
- 2
- 3

3. Complete Part A of Laboratory Report 37.
4. Obtain a microscope slide of an artery cross section, and examine it using low-power and high-power magnification. Identify the three distinct layers (tunics) of the arterial wall. The inner layer **(tunica interna)** is composed of an endothelium (simple squamous epithelium) and appears as a wavy line due to an abundance of elastic fibers that have recoiled just beneath it. The middle layer **(tunica media)** consists of numerous concentrically arranged smooth muscle cells with elastic fibers scattered among them. The outer layer **(tunica externa)** contains connective tissue rich in collagenous fibers (fig. 37.2).
5. Prepare a labeled sketch of the arterial wall in Part B of the laboratory report.
6. Obtain a slide of a vein cross section and examine it as you did the artery cross section. Note the thinner wall and larger lumen relative to an artery of comparable size. Identify the three layers of the wall, and prepare a labeled sketch in Part B of the laboratory report.
7. Complete Part B of the laboratory report.
8. Observe the blood vessels in the webbing of a frog's foot. (As a substitute for a frog, a live goldfish could be used to observe circulation in the tail.) To do this, follow these steps:

a. Obtain a live frog. Wrap its body in a moist paper towel, leaving one foot extending outward. Secure the towel with rubber bands, but be careful not to wrap the animal so tightly that it could be injured. Try to keep the nostrils exposed.

b. Place the frog on a frog board or on a piece of heavy cardboard with the foot near the hole in one corner.

c. Fasten the wrapped body to the board with masking tape.

d. Carefully spread the web of the foot over the hole and secure it to the board with dissecting pins and thread (fig. 37.3). Keep the web moist with frog Ringer's solution.

e. Secure the board on the stage of a microscope with heavy rubber bands, and position it so that the web is beneath the objective lens.

f. Focus on the web, using low-power magnification, and locate some blood vessels. Note the movement of the blood cells and the direction of the blood flow. You might notice that red blood cells of frogs are nucleated. Identify an arteriole, a capillary, and a venule.

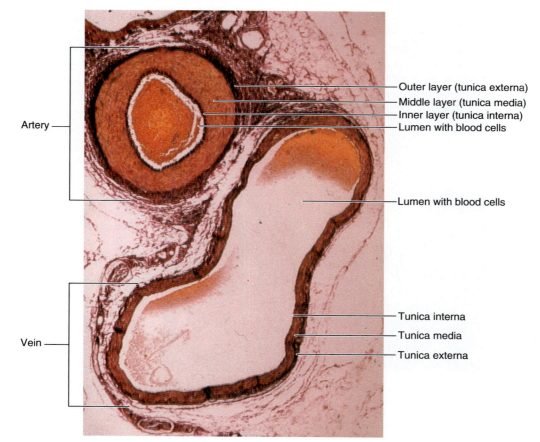

Figure 37.2 Cross section of an artery and a vein (5×).

Labels on figure:
- Artery
- Outer layer (tunica externa)
- Middle layer (tunica media)
- Inner layer (tunica interna)
- Lumen with blood cells
- Lumen with blood cells
- Vein
- Tunica interna
- Tunica media
- Tunica externa

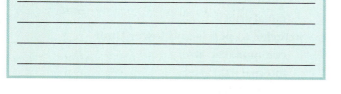

Learning Extension

Investigate the effect of temperature change on the blood vessels of the frog's foot by flooding the web with a small quantity of ice water. Observe the blood vessels with low-power magnification and note any changes in their diameters or the rate of blood flow. Remove the ice water and replace it with water heated to about 35°C (95°F). Repeat your observations. What do you conclude from this experiment?

_____

_____

_____

_____

g. Examine each of these vessels with high-power magnification.

h. When finished, return the frog to the location indicated by your instructor. The microscope lenses and stage will likely need cleaning after the experiment.

9. Complete Part C of the laboratory report.

Procedure B—Paths of Circulation

1. Review the sections entitled "Pulmonary Circuit" and "Systemic Circuit" in chapter 13 of the textbook.
2. As a review activity, label figure 37.4.
3. Locate the following blood vessels on the available anatomical charts and the human torso model:

   pulmonary trunk 1
   pulmonary arteries 2
   pulmonary veins 4
   aorta 1
   superior vena cava 1
   inferior vena cava 1

Critical Thinking Application

Why is the left ventricle wall thicker than the right ventricle wall?

_____

_____

_____

_____

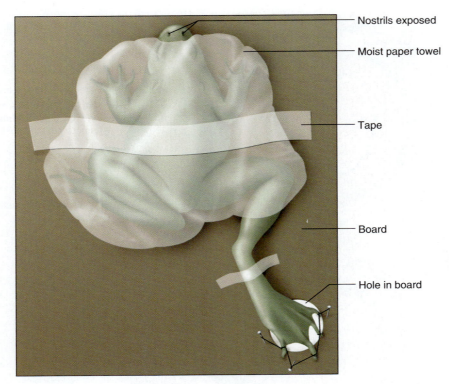

Figure 37.3 Spread the web of the foot over the hole and secure it to the board with pins and thread.

Labels on figure:
- Nostrils exposed
- Moist paper towel
- Tape
- Board
- Hole in board

Procedure C—Arterial System

1. Review the section entitled "Arterial System" in chapter 13 of the textbook.
2. As a review activity, label figures 37.5, 37.6, 37.7, and 37.8.
3. Locate the following arteries of the systemic circuit on the charts and human torso model:

   aorta
   - ascending aorta
   - aortic arch (arch of aorta)
   - thoracic aorta
   - abdominal aorta

   branches of the aorta
   - coronary artery
   - brachiocephalic trunk (artery)
   - left common carotid artery
   - left subclavian artery
   - celiac trunk (artery)
   - superior mesenteric artery
   - renal artery
   - inferior mesenteric artery
   - common iliac artery

   arteries to neck, head, and brain
   - vertebral artery
   - common carotid artery
   - external carotid artery
   - internal carotid artery
   - facial artery

   arteries to shoulder and upper limb
   - subclavian artery
   - axillary artery
   - brachial artery
   - deep brachial artery
   - ulnar artery
   - radial artery

   arteries to pelvis and lower limb
   - common iliac artery
   - internal iliac artery
   - external iliac artery
   - femoral artery
   - popliteal artery
   - anterior tibial artery
   - dorsalis pedis artery (dorsal artery of foot)
   - posterior tibial artery

4. Complete Part D of the laboratory report.

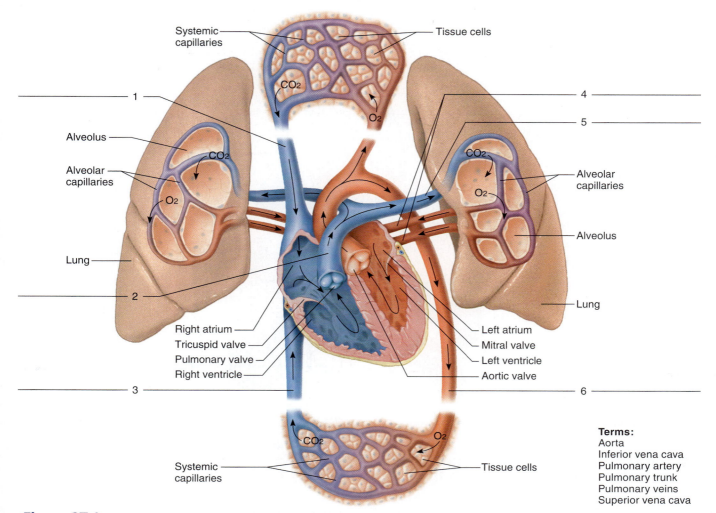

Systemic capillaries

Tissue cells

1

Alveolus

Alveolar capillaries

CO_2

O_2

Lung

2

4

5

CO_2

O_2

Alveolar capillaries

Alveolus

Lung

Right atrium

Tricuspid valve

Pulmonary valve

Right ventricle

3

Left atrium

Mitral valve

Left ventricle

Aortic valve

6

CO_2

Systemic capillaries

O_2

Tissue cells

Terms:
Aorta
Inferior vena cava
Pulmonary artery
Pulmonary trunk
Pulmonary veins
Superior vena cava

Figure 37.4   Using the terms provided, label the major blood vessels associated with the pulmonary and systemic circuits. **3**

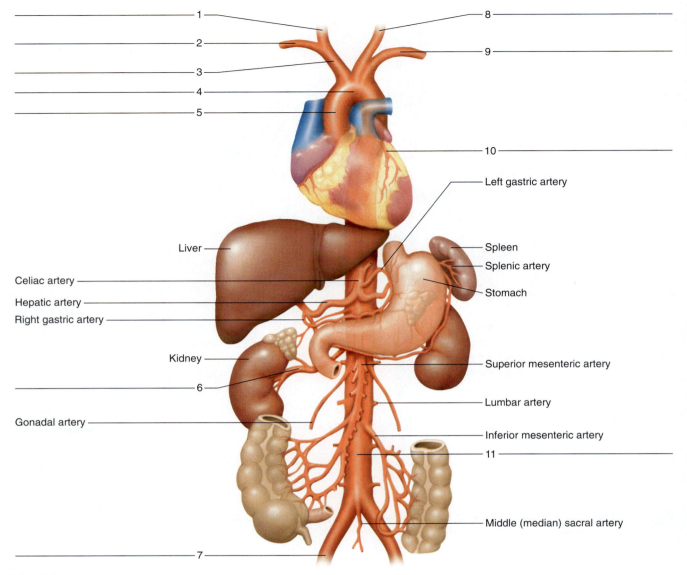

Figure 37.5 Label the portions of the aorta and its principal branches. ◄◄

Liver

Celiac artery

Hepatic artery

Right gastric artery

Kidney

Gonadal artery

Left gastric artery

Spleen

Splenic artery

Stomach

Superior mesenteric artery

Lumbar artery

Inferior mesenteric artery

Middle (median) sacral artery

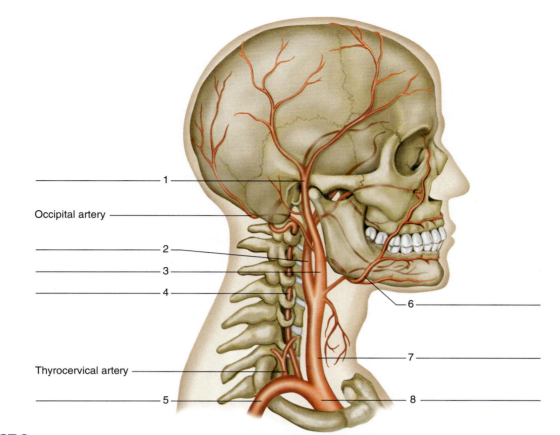

Figure 37.6   Label the arteries supplying the right side of the neck and head. (The clavicle has been removed.) ◄

Occipital artery —————————

Thyrocervical artery —————————

1
2
3
4
5
6
7
8

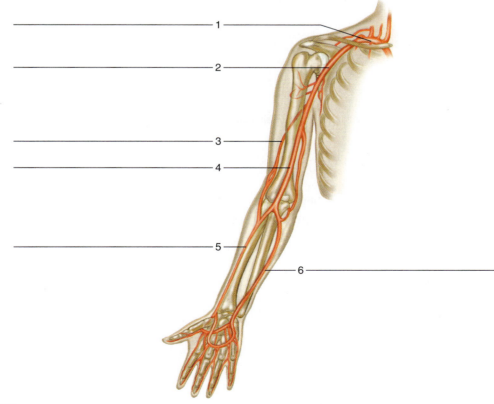

Figure 37.7   Label the major arteries of the shoulder and upper limb. ◄

1
2
3
4
5
6

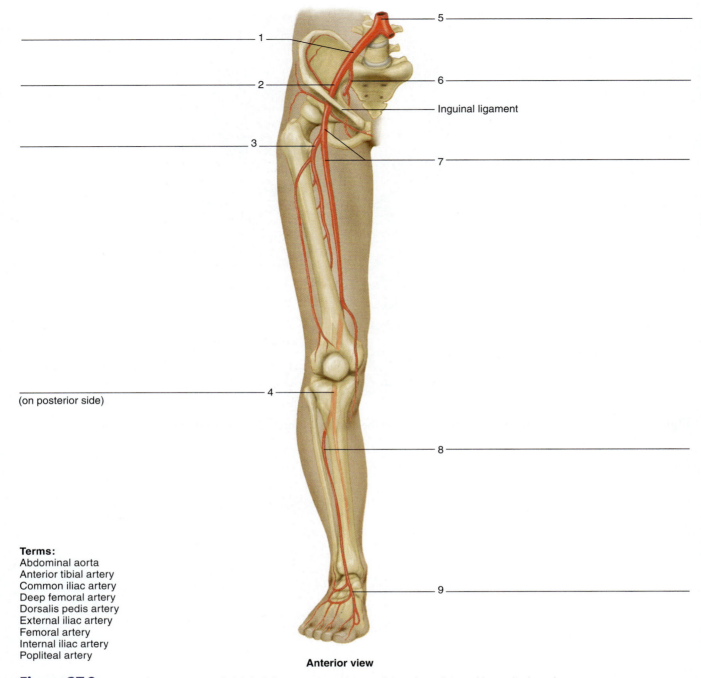

5 ⸻

1 ⸻

2 ⸻

Inguinal ligament

3 ⸻

7 ⸻

6 ⸻

4 ⸻

(on posterior side)

8 ⸻

9 ⸻

Terms:
Abdominal aorta
Anterior tibial artery
Common iliac artery
Deep femoral artery
Dorsalis pedis artery
External iliac artery
Femoral artery
Internal iliac artery
Popliteal artery

Anterior view

Figure 37.8   Using the terms provided, label the major arteries supplying the pelvis and lower limb.

Procedure D—Venous System

1. Review the section entitled "Venous System" in chapter 13 of the textbook.
2. As a review activity, label figures 37.9, 37.10, 37.11, and 37.12.
3. Locate the following veins of the systemic circuit on the charts and human torso model:

veins from brain, head, and neck
    external jugular vein
    internal jugular vein
    vertebral vein
    subclavian vein
    brachiocephalic vein
    superior vena cava

veins from upper limb and shoulder
    radial vein
    ulnar vein
    brachial vein
    basilic vein
    cephalic vein
    median cubital vein (antecubital vein)
    axillary vein
    subclavian vein

veins of abdominal viscera
    hepatic portal vein
    gastric vein
    superior mesenteric vein
    splenic vein
    inferior mesenteric vein
    hepatic vein
    renal vein

veins from lower limb and pelvis
    anterior tibial vein
    posterior tibial vein
    popliteal vein
    femoral vein
    great (long) saphenous vein
    small (short) saphenous vein
    external iliac vein
    internal iliac vein
    common iliac vein
    inferior vena cava

4. Complete Parts E and F of the laboratory report.

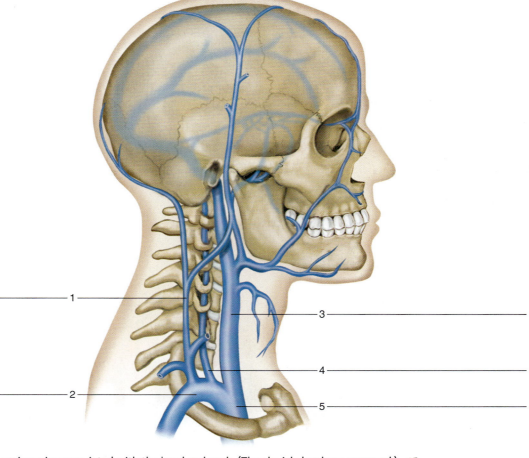

Figure 37.9   Label the major veins associated with the head and neck. (The clavicle has been removed.) ◀5

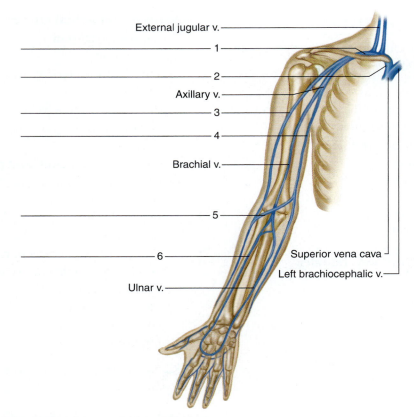

External jugular v.

1

2

Axillary v.

3

4

Brachial v.

5

Superior vena cava

6

Left brachiocephalic v.

Ulnar v.

Figure 37.10   Label the veins of the upper limb and shoulder.   5

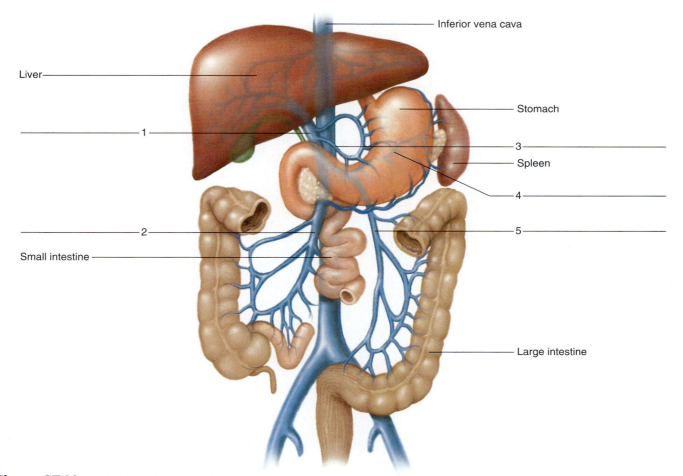

Liver

Inferior vena cava

Stomach

1

3

Spleen

4

2

5

Small intestine

Large intestine

Figure 37.11 Label the veins that drain the abdominal viscera (hepatic portal system). **5**

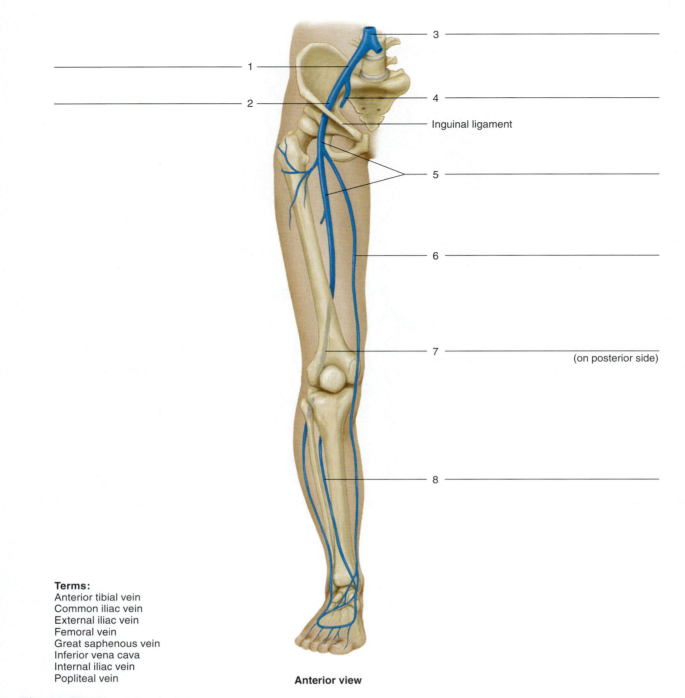

3 _____

1 _____

2 _____

4 _____

Inguinal ligament

5 _____

6 _____

7 _____   (on posterior side)

8 _____

Terms:
Anterior tibial vein
Common iliac vein
External iliac vein
Femoral vein
Great saphenous vein
Inferior vena cava
Internal iliac vein
Popliteal vein

Anterior view

Figure 37.12 Using the terms provided, label the veins of the pelvis and lower limb. ◄**5**

Name _____

Date _____

Section _____

The ← corresponds to the indicated outcome(s) found at the beginning of the laboratory exercise.

Blood Vessel Structure, Arteries, and Veins

Part A Assessments

Complete the following statements:

1. Simple squamous epithelial tissue called _____ forms the inner linings of the tunica interna of blood vessels. ◄**1**

2. The _____ of an arterial wall contains many smooth muscle cells. ◄**1**

3. The _____ of an arterial wall is largely composed of connective tissue. ◄**1**

4. When contraction of the smooth muscle in a blood vessel wall occurs, the vessel is referred to as being in a condition of _____ . ◄**1**

5. Relaxation of the smooth muscle in a blood vessel wall results in the vessel being in a condition of _____ . ◄**1**

6. The smallest blood vessels are called _____ . ◄**1**

7. Filtration results when substances are forced through capillary walls by _____ pressure. ◄**1**

8. The presence of plasma proteins in the blood increases its _____ pressure as compared to tissue fluids. ◄**1**

9. _____ in certain veins close if the blood begins to back up in that vein. ◄**1**

Part B Assessments

1. Sketch and label a section of an artery wall and a vein wall. ◄**1**

        Artery wall   **Vein wall**

2. Describe the differences you noted in the structures of the walls of an artery and a vein. Mention each of the three layers of the wall.  _____

_____

_____

Critical Thinking Application

Explain the functional significance of the differences you noted in the structures of the arterial and venous walls. ◀1 _____

_____

_____

Part C Assessments

1. How did you distinguish between arterioles and venules when you observed the vessels in the web of the frog's foot? ◀2 _____

_____

2. How did you recognize capillaries in the web? ◀2 _____

_____

3. What differences did you note in the rate of blood flow through the arterioles, capillaries, and venules? ◀2

_____

_____

Part D Assessments

Match the arteries in column A with the descriptions in column B. Place the letter of your choice in the space provided. ◀4

Column A

a. Anterior tibial
b. Brachial
c. Celiac
d. External carotid
e. Internal carotid
f. Internal iliac
g. Phrenic
h. Popliteal
i. Renal
j. Suprarenal
k. Ulnar

Column B

_____ 1. Jaw, teeth, and face

_____ 2. Kidney

_____ 3. Upper digestive tract, spleen, and liver

_____ 4. Foot and toes

_____ 5. Gluteal muscles

_____ 6. Biceps muscle

_____ 7. Knee joint

_____ 8. Adrenal gland

_____ 9. Diaphragm

_____ 10. Brain

_____ 11. Forearm muscles

Part E Assessments

Match each vein in column A with the vein it drains into from column B. Place the letter of your choice in the space provided. ◀5

Column A

a. Anterior tibial
b. Basilic
c. Brachiocephalic
d. Common iliac
e. External jugular
f. Femoral
g. Popliteal
h. Radial

Column B

_____ 1. Popliteal

_____ 2. Axillary

_____ 3. Inferior vena cava

_____ 4. Subclavian

_____ 5. Brachial

_____ 6. Superior vena cava

_____ 7. Femoral

_____ 8. External iliac

Label the major arteries and veins indicated in figure 37.13. **3** **4** **5**

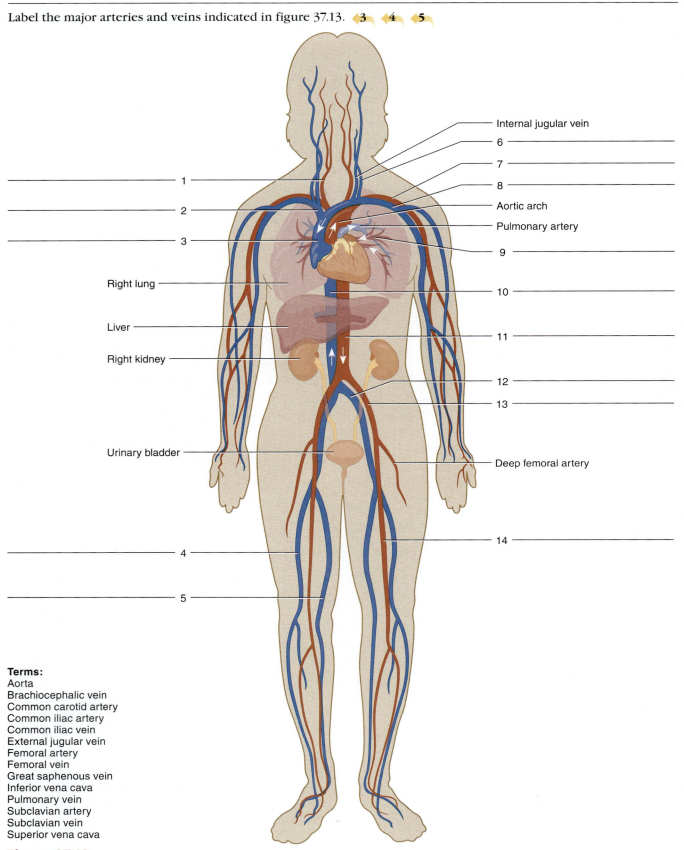

Internal jugular vein

6

7

8

Aortic arch

Pulmonary artery

9

1

2

3

Right lung

Liver

Right kidney

10

11

12

13

Urinary bladder

Deep femoral artery

14

4

5

Terms:
Aorta
Brachiocephalic vein
Common carotid artery
Common iliac artery
Common iliac vein
External jugular vein
Femoral artery
Femoral vein
Great saphenous vein
Inferior vena cava
Pulmonary vein
Subclavian artery
Subclavian vein
Superior vena cava

Figure 37.13   Uing the terms provided, label the major arteries and veins of the systemic and pulmonary circuits.

Laboratory Exercise 38

Pulse Rate and Blood Pressure

Materials Needed

Textbook
Clock with second hand
Sphygmomanometer
Stethoscope
Alcohol swabs

The surge of blood that enters the arteries each time the ventricles of the heart contract causes the elastic walls of these vessels to swell. Then, as the ventricles relax, the walls recoil. This alternate expanding and recoiling of an arterial wall can be felt as a pulse in vessels that run close to the surface of the body.

The force exerted by the blood pressing against the inner walls of arteries also creates blood pressure. This pressure reaches a maximum during ventricular contraction and then drops to its lowest level while the ventricles are relaxed.

Purpose of the Exercise

To examine the pulse, determine the pulse rate, measure blood pressure, and investigate the effects of body position and exercise on pulse rate and blood pressure.

Learning Outcomes

After completing this exercise, you should be able to

1. Determine pulse rate and correlate pulse characteristics.
2. Test the effects of various factors on pulse rate.
3. Describe and measure blood pressure using a sphygmomanometer.
4. Test the effects of various factors on blood pressure.

Procedure A—Pulse Rate

1. Review the section entitled "Blood Pressure" in chapter 13 of the textbook.
2. Complete Part A of Laboratory Report 38.
3. Examine your laboratory partner's radial pulse. To do this, follow these steps:
   a. Have your partner sit quietly, remaining as relaxed as possible.
   b. Locate the pulse by placing your index and middle fingers over the radial artery on the anterior surface of the wrist. Do not use your thumb for sensing the pulse because you may feel a pulse coming from an artery in the thumb. The radial artery is most often used because it is easy and convenient to locate.
   c. Note the characteristics of the pulse. That is, could it be described as regular or irregular, strong or weak, hard or soft? The pulse should be regular and the amplitude (magnitude) will decrease as the distance from the left ventricle increases. The strength of the pulse reveals some indications of blood pressure. Under high blood pressure the pulse feels very hard and strong; under low blood pressure the pulse feels weak and can be easily compressed.
   d. To determine the pulse rate, count the number of pulses that occur in 1 minute. This can be accomplished by counting pulses in 30 seconds and multiplying that number by 2. (A pulse count for a full minute, although taking a little longer, would give a better opportunity to detect irregularities.)
   e. Record the pulse rate in Part B of the laboratory report.
4. Repeat the procedure and determine the pulse rate in each of the following conditions:
   a. immediately after lying down;
   b. 3–5 minutes after lying down;
   c. immediately after standing;
   d. 3–5 minutes after standing quietly;
   e. immediately after 3 minutes of moderate exercise (*omit if the person has health problems*);
   f. 3–5 minutes after exercise has ended.
5. Complete Part B of the laboratory report.

Learning Extension

Determine the pulse rate and pulse characteristics in two additional locations on your laboratory partner. Locate and record the pulse rate from the common carotid artery and the dorsalis pedis artery. ◀1

Common carotid artery pulse rate _____

Dorsalis pedis artery pulse rate _____

Compare the amplitude of the pulse characteristics between these two pulse locations and interpret any of the variations noted. _____

_____

_____

Procedure B—Blood Pressure

1. Although a blood pressure is present in all blood vessels, the standard location to record blood pressure is the brachial artery.
2. Measure your laboratory partner's arterial blood pressure. To do this, follow these steps:
   a. Obtain a sphygmomanometer and a stethoscope.
   b. Clean the earpieces and the diaphragm of the stethoscope with alcohol swabs.
   c. Have your partner sit quietly with bare upper limb resting on a table at heart level. Have the person remain as relaxed as possible.
   d. Locate the brachial artery at the antecubital space. Wrap the cuff of the sphygmomanometer around the arm so that its lower border is about 2.5 cm above the bend of the elbow. Center the bladder of the cuff in line with the *brachial pulse* (fig. 38.1).
   e. Palpate the *radial pulse*. Close the valve on the neck of the rubber bulb connected to the cuff, and pump air from the bulb into the cuff. Inflate the cuff while watching the sphygmomanometer, and note the pressure when the pulse disappears. (This is a rough estimate of the systolic pressure.) Immediately deflate the cuff.
   f. Position the stethoscope over the brachial artery. Reinflate the cuff to a level 30 mm Hg higher than the point where the pulse disappeared during palpation.
   g. Slowly open the valve of the bulb until the pressure in the cuff drops at a rate of about 2 or 3 mm Hg per second.
   h. Listen for sounds (Korotkoff sounds) from the brachial artery. When the first loud tapping sound is heard, record the reading as the systolic pressure.

This indicates the pressure exerted against the arterial wall during systole.
   i. Continue to listen to the sounds as the pressure drops, and note the level when the last sound is heard. Record this reading as the diastolic pressure, which measures the constant arterial resistance.
   j. Release all of the pressure from the cuff.
   k. Repeat the procedure until you have two blood pressure measurements from each arm, allowing 2–3 minutes of rest between readings.
   l. Average your readings and enter them in the table in Part C of the laboratory report.

3. Measure and record your partner's blood pressure in each of the following conditions:
   a. 3–5 minutes after lying down;
   b. 3–5 minutes after standing quietly;
   c. immediately after 3 minutes of moderate exercise *(omit if the person has health problems)*;
   d. 3–5 minutes after exercise has ended.

4. Complete Part C of the laboratory report.

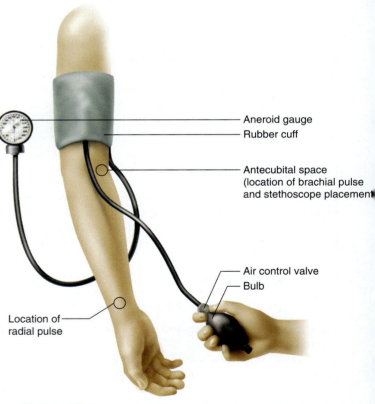

Figure 38.1   Blood pressure is commonly measured by using a sphygmomanometer (blood pressure cuff). The use of a column of mercury is the most accurate measurement, but due to environmental concerns, it has been replaced by alternative gauges and digital readouts.

Name _____

Date _____

Section _____

The ◀ corresponds to the indicated outcome(s) found at the beginning of the laboratory exercise.

Pulse Rate and Blood Pressure

Part A Assessments

Complete the following statements:

1. Blood pressure is the force exerted against the _____ . ◀3

2. The term *blood pressure* is most commonly used to refer to systemic _____ pressure. ◀3

3. The maximum pressure achieved during ventricular contraction is called_____ pressure. ◀3

4. The lowest pressure that remains in the arterial system during ventricular relaxation is called _____ pressure. ◀3

5. Blood pressure rises and falls in a pattern corresponding to the phases of the _____ . ◀3

6. A pulse is caused by the _____ . ◀1

7. The _____ artery in the arm is the standard systemic artery in which blood pressure is measured. ◀3

Part B Assessments

1. Enter your observations of pulse characteristics and pulse rates in the table. ◀1 ◀2

Test Subject	Pulse Characteristics	Pulse Rate (beats/min)
Sitting		
Lying down		
3–5 minutes later		
Standing		
3–5 minutes later		
After exercise		
3–5 minutes later		

2. Summarize the effects of body position and exercise on the characteristics and rates of the pulse. ◀2 _____

_____

_____

Part C Assessments

1. Enter the initial measurements of blood pressure in the table. ◀3

Reading	Blood Pressure in Right Arm	Blood Pressure in Left Arm
First		
Second		
Average		

2. Enter your test results in the table. ◀4

Test Subject	Blood Pressure
3–5 minutes after lying down	
3–5 minutes after standing	
After 3 minutes of moderate exercise	
3–5 minutes later	

3. Summarize the effects of body position and exercise on blood pressure. ◀4 _____

_____

_____

4. Summarize any correlations between pulse rate and blood pressure from any of the experimental conditions.

_____

_____

Critical Thinking Application

When a pulse is palpated and counted, which blood pressure (systolic or diastolic) would be characteristic at that moment? Explain your answer.

_____

_____

Laboratory Exercise

Lymphatic System

Materials Needed

Textbook
Human torso model
Anatomical chart of the lymphatic system
Compound light microscope
Prepared microscope slides:
    Lymph node section
    Human thymus section
    Human spleen section

The lymphatic system is a vast collection of cells and biochemicals that travel in lymphatic capillaries and vessels and the organs and glands that produce them. The system is closely associated with the cardiovascular system and includes a network of vessels that assist in the circulation of body fluids. These vessels provide pathways through which excess fluid can be transported away from interstitial spaces within most tissues and returned to the bloodstream. Without the lymphatic system, this fluid would accumulate in tissue spaces, producing edema.

The organs of the lymphatic system also help defend the tissues against infections by filtering particles from lymph and by supporting the activities of lymphocytes that furnish immunity against specific disease-causing agents or pathogens.

Purpose of the Exercise

To review the structure of the lymphatic system and to observe the microscopic structure of a lymph node, thymus, and spleen.

Learning Outcomes

After completing this exercise, you should be able to

1. Trace the major lymphatic pathways in an anatomical chart or model.
2. Locate and identify the major clusters of lymph nodes (lymph glands) in an anatomical chart or model.
3. Describe the structure and function of a lymph node, the thymus, and the spleen.
4. Locate and sketch the major microscopic structures of a lymph node, the thymus, and the spleen.

Procedure A—Lymphatic Pathways

1. Review the section entitled "Lymphatic Pathways" in chapter 14 of the textbook.
2. As a review activity, label figure 39.1.
3. Complete Part A of Laboratory Report 39.
4. Observe the human torso model and the anatomical chart of the lymphatic system, and locate the following features:

lymphatic vessels
lymph nodes
lymphatic trunks
collecting ducts
    thoracic duct
    right lymphatic duct
subclavian veins
internal jugular veins

Procedure B—Lymph Nodes

1. Review the section entitled "Lymph Nodes" in chapter 14 of the textbook.
2. As a review activity, label figure 39.2.
3. Complete Part B of the laboratory report.
4. Observe the anatomical chart of the lymphatic system and the human torso model and locate the clusters of lymph nodes in the following regions:

cervical region
axillary region
inguinal region
pelvic cavity
abdominal cavity
thoracic cavity

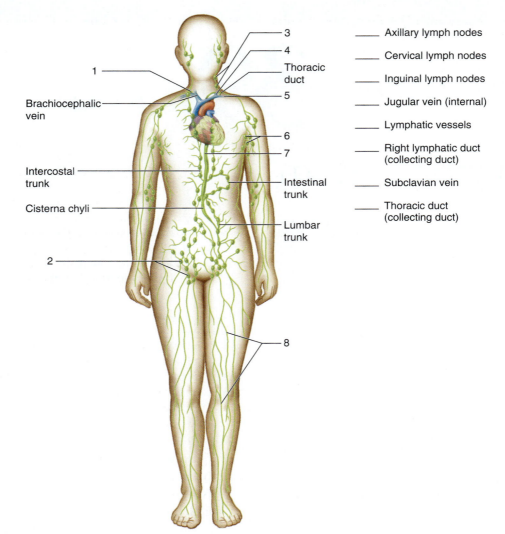

	Axillary lymph nodes
_____	Cervical lymph nodes
_____	Inguinal lymph nodes
_____	Jugular vein (internal)
_____	Lymphatic vessels
_____	Right lymphatic duct (collecting duct)
_____	Subclavian vein
_____	Thoracic duct (collecting duct)

Figure 39.1   Label the diagram by placing the correct numbers in the spaces provided.   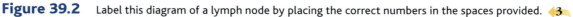 1

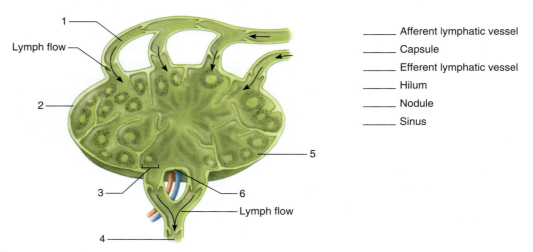

_____	Afferent lymphatic vessel
_____	Capsule
_____	Efferent lymphatic vessel
_____	Hilum
_____	Nodule
_____	Sinus

Figure 39.2   Label this diagram of a lymph node by placing the correct numbers in the spaces provided.   3

5. Palpate the lymph nodes in your cervical region. They are located along the lower border of the mandible and between the mandible and the sternocleidomastoid muscle. They feel like small, firm lumps.
6. Study figure. 39.3.
7. Obtain a prepared microscope slide of a lymph node and observe it, using low-power magnification (fig. 39.3[*b*]). Identify the *capsule* that surrounds the node and is mainly composed of collagenous fibers, the *lymph nodules* that appear as dense masses near the surface of the node, and the *lymph sinus* that appears as a narrow space between the nodules and the capsule.
8. Using high-power magnification, examine a nodule within the lymph node. The nodule contains densely packed *lymphocytes.*
9. Prepare a labeled sketch of a representative section of a lymph node in Part D of the laboratory report.

Procedure C—Thymus and Spleen

1. Review the section entitled "Thymus and Spleen" in chapter 14 of the textbook.
2. Locate the thymus and spleen in the anatomical chart of the lymphatic system and on the human torso model.
3. Complete Part C of the laboratory report.
4. Obtain a prepared microscope slide of human thymus and observe it, using low-power magnification (fig. 39.4). Note how the thymus is subdivided into *lobules* by *septa* of connective tissue that contain blood vessels. Identify the *capsule* of loose connective tissue that surrounds the thymus; the outer *cortex* of a lobule composed of densely packed cells and deeply stained; and the inner *medulla* of a lobule composed of more loosely packed lymphocytes and epithelial cells and lightly stained.

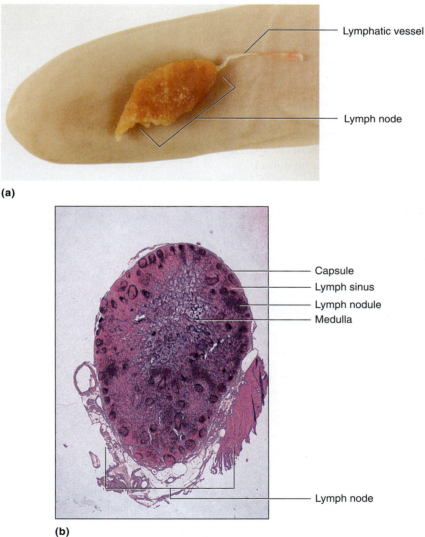

(a)

(b)

Figure 39.3 Lymph nodes: (*a*) photograph of a human lymph node on tip of finger; (*b*) micrograph of a lymph node (5×).

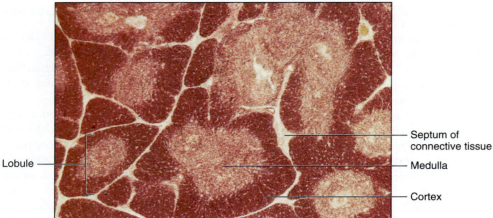

Figure 39.4 A section of the thymus (10× micrograph enlarged to 20×).

5. Examine the cortex tissue of a lobule using high-power magnification. The cells of the cortex are composed of densely packed *lymphocytes* among some epithelial cells and *macrophages*. Some of these cortical cells may be undergoing mitosis, so their chromosomes may be visible.
6. Prepare a labeled sketch of a representative section of the thymus in Part D of the laboratory report.
7. Obtain a prepared slide of the human spleen and observe it, using low-power magnification (fig. 39.5). Identify the *capsule* of dense connective tissue that surrounds the spleen. The tissues of the spleen include circular *nodules* of *white* (in unstained tissue) *pulp* enclosed in a matrix of *red pulp.*
8. Using high-power magnification, examine a nodule of white pulp and red pulp. The cells of the white pulp are mainly lymphocytes. Also, there may be an arteriole centrally located in the nodule. The cells of the red pulp are mostly red blood cells with many lymphocytes and macrophages.
9. Prepare a labeled sketch of a representative section of the spleen in Part D of the laboratory report.

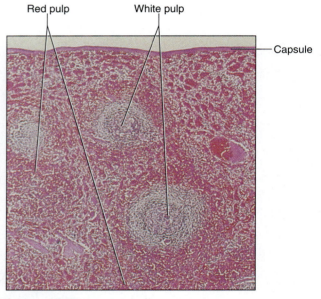

Figure 39.5 Micrograph of a section of the spleen (15×).

Name _____

Date _____

Section _____

The ⬅ corresponds to the indicated outcome(s) found at the beginning of the laboratory exercise.

Lymphatic System

Part A Assessments

Complete the following statements:

1. Lymphatic pathways begin as _____ that merge to form lymphatic vessels. ⬅**1**

2. The wall of a lymphatic capillary consists of a single layer of _____ cells.

3. Once tissue (interstitial) fluid is inside a lymph capillary, the fluid is called _____ .

4. Lymphatic vessels have walls similar to those of _____ .

5. Lymphatic vessels contain _____ that help prevent the backflow of lymph.

6. Lymphatic vessels usually lead to _____ that filter the fluid being transported. ⬅**1**

7. The _____ is the larger and longer of the two lymphatic collecting ducts. ⬅**1**

Part B Assessments

Complete the following statements:

1. Lymph nodes contain large numbers of white blood cells called _____ and macrophages that fight invading microorganisms. ⬅**3**

2. The indented region where blood vessels and nerves join a lymph node is called the _____ . ⬅**3**

3. _____ that contain germinal centers are the structural units of a lymph node. ⬅**3**

4. The spaces within a lymph node are called _____ through which lymph circulates. ⬅**3**

5. Lymph enters a node through a(n) _____ lymphatic vessel. ⬅**3**

6. The partially encapsulated lymph nodes in the pharynx are called _____ .

7. The aggregations of lymph nodules found within the mucosal lining of the small intestine are called _____ . ⬅**2**

Part C Assessments

Complete the following statements:

1. The _____ gland is located in the mediastinum, anterior to the aortic arch.

2. The thymus is composed of lymphatic tissue, subdivided into _____ . ◀3

3. The lymphocytes that leave the thymus function in providing _____ . ◀3

4. The hormones secreted by the thymus are called _____ . ◀3

5. The _____ is the largest lymphatic organ. ◀3

6. The sinuses within the spleen contain _____ . ◀3

7. The tiny islands (nodules) of tissue within the spleen that contain many lymphocytes comprise the _____ pulp. ◀3

8. The _____ pulp of the spleen contain large numbers of red blood cells, lymphocytes, and macrophages. ◀3

9. _____ within the spleen engulf and destroy foreign particles and cellular debris. ◀3

Part D Assessments

Sketch and label structural features of the following lympatic organs: ◀4

Lymph node sketch (_____ ×)	Thymus sketch (_____ ×)
Spleen sketch (_____ ×)	

Digestive Organs

Materials Needed

Textbook
Human torso model
Skull with teeth
Teeth, sectioned
Tooth model, sectioned
Paper cup
Compound light microscope
Prepared microscope slides of the following:
    Salivary gland
    Esophagus
    Stomach (fundus)
    Small intestine (jejunum)
    Large intestine

The digestive system includes the organs associated with the alimentary canal and several accessory structures. The alimentary canal, a muscular tube, passes through the body from the opening of the mouth to the anus. It includes the mouth, pharynx, esophagus, stomach, small intestine, and large intestine. The function of the canal is to move substances throughout its length. It is specialized in various regions to store, digest, and absorb food materials and to eliminate the residues. The accessory organs, which include the salivary glands, liver, gallbladder, and pancreas, secrete products into the alimentary canal that aid digestive functions.

Purpose of the Exercise

To review the structure and function of the digestive organs and to microscopically examine the tissues of these organs.

Learning Outcomes

After completing this exercise, you should be able to

1. Locate and label the major digestive organs and their major structures.
2. Describe the functions of these organs.
3. Examine and sketch the structures of a section of the small intestine.

Procedure A—Mouth and Salivary Glands

1. Review the sections entitled "Mouth" and "Salivary Glands" in chapter 15 of the textbook.
2. As a review activity, label figures 40.1, 40.2, and 40.3.
3. Examine the mouth of the human torso model and a skull. Locate the following structures:

   oral cavity (mouth)

   vestibule

   tongue

       lingual frenulum

       papillae

   palate

       hard palate

       soft palate

       uvula

   palatine tonsils

   gums (gingivae)

   teeth

       incisors

       canines (cuspids)

       premolars (bicuspids)

       molars

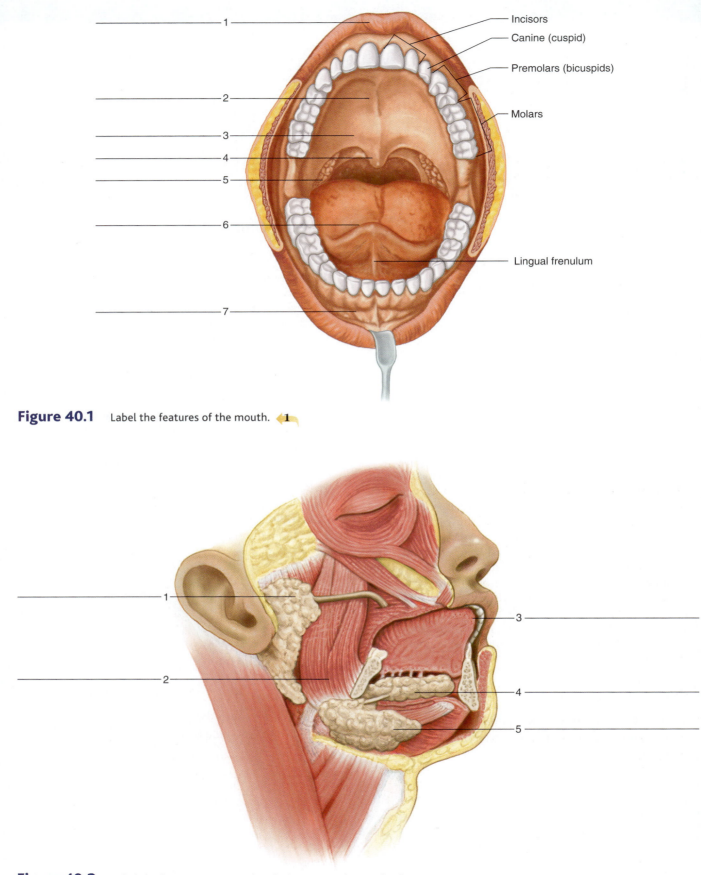

Incisors

Canine (cuspid)

Premolars (bicuspids)

Molars

Lingual frenulum

Figure 40.1   Label the features of the mouth. ◀**1**

Figure 40.2   Label the features associated with the major salivary glands. ◀**1**

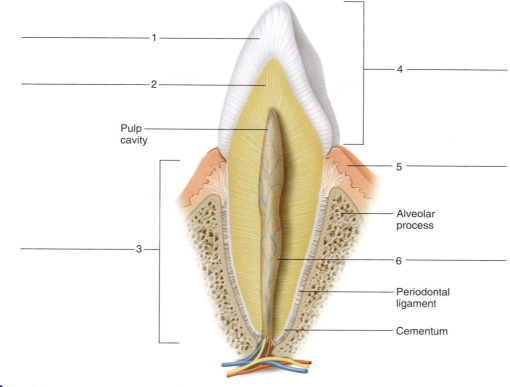

Pulp
cavity

Alveolar
process

Periodontal
ligament

Cementum

Figure 40.3 Label the features of this canine (cuspid) tooth. ◀1

4. Examine a sectioned tooth and a tooth model. Locate the following features:

crown
    enamel
    dentin
neck
root
    pulp cavity
    cementum
    root canal

5. Observe the head of the human torso model and locate the following:

parotid salivary gland
submandibular salivary gland
sublingual salivary gland

6. Examine a microscopic section of a salivary gland, using low- and high-power magnification. Note the mucous cells that produce mucus and serous cells that produce enzymes are clustered around small ducts. Also note a larger secretory duct surrounded by lightly stained cuboidal epithelial cells (fig. 40.4).

7. Complete Part A of Laboratory Report 40.

Procedure B—Pharynx and Esophagus

1. Review the section entitled "Pharynx and Esophagus" in chapter 15 of the textbook.
2. As a review activity, label figure 40.5.
3. Observe the human torso model and locate the following features:

pharynx
    nasopharynx
    oropharynx
    laryngopharynx
esophagus
lower esophageal sphincter (cardiac sphincter)

4. Have your partner take a drink from a cup of water. Carefully watch the movements in the anterior region of the neck. What steps in the swallowing process did you observe?

_____

_____

_____

_____

_____

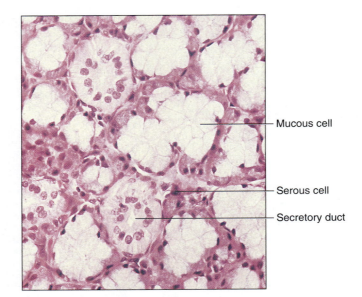

Figure 40.4 Micrograph of a salivary gland (300×).

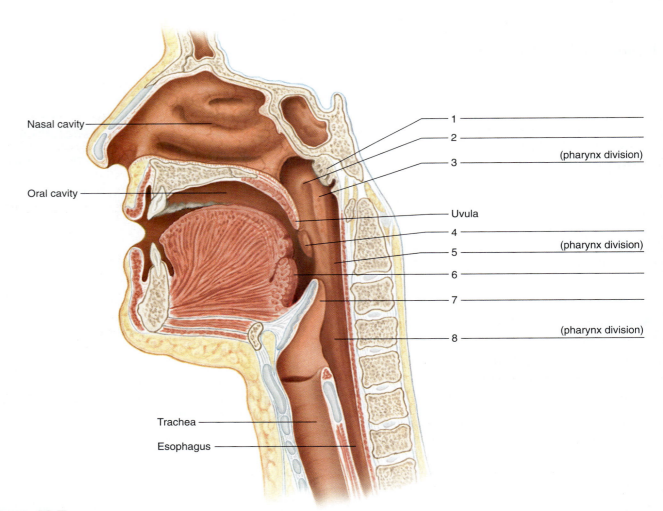

Mucous cell

Serous cell

Secretory duct

Nasal cavity

Oral cavity

1

2

(pharynx division)

3

Uvula

4

(pharynx division)

5

6

7

(pharynx division)

8

Trachea

Esophagus

Figure 40.5 Label the features associated with the pharynx. ◀1

5. Examine a microscopic section of the esophagus wall, using low-power magnification (fig. 40.6). The inner lining is composed of stratified squamous epithelium, and there are layers of muscle tissue in the wall. Locate some mucous glands in the submucosa. They appear as clusters of lightly stained cells.

6. Complete Part B of the laboratory report.

Procedure C—The Stomach

1. Review the section entitled "Stomach" in chapter 15 of the textbook.

2. As a review activity, label figures 40.7 and 40.8.

3. Observe the human torso model and locate the following features of the stomach:

cardiac region

fundic region

body region

pyloric region

pyloric canal

rugae (gastric folds)

pyloric sphincter (valve)

4. Examine a microscopic section of the stomach wall, using low-power magnification (fig. 40.9). Note how the inner lining of simple columnar epithelium dips

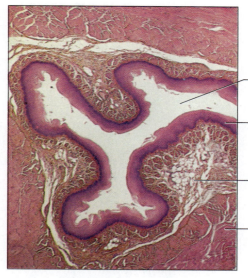

Lumen

Mucosa composed of stratified squamous epithelium

Submucosa with mucous glands

Muscular layer

Figure 40.6 Micrograph of a cross section of the esophagus (10×).

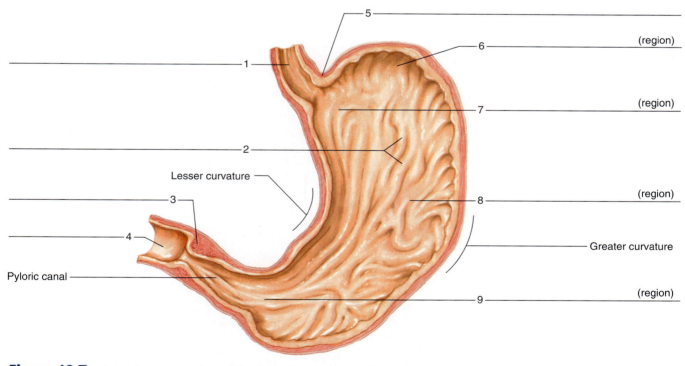

5

6 _____ (region)

1

7 _____ (region)

2

Lesser curvature

3

8 _____ (region)

4

Greater curvature

Pyloric canal

9 _____ (region)

Figure 40.7 Label the major regions of the stomach and associated structures.

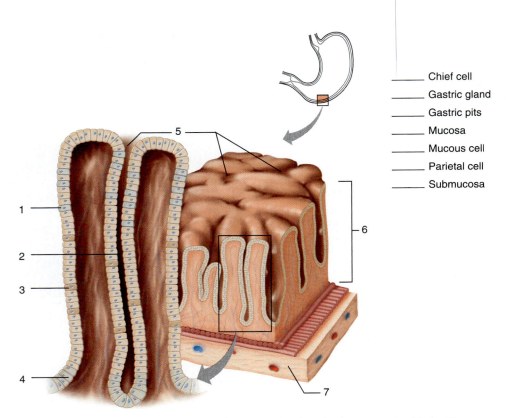

Chief cell
_____ Gastric gland
_____ Gastric pits
_____ Mucosa
_____ Mucous cell
_____ Parietal cell
_____ Submucosa

1
2
3
4
5
6
7

Figure 40.8   Label the lining of the stomach by placing the correct numbers in the spaces provided. ◀ **1**

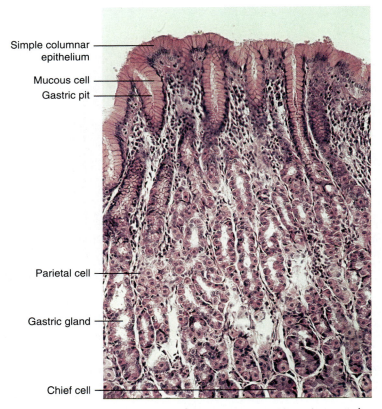

Simple columnar epithelium
Mucous cell
Gastric pit
Parietal cell
Gastric gland
Chief cell

Figure 40.9   Micrograph of the mucosa of the stomach wall (50× micrograph enlarged to 100×).

inward to form gastric pits. The gastric glands are tubular structures that open into the gastric pits. Near the deep ends of these glands, you should be able to locate some intensely stained (bluish) chief cells and some lightly stained (pinkish) parietal cells. What are the functions of these cells? _____

_____

_____

_____

_____

5. Complete Part C of the laboratory report.

Procedure D—Pancreas and Liver

1. Review the sections entitled "Pancreas" and "Liver" in chapter 15 of the textbook.
2. As a review activity, label figure 40.10.

3. Observe the human torso model and locate the following structures:

> **pancreas**
> **pancreatic duct**
> **liver**
> **gallbladder**
> **hepatic ducts**
> **cystic duct**
> **common bile duct**
> **hepatopancreatic sphincter (sphincter of Oddi)**

Procedure E—Small and Large Intestines

1. Review the sections entitled "Small Intestine" and "Large Intestine" in chapter 15 of the textbook.

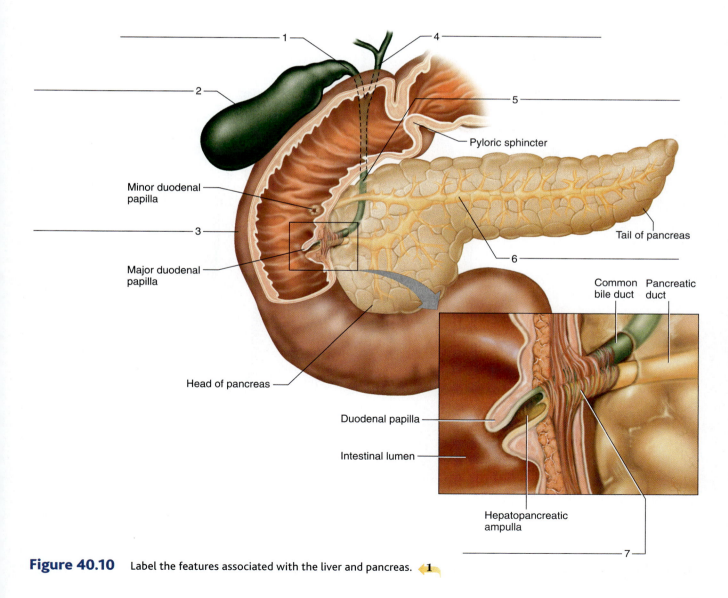

Figure 40.10 Label the features associated with the liver and pancreas. ◀**1**

2. As a review activity, label figure 40.11 and examine figure 40.12.
3. Observe the human torso model and locate each of the following features:

small intestine
   duodenum
   jejunum
   ileum
mesentery
ileocecal sphincter (valve)
large intestine
   cecum
   appendix (vermiform appendix)

ascending colon
transverse colon
descending colon
sigmoid colon
rectum
anal canal
anal columns
anal sphincter muscles
   external anal sphincter
   internal anal sphincter
anus

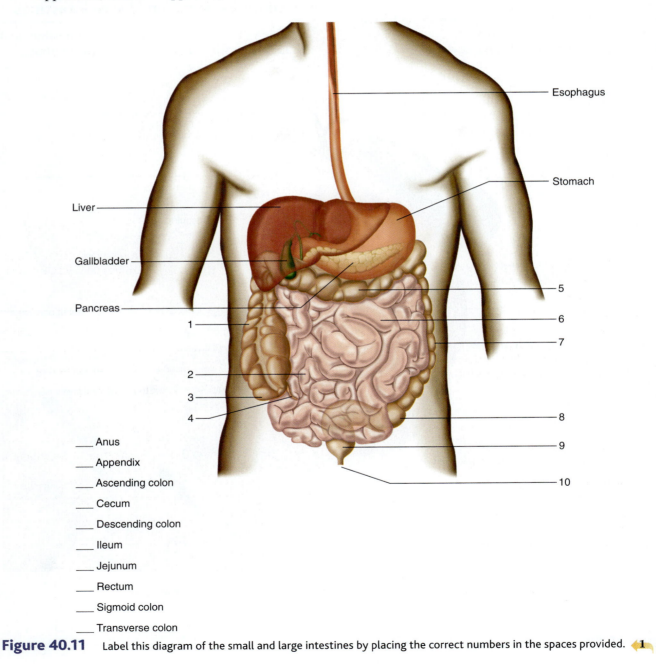

___ Anus
___ Appendix
___ Ascending colon
___ Cecum
___ Descending colon
___ Ileum
___ Jejunum
___ Rectum
___ Sigmoid colon
___ Transverse colon

Figure 40.11   Label this diagram of the small and large intestines by placing the correct numbers in the spaces provided.

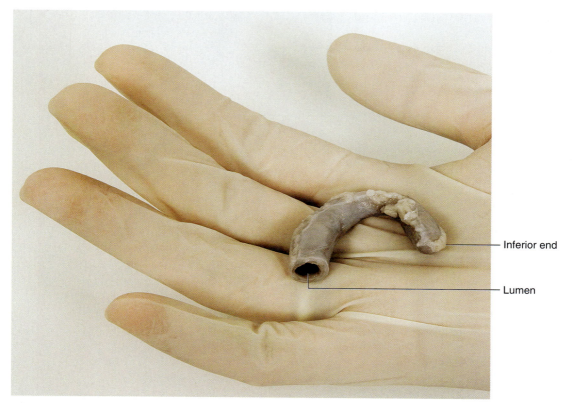

Figure 40.12   Normal appendix.

4. Using low-power magnification, examine a microscopic section of the small intestine wall. Identify the mucosa, submucosa, muscular layer, and serosa. Note the villi that extend into the lumen of the tube. Study a single villus, using high-power magnification. Note the core of connective tissue and the covering of simple columnar epithelium that contains some lightly stained goblet cells (fig. 40.13). What is the function of these villi?

_____

_____

_____

5. Prepare a labeled sketch of the wall of the small intestine in Part D of the laboratory report.
6. Examine a microscopic section of the large intestine wall. Note the lack of villi. Locate the four layers of the wall. Also note the tubular mucous glands that open on the surface of the inner lining and the

numerous lightly stained goblet cells (fig. 40.14). What is the function of the mucus secreted by these glands?

_____

_____

_____

_____

7. Complete Part E of the laboratory report.

Critical Thinking Application

How is the structure of the small intestine better adapted for absorption than the large intestine?

_____

_____

_____

_____

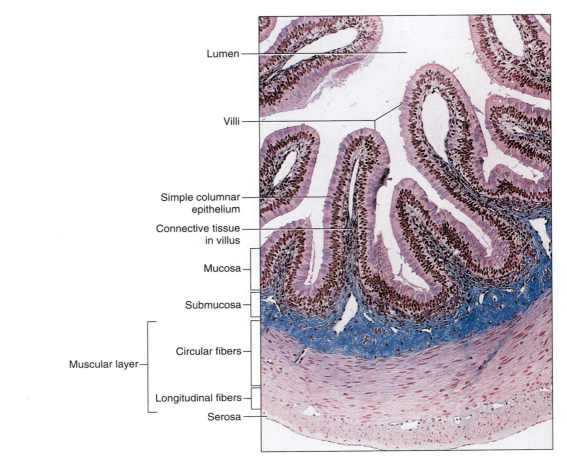

Figure 40.13   Micrograph of the small intestine wall (40×).

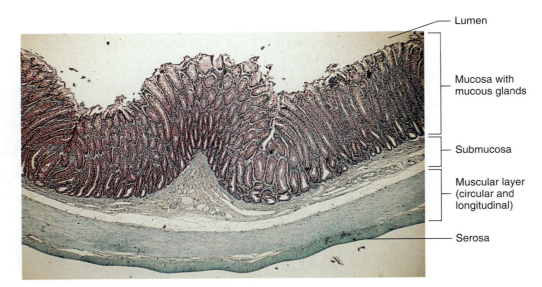

Figure 40.14   Micrograph of the large intestine wall (64×).

Name _____

Date _____

Section _____

The ⬅ corresponds to the indicated outcome(s) found at the beginning of the laboratory exercise.

Digestive Organs

Part A Assessments

Match the terms in column A with the descriptions in column B. Place the letter of your choice in the space provided. ⬅1

Column A	Column B
a. Adenoids (pharyngeal tonsils)	_____ **1.** Bonelike substance beneath tooth enamel
b. Crown	
c. Dentin	_____ **2.** Smallest of major salivary glands
d. Incisor	_____ **3.** Tooth specialized for grinding
e. Lingual frenulum	
f. Molar	_____ **4.** Chamber between tongue and palate
g. Oral cavity	_____ **5.** Projections on tongue surface
h. Papillae	_____ **6.** Cone-shaped projection of soft palate
i. Periodontal ligament	
j. Sublingual gland	_____ **7.** Attaches tooth to jaw
k. Uvula	_____ **8.** Chisel-shaped tooth
l. Vestibule	
	_____ **9.** Space between the teeth, cheeks, and lips
	_____ **10.** Anchors tongue to floor of mouth
	_____ **11.** Lymphatic tissue in posterior wall of pharynx near auditory tubes
	_____ **12.** Portion of tooth projecting beyond gum

Part B Assessments

Complete the following:

1. The part of the pharynx superior to the soft palate is called the _____ . ⬅1

2. The middle part of the pharynx is called the _____ . ⬅1

3. The inferior portion of the pharynx is called the_____ . ⬅1

4. _____ is the main secretion of the esophagus. ⬅2

5. Summarize the functions of the esophagus. ⬅2 _____

_____

_____

Part C Assessments

Complete the following:

1. Name the four regions of the stomach. **1** _____

2. Name the valve that prevents the regurgitation of food from the small intestine back into the stomach. **2**

   _____

3. Name the gastric cells that secrete digestive enzymes. **2** _____

4. Name the gastric cells that secrete hydrochloric acid. **2** _____

5. Name the most important digestive enzyme in gastric juice. **2** _____

6. Name the substance needed for the efficient absorption of vitamin B_{12}. **2** _____

7. Summarize the functions of the stomach. **2** _____

   _____

   _____

Part D Assessments

In the space that follows, sketch a representative area of the wall of the small intestine. Label the four layers, the lumen, and a villus. Indicate the magnification used for the sketch. **3**

Part E Assessments

Complete the following:

1. Name the three portions of the small intestine. **1** _____

2. Describe the function of the mesentery. **2** _____

   _____

3. Name five digestive enzymes secreted by the intestinal glands. **2** _____

   _____

4. Name the valve located between the small and large intestines. **1** _____

5. Name the small projection that contains lymphatic tissue attached to the cecum. **1** _____

6. Summarize the functions of the small intestine. **2** _____

   _____

   _____

7. Summarize the functions of the large intestine. **2** _____

   _____

Action of a Digestive Enzyme

Materials Needed

0.5% amylase solution*
Beakers (50 and 500 mL)
Distilled water
Funnel
Pipets (1 and 10 mL)
Pipet rubber bulbs
0.5% starch solution
Graduated cylinder (10 mL)
Test tubes
Test-tube clamps
Wax marker
Iodine-potassium-iodide solution
Medicine dropper
Ice
Water bath, 37°C (98.6°F)
Porcelain test plate
Benedict's solution
Hot plates
Test-tube rack
Thermometer

*The amylase must be free of sugar. See Appendix 1.

⚠ Safety

- Wear safety glasses when working with acids and when heating test tubes.
- Use test-tube clamps when handling hot test tubes.
- If an open flame is used for heating the test solutions, keep clothes and hair away from the flame.
- Use only a mechanical pipetting device (never your mouth). Use pipets with rubber bulbs or dropping pipets.

The digestive enzyme in salivary secretions is called *salivary amylase.* Pancreatic amylase is secreted among several other pancreatic enzymes. A bacterial extraction of amylase is available for laboratory experiments. This enzyme splits starch molecules into sugar (disaccharide) molecules, which is the first step in the digestion of complex carbohydrates.

As in the case of other enzymes, amylase is a protein catalyst. Its activity is affected by exposure to certain environmental factors, including various temperatures, pH, radiation, and electricity. As temperatures increase, faster chemical reactions occur as the collisions of molecules happen at a greater frequency. Eventually temperatures increase to a point that the enzyme is denatured and the rate of the enzyme activity rapidly declines. As temperatures decrease, enzyme activity also decreases due to fewer collisions of the molecules, however the colder temperatures do not denature the enzyme. Normal body temperature provides an environment for enzyme activity near the optimum for enzymatic reactions.

The pH of the environment where enzymes are secreted also has a major influence on the activity of the enzyme reactions. The optimum pH for amylase activity is between 6.8 and 7.0, typical of salivary secretions. When salivary amylase arrives within the stomach, hydrochloric acid deactivates the enzyme, diminishing any further chemical digestion of remaining starch located in the stomach. Pancreatic amylase is secreted into the small intestine where an optimum pH for amylase is once more provided. Other enzymes in the digestive system have different optimum pH ranges of activity compared to amylase. For example, pepsin from stomach secretions has an optimum activity around pH 2, while trypsin from pancreatic secretions operate best around pH 7–8.

Purpose of the Exercise

To investigate the action of amylase and the effect of heat on its enzymatic activity.

Learning Outcomes

After completing this exercise, you should be able to

1. Explain the action of amylase.
2. Test a solution for the presence of starch or the presence of sugar.
3. Test the effects of varying temperatures on the activity of amylase.

Procedure A—Amylase Activity

1. Study the lock-and-key model of amylase action on starch digestion (fig. 41.1).
2. Examine the digestive locations of starch to glucose (fig. 41.2).
3. Mark three clean test tubes as *tubes 1, 2,* and *3,* and prepare the tubes as follows:

   Tube 1: **Add 6 mL of amylase solution.**

   Tube 2: **Add 6 mL of starch solution.**

   Tube 3: **Add 5 mL of starch solution and 1 mL of amylase solution.**

4. Shake the tubes well to mix the contents, and place them in a warm water bath, 37°C (98.6°F), for 10 minutes.
5. At the end of the 10 minutes, test the contents of each tube for the presence of starch. To do this, follow these steps:
   a. Place 1 mL of the solution to be tested in a depression of a porcelain test plate.
   b. Next add one drop of iodine-potassium-iodide solution, and note the color of the mixture. If the solution becomes blue-black, starch is present.

   c. Record the results in Part A of Laboratory Report 41.

6. Test the contents of each tube for the presence of sugar (disaccharides in this instance). To do this, follow these steps:
   a. Place 1 mL of the solution to be tested in a clean test tube.
   b. Add 1 mL of Benedict's solution.
   c. Place the test tube with a test-tube clamp in a beaker of boiling water for 2 minutes.
   d. Note the color of the liquid. If the solution becomes green, yellow, orange, or red, sugar is present. Blue indicates a negative test, whereas green indicates a positive test with the least amount of sugar, and red indicates the greatest amount of sugar present.
   e. Record the results in Part A of the laboratory report.

7. Complete Part A of the laboratory report.

Procedure B—Effect of Heat

1. Mark three clean test tubes as *tubes 4, 5,* and *6.*
2. Add 1 mL of amylase solution to each of the tubes, and expose each solution to a different test temperature for 3 minutes as follows:

   Tube 4: **Place in beaker of ice water (about 0°C [32°F]).**

   Tube 5: **Place in warm water bath (about 37°C [98.6°F]).**

   Tube 6: **Place in beaker of boiling water (about 100°C [212°F]). Use a test-tube clamp.**

3. Add 5 mL of starch solution to each tube, shake to mix the contents, and return the tubes to their respective test temperatures for 10 minutes. It is important that the 5 mL of starch solution added to tube 4 be at ice-water temperature before it is added to the 1 mL of amylase solution.
4. At the end of the 10 minutes, test the contents of each tube for the presence of starch and the presence of sugar by following the directions in Procedure A.
5. Complete Part B of the laboratory report.

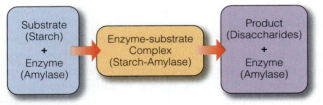

Figure 41.1   Lock-and-key model of enzyme action of amylase on starch digestion.

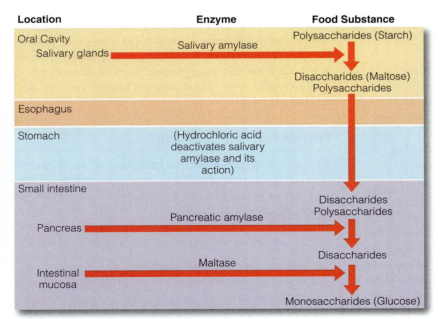

Figure 41.2   Flowchart of digestion of starch.

The ⬅ corresponds to the indicated outcome(s) found at the beginning of the laboratory exercise.

Action of a Digestive Enzyme

Part A—Amylase Activity Assessments

1. Test results: ⬅2

Tube		Starch	Sugar
1	Amylase solution		
2	Starch solution		
3	Starch-amylase solution		

2. Complete the following:

   a. Explain the reason for including tube 1 in this experiment. ⬅1

   _____

   _____

   _____

   b. What is the importance of tube 2? ⬅1

   _____

   _____

   _____

   c. What do you conclude from the results of this experiment? ⬅2

   _____

   _____

   _____

Part B—Effect of Heat Assessments

1. Test results: ◀**3**

Tube		Starch	Sugar
4	0°C (32°F)		
5	37°C (98.6°F)		
6	100°C (212°F)		

2. Complete the following:

   a. What do you conclude from the results of this experiment? ◀**1**

   _____

   _____

   _____

   _____

   _____

   b. If digestion failed to occur in one of the tubes in this experiment, how can you tell if the amylase was destroyed by the factor being tested or if the amylase activity was simply inhibited by the test treatment? ◀**1**

   _____

   _____

   _____

   _____

   _____

Critical Thinking Application

What test result would occur if the amylase you used contained sugar? _____ Would your results be valid? Explain your answer.

_____

_____

_____

_____

Laboratory Exercise  **42**

Respiratory Organs

Materials Needed

Textbook
Human skull (sagittal section)
Human torso model
Larynx model
Thoracic organs model
Compound light microscope
Prepared microscope slides of the following:
    Trachea (cross section)
    Lung, human (normal)

For Demonstrations:
Animal lung with trachea (fresh or preserved)
Bicycle pump
Prepared microscope slides of the following:
    Lung tissue (smoker)
    Lung tissue (emphysema)

⚠ Safety

- Wear disposable gloves when working on the fresh or preserved animal lung demonstration.
- Wash your hands before leaving the laboratory.

The organs of the respiratory system include the nose, nasal cavity, sinuses, pharynx, larynx, trachea, bronchial tree, and lungs. They mainly function to process incoming air and to transport it to and from the atmosphere outside the body and the air sacs of the lungs.

In the air sacs, gas exchanges take place between the air and the blood of nearby capillaries. The blood, in turn, transports gases to and from the air sacs and the body cells. This entire process of transporting and exchanging gases between the atmosphere and the body cells is called *respiration*.

Purpose of the Exercise

To review the structure and function of the respiratory organs and to microscopically examine the tissues of some of these organs.

Learning Outcomes

After completing this exercise, you should be able to

1. Locate the major organs and structural features of the respiratory system.
2. Describe the functions of these organs.
3. Sketch and label features of tissue sections of the trachea and lung.

Procedure A—Respiratory Organs

1. Review the section entitled "Organs of the Respiratory System" in chapter 16 of the textbook.
2. As a review activity, label figures 42.1 and 42.2.
3. Examine the sagittal section of the human skull and locate the following features:

nasal cavity
    nostril
    nasal septum
    nasal conchae
paranasal sinuses
    maxillary sinus
    frontal sinus
    ethmoidal sinus
    sphenoidal sinus

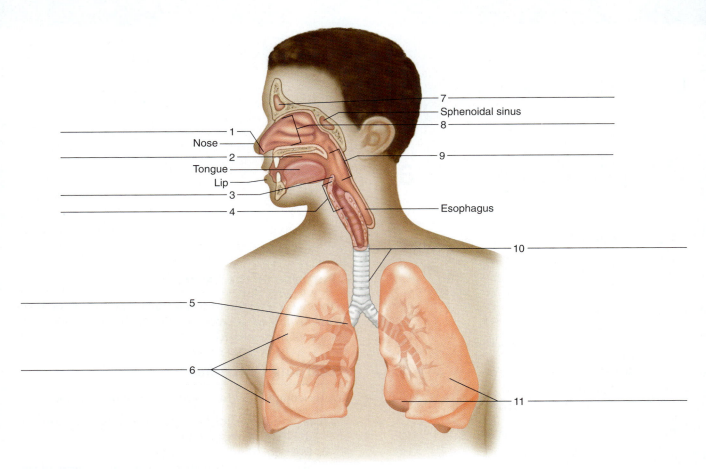

Figure 42.1 Label the major features of the respiratory system.

7 ————
Sphenoidal sinus
8 ————
1
Nose
2
9 ————
Tongue
Lip
3
Esophagus
4
10 ————
5
6
11 ————

4. Observe the larynx model, the thoracic organs model, and the human torso model. Locate the features listed in step 3. Also locate the following features:

pharynx
    nasopharynx
    oropharynx
    laryngopharynx
larynx (palpate your larynx)
    vocal cords
        false vocal cords (vestibular folds)
        true vocal cords (vocal folds)
    thyroid cartilage ("Adam's apple")
    cricoid cartilage
    epiglottic cartilage
    glottis

trachea (palpate your trachea)
bronchi
    primary (right and left main) bronchi
    secondary (lobar) bronchi
    tertiary (segmental) bronchi
bronchioles
lung
    lobes (right lung, three; left lung, two)
visceral pleura
parietal pleura
pleural cavity

5. Complete Part A of Laboratory Report 42.

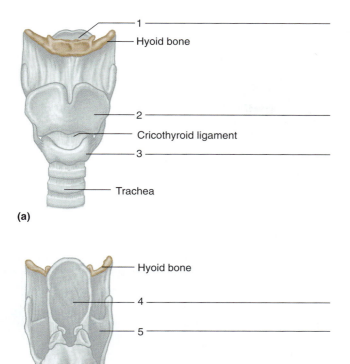

(a)

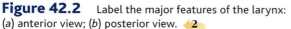

(b)

Figure 42.2 Label the major features of the larynx:
(a) anterior view; (b) posterior view. 2

1 ———
——— Hyoid bone

2 ———
——— Cricothyroid ligament
3 ———

——— Trachea

Hyoid bone

4 ———

5 ———

6 ———

——— Trachea

Procedure B—Respiratory Tissues

1. Obtain a prepared microscope slide of a trachea, and use low-power magnification to examine it. Notice the inner lining of ciliated pseudostratified columnar epithelium and the deep layer of hyaline cartilage, which represents a portion of an incomplete (C-shaped) tracheal ring (fig. 42.3).
2. Use high-power magnification to observe the cilia on the free surface of the epithelial lining. Locate the wineglass-shaped goblet cells, which secrete a protective mucus, in the epithelium (fig. 42.4).
3. Prepare a labeled sketch of a representative portion of the tracheal wall in Part B of the laboratory report.
4. Obtain a prepared microscope slide of human lung tissue. Examine it, using low-power magnification, and note the many open spaces of the air sacs (alveoli). Look for a bronchiole—a tube with a relatively thick wall and a wavy inner lining. Locate the smooth muscle tissue in the wall of this tube (fig. 42.5). You also may see a section of cartilage as part of the bronchiole wall.

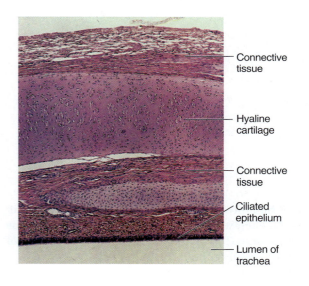

Connective tissue

Hyaline cartilage

Connective tissue

Ciliated epithelium

Lumen of trachea

Figure 42.3   Micrograph of a section of the tracheal wall (63×).

Demonstration

Observe the animal lung and the attached trachea. Identify the larynx, major laryngeal cartilages, trachea, and the incomplete cartilaginous rings of the trachea. Open the larynx and locate the vocal folds. Examine the visceral pleura on the surface of a lung. A bicycle pump could be used to demonstrate lung inflation. Section the lung and locate some bronchioles and alveoli. Squeeze a portion of a lung between your fingers. How do you describe the texture of the lung?

_____

_____

_____

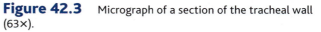

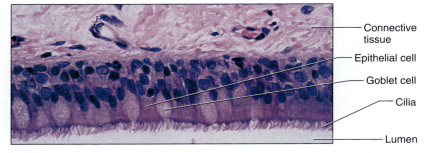

Figure 42.4   Micrograph of ciliated epithelium in the respiratory tract (275×).

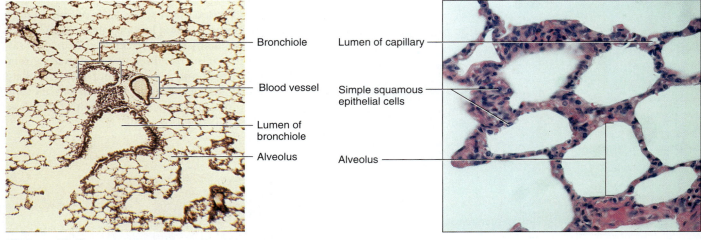

Figure 42.5   Micrograph of human lung tissue (35×).

Figure 42.6   Micrograph of human lung tissue (250×).

5. Use high-power magnification to examine the alveoli. Their walls are composed of simple squamous epithelium (fig. 42.6). You also may see sections of blood vessels containing blood cells.
6. Prepare a labeled sketch of a representative portion of the lung in Part B of the laboratory report.
7. Complete Part C of the laboratory report.

Demonstration

Examine the prepared microscope slides of the lung tissue of a smoker and a person with emphysema, using low-power magnification. How does the smoker's lung tissue compare with that of the normal lung tissue that you examined previously?

_____

_____

_____

How does the emphysema patient's lung tissue compare with the normal lung tissue?

_____

_____

_____

The ⬅ corresponds to the indicated outcome(s) found at the beginning of the laboratory exercise.

Respiratory Organs

Part A Assessments

Match the terms in column A with the descriptions in column B. Place the letter of your choice in the space provided. ⬅**1** ⬅**2**

Column A	Column B
a. Alveolus	_____ **1.** Potential space between visceral and parietal pleurae
b. Cricoid cartilage	_____ **2.** Most inferior portion of larynx
c. Epiglottis	_____ **3.** Serves as resonant chamber and reduces weight of skull
d. Glottis	_____ **4.** Microscopic air sac for gas exchange
e. Lung	_____ **5.** Consists of large lobes
f. Nasal concha	_____ **6.** Opening between vocal cords
g. Pharynx	_____ **7.** Fold of mucous membrane containing elastic fibers responsible for sounds
h. Pleural cavity	
i. Sinus (paranasal sinus)	_____ **8.** Increases surface area of nasal mucous membrane
j. Vocal cord (true)	_____ **9.** Passageway for air and food
	_____ **10.** Partially covers opening of larynx during swallowing

Part B Assessments

1. Sketch and label a portion of the tracheal wall. ⬅**3**

2. Sketch and label a portion of lung tissue. ◀**3**

Part C Assessments

Complete the following:

1. What is the function of the mucus secreted by the goblet cells? ◀**2** _____

_____

_____

2. Describe the function of the cilia in the respiratory tubes. ◀**2** _____

_____

_____

3. How is breathing affected if the smooth muscle of the bronchial tree relaxes? ◀**2** _____

_____

_____

Critical Thinking Application

Why are the alveolar walls so thin?

_____

_____

_____

_____

Breathing and Respiratory Volumes and Capacities

Breathing involves the movement of air from outside the body through the bronchial tree and into the alveoli and the reversal of this air movement to allow gas (oxygen and carbon dioxide) exchange between air and blood. These movements are caused by changes in the size of the thoracic cavity that result from skeletal muscle contractions and from the elastic recoil of stretched tissues.

The volumes of air that move in and out of the lungs during various phases of breathing are called *respiratory air volumes* and *capacities.* These volumes can be measured by using an instrument called a *spirometer.* However, the values obtained vary with a person's age, sex, height, weight, stress, and physical fitness. Various respiratory capacities can be calculated by combining two or more of the respiratory volumes.

Purpose of the Exercise

To review the mechanisms of breathing and to measure or calculate certain respiratory air volumes and respiratory capacities.

Learning Outcomes

After completing this exercise, you should be able to

1. Differentiate between the mechanisms responsible for inspiration and expiration.
2. Match the respiratory air volumes and respiratory capacities with their definitions.
3. Measure respiratory air volumes using a spirometer.
4. Calculate respiratory capacities using the data obtained from respiratory air volumes.

Procedure A—Breathing Mechanisms

1. Review the sections entitled "Inspiration" and "Expiration" in chapter 16 of the textbook.
2. Complete Part A of Laboratory Report 43.

WARNING: *If the subject begins to feel dizzy or light-headed while performing Procedure B, stop the exercise and breathe normally.*

Procedure B—Respiratory Air Volumes and Capacities

1. Review the section entitled "Respiratory Air Volumes and Capacities" in chapter 16 of the textbook.
2. Complete Part B of the laboratory report.
3. Get a handheld spirometer (fig. 43.2). Point the needle to zero by rotating the adjustable dial. Before using the instrument, clean it with an alcohol swab and place a new disposable mouthpiece over its stem. The instrument should be held with the dial upward, and air should be blown only into the disposable mouthpiece (fig. 43.3). Movement of the needle indicates the air volume that leaves the lungs. The exhalations should be slowly and forcefully performed. Too rapid an exhalation could result in erroneous data or damage to the spirometer.
4. *Tidal volume* (TV) (about 500 mL) is the volume of air that enters (or leaves) the lungs during a *respiratory cycle* (one inspiration plus the following expiration). *Resting tidal volume* is the volume of air that enters (or leaves) the lungs during normal, quiet breathing (fig. 43.4). To measure this volume, follow these steps:
   a. Sit quietly for a few moments.
   b. Position the spirometer dial so that the needle points to zero.
   c. Place the mouthpiece between your lips and exhale three ordinary expirations into it after inhaling through the nose each time. *Do not force air out of your lungs; exhale normally.*
   d. Divide the total value indicated by the needle by 3, and record this amount as your resting tidal volume on the table in Part C of the laboratory report.

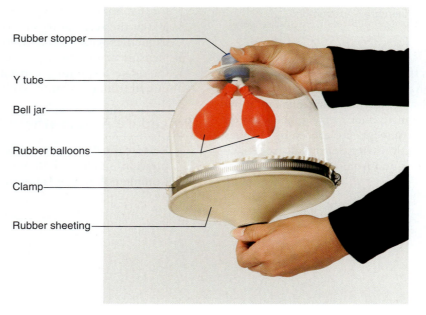

Rubber stopper
Y tube
Bell jar
Rubber balloons
Clamp
Rubber sheeting

Figure 43.1   A lung function model.

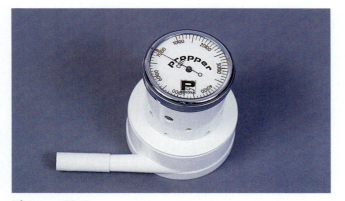

Figure 43.2 A handheld spirometer can be used to measure respiratory air volumes.

Figure 43.3 Demonstration of a handheld spirometer. Air should only be blown slowly and forcefully into a disposable mouthpiece. Use a nose clip or pinch your nostrils when measuring expiratory reserve volume and vital capacity volume.

5. *Expiratory reserve volume* (ERV) (about 1,100 mL) is the volume of air in addition to the tidal volume that leaves the lungs during forced expiration. To measure this volume, follow these steps:
   a. Breathe normally for a few moments. Set the needle to zero.
   b. At the end of an ordinary expiration, place the mouthpiece between your lips and exhale all of the air you can force from your lungs through the spirometer. Use a nose clip or pinch your nose to prevent air from exiting your nostrils.
   c. Record the results as your expiratory reserve volume in Part C.

6. *Vital capacity* (VC) (about 4,600 mL) is the maximum volume of air that can be exhaled after taking the deepest breath possible. To measure this volume, follow these steps:
   a. Breathe normally for a few moments. Set the needle at zero.
   b. Breathe in and out deeply a couple of times, then take the deepest breath possible.
   c. Place the mouthpiece between your lips and exhale all the air out of your lungs, slowly and forcefully. Use a nose clip or pinch your nose to prevent air from exiting your nostrils.
   d. Record the value as your vital capacity in Part C. Compare your result with that expected for a person of your sex, age, and height as listed in tables 43.1 and 43.2. Use the meterstick to determine your height in centimeters if necessary or multiply your height in inches times 2.54 to calculate your height in centimeters. Considerable individual variations from the expected will be noted due to parameters other than sex, age, and height, which could include physical shape, health, medications, and others.

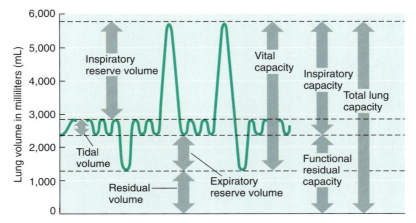

Figure 43.4 Graphic representation of respiratory volumes and capacities.

It can be noted from the data in tables 43.1 and 43.2 that vital capacities gradually decrease with age. Propose an explanation for this normal correlation.

_____

_____

_____

_____

7. *Inspiratory reserve volume* (IRV) (about 3,000 mL) is the volume of air in addition to the tidal volume that enters the lungs during forced inspiration. Calculate your inspiratory reserve volume by subtracting your tidal volume (TV) and your expiratory reserve volume (ERV) from your vital capacity (VC):

$$IRV = VC - (TV + ERV)$$

8. *Inspiratory capacity* (IC) (about 3,500 mL) is the maximum volume of air a person can inhale following exhalation of the tidal volume. Calculate your inspiratory capacity by adding your tidal volume (TV) and your inspiratory reserve volume (IRV):

$$IC = TV + IRV$$

9. *Functional residual capacity* (FRC) (about 2,300 mL) is the volume of air that remains in the lungs following exhalation of the tidal volume. Calculate your functional residual capacity (FRC) by adding your expiratory reserve volume (ERV) and your residual volume (RV), which you can assume is 1,200 mL:

$$FRC = ERV + 1,200$$

10. *Residual volume* (RV) (about 1,200 mL) is the volume of air that always remains in the lungs after the most forceful expiration. Although it is part of *total lung capacity* (about 5,800 mL), it cannot be measured with a spirometer. The residual air allows gas exchange and the alveoli to remain open during the respiratory cycle.

11. Complete Part C of the laboratory report.

✏️ **Learning Extension**

Determine your *minute respiratory volume*. To do this, follow these steps:

1. Sit quietly for a while, and then to establish your breathing rate, count the number of times you breathe in 1 minute. This might be inaccurate because conscious awareness of breathing rate can alter the results. You might ask a laboratory partner to record your breathing rate at some time when you are not expecting it to be recorded. A normal breathing rate is about 12–15 breaths per minute.

2. Calculate your minute respiratory volume by multiplying your breathing rate by your tidal volume:

_____ × _____ = _____
(breathing   (tidal   (minute respiratory
rate)   volume)   volume)

3. This value indicates the total volume of air that moves into your respiratory passages during each minute of ordinary breathing.

Table 43.1   Predicted Vital Capacities (in Milliliters) for Females

Height in Centimeters

Age	146	148	150	152	154	156	158	160	162	164	166	168	170	172	174	176	178	180	182	184	186	188	190	192	194
16	2950	2990	3030	3070	3110	3150	3190	3230	3270	3310	3350	3390	3430	3470	3510	3550	3590	3630	3670	3715	3755	3800	3840	3880	3920
18	2920	2960	3000	3040	3080	3120	3160	3200	3240	3280	3320	3360	3400	3440	3480	3520	3560	3600	3640	3680	3720	3760	3800	3840	3880
20	2890	2930	2970	3010	3050	3090	3130	3170	3210	3250	3290	3330	3370	3410	3450	3490	3525	3565	3605	3645	3695	3720	3760	3800	3840
22	2860	2900	2940	2980	3020	3060	3095	3135	3175	3215	3255	3290	3330	3370	3410	3450	3490	3530	3570	3610	3650	3685	3725	3765	3800
24	2830	2870	2910	2950	2985	3025	3065	3100	3140	3180	3220	3260	3300	3335	3375	3415	3455	3490	3530	3570	3610	3650	3685	3725	3765
26	2800	2840	2880	2920	2960	3000	3035	3070	3110	3150	3190	3230	3265	3300	3340	3380	3420	3455	3490	3530	3570	3610	3650	3685	3725
28	2775	2810	2850	2890	2930	2965	3000	3040	3070	3115	3155	3190	3230	3270	3305	3345	3380	3420	3460	3495	3535	3570	3610	3650	3685
30	2745	2780	2820	2860	2895	2935	2970	3010	3045	3085	3120	3160	3195	3235	3270	3310	3345	3385	3420	3460	3495	3535	3570	3610	3645
32	2715	2750	2790	2825	2865	2900	2940	2975	3015	3050	3090	3125	3160	3200	3235	3275	3310	3350	3385	3425	3460	3495	3535	3570	3610
34	2685	2725	2760	2795	2835	2870	2910	2945	2980	3020	3055	3090	3130	3165	3200	3240	3275	3310	3350	3385	3425	3460	3495	3535	3570
36	2655	2695	2730	2765	2805	2840	2875	2910	2950	2985	3020	3060	3095	3130	3165	3205	3240	3275	3310	3350	3385	3420	3460	3495	3530
38	2630	2665	2700	2735	2770	2810	2845	2880	2915	2950	2990	3025	3060	3095	3130	3170	3205	3240	3275	3310	3350	3385	3420	3455	3490
40	2600	2635	2670	2705	2740	2775	2810	2850	2885	2920	2955	2990	3025	3060	3095	3135	3170	3205	3240	3275	3310	3345	3380	3420	3455
42	2570	2605	2640	2675	2710	2745	2780	2815	2850	2885	2920	2955	2990	3025	3060	3100	3135	3170	3205	3240	3275	3310	3345	3380	3415
44	2540	2575	2610	2645	2680	2715	2750	2785	2820	2855	2890	2925	2960	2995	3030	3060	3095	3130	3165	3200	3235	3270	3305	3340	3375
46	2510	2545	2580	2615	2650	2685	2715	2750	2785	2820	2855	2890	2925	2960	2995	3030	3060	3095	3130	3165	3200	3235	3270	3305	3340
48	2480	2515	2550	2585	2620	2650	2685	2715	2750	2785	2820	2855	2890	2925	2960	2995	3030	3060	3095	3130	3160	3195	3230	3265	3300
50	2455	2485	2520	2555	2590	2625	2655	2690	2720	2755	2785	2820	2855	2890	2925	2955	2990	3025	3060	3090	3125	3155	3190	3225	3260
52	2425	2455	2490	2525	2555	2590	2625	2655	2690	2720	2755	2790	2820	2855	2890	2925	2955	2990	3020	3055	3090	3125	3155	3190	3220
54	2395	2425	2460	2495	2530	2560	2590	2625	2655	2690	2720	2755	2790	2820	2855	2885	2920	2950	2985	3020	3050	3085	3115	3150	3180
56	2365	2400	2430	2460	2495	2525	2560	2590	2625	2655	2690	2720	2755	2790	2820	2855	2885	2920	2950	2980	3015	3045	3080	3110	3145
58	2335	2370	2400	2430	2460	2495	2525	2560	2590	2625	2655	2690	2720	2750	2785	2815	2850	2880	2920	2945	2975	3010	3040	3075	3105
60	2305	2340	2370	2400	2430	2460	2495	2525	2560	2590	2625	2655	2685	2720	2750	2780	2810	2845	2875	2915	2940	2970	3000	3035	3065
62	2280	2310	2340	2370	2405	2435	2465	2495	2525	2560	2590	2620	2655	2685	2715	2745	2775	2810	2840	2870	2900	2935	2965	2995	3025
64	2250	2280	2310	2340	2370	2400	2430	2465	2495	2525	2555	2585	2620	2650	2680	2710	2740	2770	2805	2835	2865	2895	2925	2955	2990
66	2220	2250	2280	2310	2340	2370	2400	2430	2460	2495	2525	2555	2585	2615	2645	2675	2705	2735	2765	2800	2825	2860	2890	2920	2950
68	2190	2220	2250	2280	2310	2340	2370	2400	2430	2460	2490	2520	2550	2580	2610	2640	2670	2700	2730	2760	2795	2820	2850	2880	2910
70	2160	2190	2220	2250	2280	2310	2340	2370	2400	2425	2455	2485	2515	2545	2575	2605	2635	2665	2695	2725	2755	2780	2810	2840	2870
72	2130	2160	2190	2220	2250	2280	2310	2335	2365	2395	2425	2455	2480	2510	2540	2570	2600	2630	2660	2685	2715	2745	2775	2805	2830
74	2100	2130	2160	2190	2220	2245	2275	2305	2335	2360	2390	2420	2450	2475	2505	2535	2565	2590	2620	2650	2680	2710	2740	2765	2795

From E. DeF. Baldwin and E.W. Richards, Jr., "Pulmonary Insufficiency 1. Physiologic Classification, Clinical Methods of Analysis, Standard Values in Normal Subjects" in Medicine 27:243, © by William & Wilkins. Used by permission.

Table 43.2 Predicted Vital Capacities (in Milliliters) for Males

Age	Height in Centimeters																								
	146	148	150	152	154	156	158	160	162	164	166	168	170	172	174	176	178	180	182	184	186	188	190	192	194
16	3765	3820	3870	3920	3975	4025	4075	4130	4180	4230	4285	4335	4385	4440	4490	4540	4590	4645	4695	4745	4800	4850	4900	4955	5005
18	3740	3790	3840	3890	3940	3995	4045	4095	4145	4200	4250	4300	4350	4405	4455	4505	4555	4610	4660	4710	4760	4815	4865	4915	4965
20	3710	3760	3810	3860	3910	3960	4015	4065	4115	4165	4215	4265	4320	4370	4420	4470	4520	4570	4625	4675	4725	4775	4825	4875	4930
22	3680	3730	3780	3830	3880	3930	3980	4030	4080	4135	4185	4235	4285	4335	4385	4435	4485	4535	4585	4635	4685	4735	4790	4840	4890
24	3635	3685	3735	3785	3835	3885	3935	3985	4035	4085	4135	4185	4235	4285	4330	4380	4430	4480	4530	4580	4630	4680	4730	4780	4830
26	3605	3655	3705	3755	3805	3855	3905	3955	4000	4050	4100	4150	4200	4250	4300	4350	4395	4445	4495	4545	4595	4645	4695	4740	4790
28	3575	3625	3675	3725	3775	3820	3870	3920	3970	4020	4070	4115	4165	4215	4265	4310	4360	4410	4460	4510	4555	4605	4655	4705	4755
30	3550	3595	3645	3695	3740	3790	3840	3890	3935	3985	4035	4080	4130	4180	4230	4275	4325	4375	4425	4470	4520	4570	4615	4665	4715
32	3520	3565	3615	3665	3710	3760	3810	3855	3905	3950	4000	4050	4095	4145	4195	4240	4290	4340	4385	4435	4485	4530	4580	4625	4675
34	3475	3525	3570	3620	3665	3715	3760	3810	3855	3905	3950	4000	4045	4095	4140	4190	4225	4285	4330	4380	4425	4475	4520	4570	4615
36	3445	3495	3540	3585	3635	3680	3730	3775	3825	3870	3920	3965	4010	4060	4105	4155	4200	4250	4295	4340	4390	4435	4485	4530	4580
38	3415	3465	3510	3555	3605	3650	3695	3745	3790	3840	3885	3930	3980	4025	4070	4120	4165	4210	4260	4305	4350	4400	4445	4495	4540
40	3385	3435	3480	3525	3575	3620	3665	3710	3760	3805	3850	3900	3945	3990	4035	4085	4130	4175	4220	4270	4315	4360	4410	4455	4500
42	3360	3405	3450	3495	3540	3590	3635	3680	3725	3770	3820	3865	3910	3955	4000	4050	4095	4140	4185	4230	4280	4325	4370	4415	4460
44	3315	3360	3405	3450	3495	3540	3585	3630	3675	3725	3770	3815	3860	3905	3950	3995	4040	4085	4130	4175	4220	4270	4315	4360	4405
46	3285	3330	3375	3420	3465	3510	3555	3600	3645	3690	3735	3780	3825	3870	3915	3960	4005	4050	4095	4140	4185	4230	4275	4320	4365
48	3255	3300	3345	3390	3435	3480	3525	3570	3615	3655	3700	3745	3790	3835	3880	3925	3970	4015	4060	4105	4150	4190	4235	4280	4325
50	3210	3255	3300	3345	3390	3430	3475	3520	3565	3610	3650	3695	3740	3785	3830	3870	3915	3960	4005	4050	4090	4135	4180	4225	4270
52	3185	3225	3270	3315	3355	3400	3445	3490	3530	3575	3620	3660	3705	3750	3795	3835	3880	3925	3970	4010	4055	4100	4140	4185	4230
54	3155	3195	3240	3285	3325	3370	3415	3455	3500	3540	3585	3630	3670	3715	3760	3800	3845	3890	3930	3975	4020	4060	4105	4145	4190
56	3125	3165	3210	3255	3295	3340	3380	3425	3465	3510	3550	3595	3640	3680	3725	3765	3810	3850	3895	3940	3980	4025	4065	4110	4150
58	3080	3125	3165	3210	3250	3290	3335	3375	3420	3460	3500	3545	3585	3630	3670	3715	3755	3800	3840	3880	3925	3965	4010	4050	4095
60	3050	3095	3135	3175	3220	3260	3300	3345	3385	3430	3470	3500	3555	3595	3635	3680	3720	3760	3805	3845	3885	3930	3970	4015	4055
62	3020	3060	3110	3150	3190	3230	3270	3310	3350	3390	3440	3480	3520	3560	3600	3640	3680	3730	3770	3810	3850	3890	3930	3970	4020
64	2990	3030	3080	3120	3160	3200	3240	3280	3320	3360	3400	3440	3490	3530	3570	3610	3650	3690	3730	3770	3810	3850	3900	3940	3980
66	2950	2990	3030	3070	3110	3150	3190	3230	3270	3310	3350	3390	3430	3470	3510	3550	3600	3640	3680	3720	3760	3800	3840	3880	3920
68	2920	2960	3000	3040	3080	3120	3160	3200	3240	3280	3320	3360	3400	3440	3480	3520	3560	3600	3640	3680	3720	3760	3800	3840	3880
70	2890	2930	2970	3010	3050	3090	3130	3170	3210	3250	3290	3330	3370	3410	3450	3480	3520	3560	3600	3640	3680	3720	3760	3800	3840
72	2860	2900	2940	2980	3020	3060	3100	3140	3180	3210	3250	3290	3330	3370	3410	3450	3490	3530	3570	3610	3650	3680	3720	3760	3800
74	2820	2860	2900	2930	2970	3010	3050	3090	3130	3170	3200	3240	3280	3320	3360	3400	3440	3470	3510	3550	3590	3630	3670	3710	3740

From E. DeF. Baldwin and E.W. Richards, Jr., "Pulmonary Insufficiency 1, Physiologic Classification, Clinical Methods of Analysis, Standard Values in Normal Subjects" in Medicine 27:243, (c) by William & Wilkins. Used by permission.

Name _____

Date _____

Section _____

The ⬅ corresponds to the indicated outcome(s) found at the beginning of the laboratory exercise.

Breathing and Respiratory Volumes and Capacities

Part A Assessments

Complete the following statements:

1. Breathing can also be called _____ .

2. The weight of air causes a force called _____ pressure.

3. The weight of air at sea level is sufficient to support a column of mercury within a tube _____ mm high.

4. If the pressure inside the lungs decreases, outside air is forced into the airways by _____ . ⬅1

5. Nerve impulses are carried to the diaphragm by the _____ nerves. ⬅1

6. When the diaphragm contracts, the size of the thoracic cavity _____ . ⬅1

7. The ribs are raised by the contraction of the _____ muscles, which increases the size of the thoracic cavity. ⬅1

8. Only a thin film of lubricating serous fluid separates the parietal pleura from the _____ of a lung.

9. A mixture of lipoproteins, called _____ , acts to reduce the tendency of alveoli to collapse.

10. The force responsible for normal expiration comes from _____ of tissues and from surface tension. ⬅1

11. Muscles that help to force out more than the normal volume of air by pulling the ribs downward and inward include the _____ . ⬅1

12. The diaphragm can be forced to move higher than normal by the contraction of the _____ muscles. ⬅1

Part B Assessments

Match the air volumes in column A with the definitions in column B. Place the letter of your choice in the space provided. ⬅2

Column A	Column B
a. Expiratory reserve volume	_____ **1.** Volume in addition to tidal volume that leaves the lungs during forced expiration
b. Functional residual capacity	
c. Inspiratory capacity	_____ **2.** Vital capacity plus residual volume
d. Inspiratory reserve volume	_____ **3.** Volume that remains in lungs after the most forceful expiration
e. Residual volume	_____ **4.** Volume that enters or leaves lungs during a respiratory cycle
f. Tidal volume	_____ **5.** Volume in addition to tidal volume that enters lungs during forced inspiration
g. Total lung capacity	
h. Vital capacity	_____ **6.** Maximum volume a person can exhale after taking the deepest possible breath
	_____ **7.** Maximum volume a person can inhale following exhalation of the tidal volume
	_____ **8.** Volume of air remaining in the lungs following exhalation of the tidal volume

Part C Assessments

1. Test results for respiratory air volumes and capacities: ◄3 ◄4

Respiratory Volume or Capacity	Expected Value* (approximate)	Test Result	Percent of Expected Value (test result/expected value × 100)
Tidal volume (resting) (TV)	500 mL		
Expiratory reserve volume (ERV)	1,100 mL		
Vital capacity (VC)	4,600 mL _ _ _ _ _ _ (enter yours from table 43.1 or 43.2)		
Inspiratory reserve volume (IRV)	3,000 mL		
Inspiratory capacity (IC)	3,500 mL		
Functional residual capacity (FRC)	2,300 mL		

*The values listed are most characteristic for a tall, young adult. If your expected value for vital capacity is considerably different than 4,600 mL, your other values would vary accordingly.

2. Complete the following:

   a. How do your test results compare with the expected values?

   _____

   _____

   _____

   _____

   b. How does your vital capacity compare with the average value for a person of your sex, age, and height?

   _____

   _____

   c. What measurement in addition to vital capacity is needed before you can calculate your total lung capacity? ◄4

   _____

   _____

3. If your experimental results are considerably different than the predicted vital capacities, propose reasons for the differences. As you write this paragraph, consider factors such as smoking, physical fitness, respiratory disorders, and medications. (Your instructor might have you make some class correlations from class data.)

   _____

   _____

   _____

   _____

   _____

Kidney Structure

Materials Needed

Textbook
Human torso model
Kidney model
Preserved pig (or sheep) kidney
Dissecting tray
Dissecting instruments
Long knife
Compound light microscope
Prepared microscope slide of a kidney section

⚠ Safety

- Wear disposable gloves when working on the kidney dissection.
- Dispose of the kidney and gloves as directed by your laboratory instructor.
- Wash the dissecting tray and instruments as instructed.
- Wash your laboratory table.
- Wash your hands before leaving the laboratory.

The two kidneys are the primary organs of the urinary system. They are located in the abdominal cavity, against the posterior wall and behind the parietal peritoneum (retroperitoneal). Masses of adipose tissue associated with the kidneys hold them in place at a vertebral level between T12 and L3. The right kidney is slightly more inferior due to the large mass of the liver near its superior border. Ureters force urine by means of peristaltic waves into the urinary bladder, which temporarily stores urine. The urethra conveys urine to the outside of the body

Each kidney harbors over 1 million nephrons serving as the basic structural and functional unit of the kidney. A glomerular capsule, proximal convoluted tubule, nephron loop, and a distal convoluted tubule compose the microscopic, multicellular structure of a relatively long nephron tubule, which drains into a collecting duct. Approximately 85% of the nephrons are cortical nephrons with short nephron loops, while the remaining represent juxtamedullary nephrons, with long nephron loops extending deeper into the renal medulla. An elaborate network of blood vessels surrounds the entire nephron. Glomerular filtration, tubular reabsorption, and tubular secretion represent three processes resulting in urine as the final product.

A variety of overall functions occur in the kidneys. They remove metabolic wastes from the blood; help regulate blood volume, blood pressure, and pH of blood; control water and electrolyte concentrations; and secrete renin and erythropoietin.

Purpose of the Exercise

To review the structure of the kidney, to dissect a kidney, and to microscopically observe the major structures of a nephron.

Learning Outcomes

After completing this exercise, you should be able to

1. Locate and identify the major structures of the urinary system.
2. Locate and identify the major structures of a kidney.
3. Identify and sketch the major structures of a nephron.
4. Trace the path of filtrate through a renal nephron.

Procedure A—Kidney Structure

1. Review the section entitled "Kidney Structure" in chapter 17 of the textbook.
2. As a review activity, label figures 44.1 and 44.2.
3. Complete Part A of Laboratory Report 44.

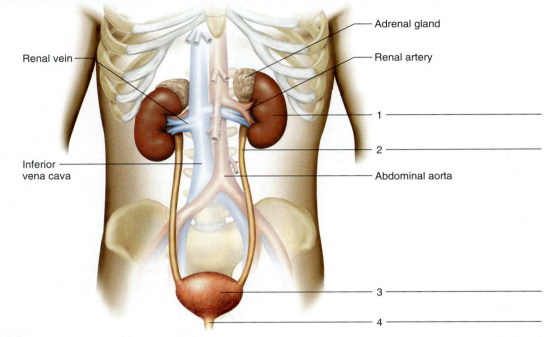

Renal vein —

Inferior
vena cava —

Adrenal gland

Renal artery

— 1 —————————

— 2 —————————

Abdominal aorta

— 3 —————————

— 4 —————————

Figure 44.1 Label the major structures of the urinary system. ◀**1**

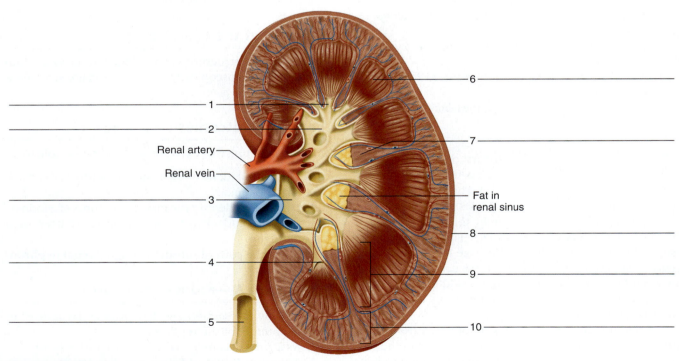

————————— 1 —

————————— 2 —

Renal artery—

Renal vein—

————————— 3 —

————————— 4 —

————————— 5 —

— 6 —————————

— 7 —————————

Fat in
renal sinus

— 8 —————————

— 9 —————————

— 10 —————————

Figure 44.2 Label the major structures in the longitudinal section of a kidney. ◀**2**

4. Observe the human torso model and the kidney model. Locate the following:

kidneys
ureters
urinary bladder
urethra
renal sinus
renal pelvis
    major calyces
    minor calyces
renal medulla
    renal pyramids
    renal papillae
renal cortex
    renal columns (extensions of renal cortex between renal pyramids)
nephrons

5. To observe the structure of a kidney:
   a. Obtain a preserved pig or sheep kidney and rinse it with water to remove as much of the preserving fluid as possible.

   b. Carefully remove any adipose tissue from the surface of the specimen.
   c. Locate the following features:

   renal (fibrous) capsule
   hilum of kidney
   renal artery
   renal vein
   ureter

   d. Use a long knife to cut the kidney in half longitudinally along the coronal plane, beginning on the convex border.
   e. Rinse the interior of the kidney with water, and using figure 44.3 as a reference, locate the following:

   renal pelvis
       major calyces
       minor calyces
   renal cortex
       renal columns (extensions of renal cortex between renal pyramids)
   renal medulla
       renal pyramids

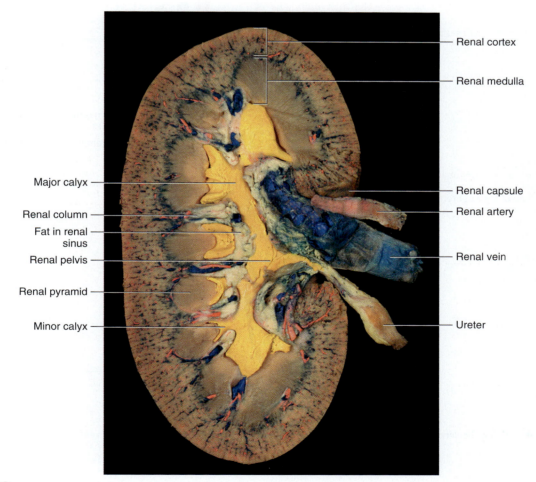

Figure 44.3 Longitudinal section of a pig kidney that has a triple injection of latex (*red* in the renal artery, *blue* in the renal vein, and *yellow* in the ureter and renal pelvis).

Procedure B—Renal Blood Vessels and Nephrons

1. Review the sections entitled "Renal Blood Vessels" and "Nephrons" in chapter 17 of the textbook.
2. As a review activity, label figure 44.4.
3. Complete Part B of the laboratory report.
4. Obtain a microscope slide of a kidney section, and examine it using low-power magnification. Locate the *renal capsule, renal cortex* (which appears somewhat granular and may be more darkly stained than the other renal tissues), and *renal medulla* (fig. 44.5).
5. Examine the renal cortex, using high-power magnification. Locate a *renal corpuscle.* These structures appear as isolated circular areas. Identify the *glomerulus,* the capillary cluster inside the corpuscle, and the *glomerular capsule,* which appears as a clear area surrounding the glomerulus. Also note the numerous sections of renal tubules that occupy the spaces between renal corpuscles (fig. 44.5a).
6. Prepare a labeled sketch of a representative section of the renal cortex in Part C of the laboratory report.
7. Examine the renal medulla using high-power magnification (fig. 44.5b). Identify longitudinal and cross sections of various collecting ducts. These ducts are lined with simple epithelial cells, which vary in shape from squamous to cuboidal.
8. Prepare a labeled sketch of a representative section of renal medulla in Part C of the laboratory report.

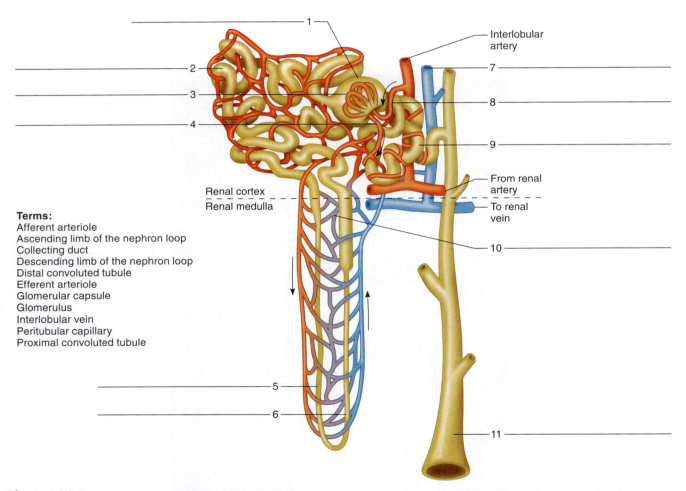

Terms:
Afferent arteriole
Ascending limb of the nephron loop
Collecting duct
Descending limb of the nephron loop
Distal convoluted tubule
Efferent arteriole
Glomerular capsule
Glomerulus
Interlobular vein
Peritubular capillary
Proximal convoluted tubule

Interlobular artery
From renal artery
To renal vein
Renal cortex
Renal medulla

Figure 44.4 Using the terms provided, label the major structures of the nephron and the blood vessels associated with it. ◀3

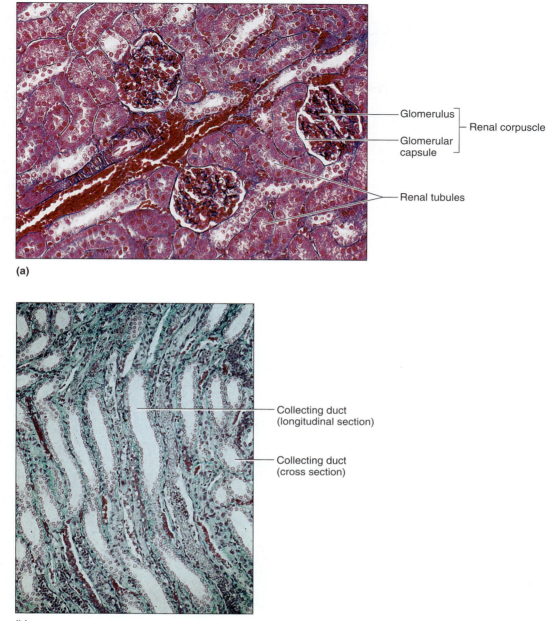

(a)

Glomerulus ⎤
⎥ Renal corpuscle
Glomerular ⎦
capsule

Renal tubules

(b)

Collecting duct
(longitudinal section)

Collecting duct
(cross section)

Figure 44.5 (a) Micrograph of a section of the renal cortex (220×). (b) Micrograph of a section of the renal medulla (80× micrograph enlarged to 200×).

Notes

Name _____

Date _____

Section _____

The ⬅ corresponds to the indicated outcome(s) found at the beginning of the laboratory exercise.

Kidney Structure

Part A Assessments

Match the terms in column A with the descriptions in column B. Place the letter of your choice in the space provided. ◀2

Column A	Column B
a. Calyces	_____ **1.** Shell around renal medulla
b. Hilum of kidney	_____ **2.** Branches of renal pelvis to renal papillae
c. Nephron	_____ **3.** Conical mass of tissue within renal medulla
d. Renal column	_____ **4.** Projection with tiny openings into minor calyx
e. Renal cortex	_____ **5.** Medial depression for blood vessels and ureter to enter kidney chamber
f. Renal papilla	_____ **6.** Hollow chamber within kidney
g. Renal pelvis	_____ **7.** Microscopic functional unit of kidney
h. Renal pyramid	_____ **8.** Cortical tissue between renal pyramids
i. Renal sinus	_____ **9.** Superior funnel-shaped end of ureter inside renal sinus

Part B Assessments

1. Distinguish between a renal corpuscle and a renal tubule. ◀3 _____

_____

_____

2. Number the following structures to indicate their respective positions in relation to the nephron.

Assign number 1 to the structure nearest the glomerulus. ◀4

_____ Ascending limb of the nephron loop

_____ Collecting duct

_____ Descending limb of the nephron loop

_____ Distal convoluted tubule

_____ Glomerular capsule

_____ Proximal convoluted tubule

_____ Renal papilla

3. Explain how the blood vessels associated with the renal corpuscle help to maintain a relatively high blood pressure within the glomerulus. _____

_____

_____

_____

_____

_____

_____

4. Describe the *juxtaglomerular apparatus.* _____

_____

_____

_____

_____

_____

Part C Assessments

Sketch a representative section of renal cortex. Label the glomerulus, glomerular capsule, and sections of renal tubules. ◀**3**

Sketch a representative section of renal medulla. Label a longitudinal section and a cross section of a collecting duct. ◀**3**

Laboratory Exercise

Urinalysis

Materials Needed

Normal and abnormal simulated urine specimens are suggested as a substitute for collected urine.

Disposable urine-collecting container
Paper towel
Urinometer cylinder
Urinometer hydrometer
Laboratory thermometer
pH test paper
Reagent strips (individual or combination strips such as Chemstrip or Multistix) to test for the presence of the following:
  Glucose
  Protein
  Ketones
  Bilirubin
  Hemoglobin/occult blood
Compound microscope
Microscope slide
Coverslip
Centrifuge
Centrifuge tube
Graduated cylinder, 10 mL
Medicine dropper
Sedi-stain

⚠ Safety

- Consider using normal and abnormal simulated urine samples available from various laboratory supply houses.
- Wear disposable gloves when working with body fluids.
- Work only with your own urine sample.
- Use an appropriate disinfectant to wash the laboratory table before and after the procedures.
- Place glassware in a disinfectant when finished.
- Dispose of contaminated items as directed by your laboratory instructor.
- Wash your hands before leaving the laboratory.

Urine is the product of three processes of the nephrons within the kidneys: glomerular filtration, tubular secretion, and tubular reabsorption. As a result of these processes, various waste substances are removed from the blood as well as maintenance of body fluid and elecyrolyte balance. Consequently, the composition of urine varies considerably because of differences in dietary intake and physical activity from day to day. The normal urinary output ranges from 1.0–1.8 liters per day. Typically, urine consists of 95% water and 5% solutes. The volume of urine produced by the kidneys varies with such factors as fluid intake, environmental temperature, relative humidity, respiratory rate, and body temperature.

An analysis of urine composition and volume often is used to evaluate the functions of the kidneys and other organs. This procedure, called *urinalysis,* is a clinical assessment and a diagnostic tool of certain pathological conditions and general overall health. A urinalysis and a complete blood analysis complement each other for an evaluation of certain diseases and general health.

A urinalysis involves three aspects: physical characteristics, chemical analysis, and a microscopic examination. Physical characteristics of urine include volume, color, transparency, and odor. The chemical analysis of solutes in urine includes urea and other nitrogenous wastes; electrolytes; pigments; as well as possible glucose, protein, ketones, bilirubin, and hemoglobin. The specific gravity and pH of urine are greatly influenced by the components and amounts of solutes. An examination of microscopic solids, including cells, casts, and crystals, assist the diagnosis of injury, various diseases, and urinary infections.

Purpose of the Exercise

To perform the observations and tests commonly used to analyze the characteristics and composition of urine.

Learning Outcomes

After completing this exercise, you should be able to

1. Evaluate the color, transparency, and specific gravity of a urine sample.
2. Measure the pH of a urine sample.
3. Test a urine sample for the presence of glucose, protein, ketones, bilirubin, and hemoglobin.
4. Perform a microscopic study of urine sediment.
5. Summarize the results of these observations and tests.

> **WARNING:** *While performing the following tests, you should wear disposable latex gloves so that skin contact with the urine is avoided. Observe all safety procedures listed for this lab. (Normal and abnormal simulated urine specimens could be used instead of real urine for this lab.)*

Procedure—Urinalysis

1. Proceed to the restroom with a clean disposable container. The first small volume of urine should not be collected because it contains abnormally high levels of microorganisms from the urethra. Collect a midstream sample of about 50 mL of urine. The best collections are the first specimen in the morning or one taken 3 hours after a meal. Refrigerate samples if they are not used immediately.

2. Place a sample of urine in a clean, transparent container. Describe the *color* of the urine. Normal urine varies from light yellow to amber, depending on the presence of urochromes, end-product pigments produced during the decomposition of hemoglobin. Dark urine indicates a high concentration of pigments.

Abnormal urine colors include yellow-brown or green, due to elevated concentrations of bile pigments, and red to dark brown, due to the presence of blood. Certain foods, such as beets or carrots, and various drug substances also may cause color changes in urine, but in such cases the colors have no clinical significance. Enter the results of this and the following tests in Part A of Laboratory Report 45.

3. Evaluate the *transparency* of the urine sample (judge whether the urine is clear, slightly cloudy, or very cloudy). Normal urine is clear enough to see through. You can read newsprint through slightly cloudy urine; you can no longer read newsprint through cloudy urine. Cloudy urine indicates the presence of various substances that may include mucus, bacteria, epithelial cells, fat droplets, or inorganic salts.

4. Determine the *specific gravity* of the urine sample, which indicates the solute concentration. Specific gravity is the ratio of the weight of something to the weight of an equal volume of pure water. For ex-

ample, mercury (at 15°C) weighs 13.6 times as much as an equal volume of water; thus, it has a specific gravity of 13.6. Although urine is mostly water, it has substances dissolved in it and is slightly heavier than an equal volume of water. Thus, urine has a specific gravity of more than 1.000. Actually, the specific gravity of normal urine varies from 1.003 to 1.035. If the specific gravity is too low, the urine contains few solutes and represents dilute urine, a likely result of excessive fluid intake or use of diuretics. A specific gravity above the normal range represents a higher concentration of solutes, likely from a limited fluid intake. Concentrated urine over an extended time increases the risk of the formation of kidney stones.

To determine the specific gravity of a urine sample, follow these steps:

a. Pour enough urine into a clean urinometer cylinder to fill it about three-fourths full. Any foam that appears should be removed with a paper towel.
b. Use a laboratory thermometer to measure the temperature of the urine.
c. Gently place the urinometer hydrometer into the urine, and *make sure that the float is not touching the sides or the bottom of the cylinder* (fig. 45.1).
d. Position your eye at the level of the urine surface. Determine which line on the stem of the hydrometer intersects the lowest level of the concave surface (meniscus) of the urine.
e. Liquids tend to contract and become denser as they are cooled, or to expand and become less dense as they are heated, so it may be necessary to make a temperature correction to obtain an accurate specific gravity measurement. To do this, add 0.001 to the hydrometer reading for each 3 degrees of urine temperature above 25°C or subtract 0.001 for each 3 degrees below 25°C. Enter this calculated value in the table of the laboratory report as the test result.

5. Reagent strips can be used to perform a variety of urine tests. In each case, directions for using the strips are found on the strip container. *Be sure to read them.*

To perform each test, follow these steps:

a. Obtain a urine sample and the proper reagent strip.
b. Read the directions on the strip container.
c. Dip the strip in the urine sample.
d. Remove the strip at an angle and let it touch the inside rim of the urine container to remove any excess liquid.
e. Wait for the length of time indicated by the directions on the container before you compare the color of the test strip with the standard color scale on the side of the container. The value or amount represented by the matching color should be used as the test result and recorded in Part A of the laboratory report.

Hydrometer

Urine in urinometer cylinder

URINE SPECIMEN

Figure 45.1 Float the hydrometer in the urine, making sure that it does not touch the sides or the bottom of the cylinder.

Alternative Procedure

If combination reagent test strips (Multistix or Chemstrip) are being used, locate the appropriate color chart for each test being evaluated. Wait the designated time for each color reaction before the comparison is made to the standard color scale.

6. Perform the *pH test.* The pH of normal urine varies from 4.6 to 8.0, but most commonly, it is near 6.0 (slightly acidic). The pH of urine may decrease as a result of a diet high in protein, or it may increase with a vegetarian diet. Significant daily variations within the broad normal range are results of concentrations of excesses from variable diets.

7. Perform the *glucose test.* Normally, there is no glucose in urine. However, glucose may appear in the urine temporarily following a meal high in carbohydrates. Glucose also may appear in the urine as a result of uncontrolled diabetes mellitus.

8. Perform the *protein test.* Normally, proteins of large molecular size are not present in urine. However, those of small molecular sizes, such as albumins, may appear in trace amounts, particularly following strenuous exercise. Increased amounts of proteins also may appear as a result of kidney diseases in

which the glomeruli are damaged or as a result of high blood pressure.

9. Perform the *ketone test.* Ketones are products of fat metabolism. Usually they are not present in urine. However, they may appear in the urine if the diet fails to provide adequate carbohydrate, as in the case of prolonged fasting or starvation or as a result of insulin deficiency (diabetes mellitus).

10. Perform the *bilirubin test.* Bilirubin, which results from hemoglobin decomposition in the liver, normally is absent in urine. It may appear, however, as a result of liver disorders that cause obstructions of the biliary tract. Urochrome, a normal yellow component of urine, is a result of additional breakdown of bilirubin.

11. Perform the *hemoglobin/occult blood test.* Hemoglobin occurs in the red blood cells, and because such cells normally do not pass into the renal tubules, hemoglobin is not found in normal urine. Its presence in urine usually indicates a disease process, a transfusion reaction, or an injury to the urinary organs.

12. Complete Part A of the laboratory report.

13. A urinalysis usually includes a study of urine sediment—the microscopic solids present in a urine sample. This sediment normally includes mucus; certain crystals; and a variety of cells, such as the epithelial cells that line the urinary tubes and an occasional white blood cell. Other types of solids, such as casts or red blood cells, may indicate a disease or injury if they are present in excess. (Casts are cylindrical masses of cells or other substances that form in the renal tubules and are flushed out by the flow of urine.)

   To observe urine sediment, follow these steps:

   a. Thoroughly stir or shake a urine sample to suspend the sediment, which tends to settle to the bottom of the container.

   b. Pour 10 mL of urine into a clean centrifuge tube and centrifuge it for 5 minutes at a slow speed (1,500 rpm). Be sure to balance the centrifuge with an even number of tubes filled to the same levels.

   c. Carefully decant 9 mL (leave 1 mL) of the liquid from the sediment in the bottom of the centrifuge tube, as directed by your laboratory instructor. Resuspend the 1 mL of sediment.

   d. Use a medicine dropper to remove some of the sediment and place it on a clean microscope slide.

   e. Add a drop of Sedi-stain to the sample, and add a coverslip.

   f. Examine the sediment with low-power (reduce the light when using low power) and high-power magnifications.

   g. With the aid of figure 45.2, identify the types of solids present.

   h. In Part B of the laboratory report, make a sketch of each type of sediment that you observed.

14. Complete Part B of the laboratory report.

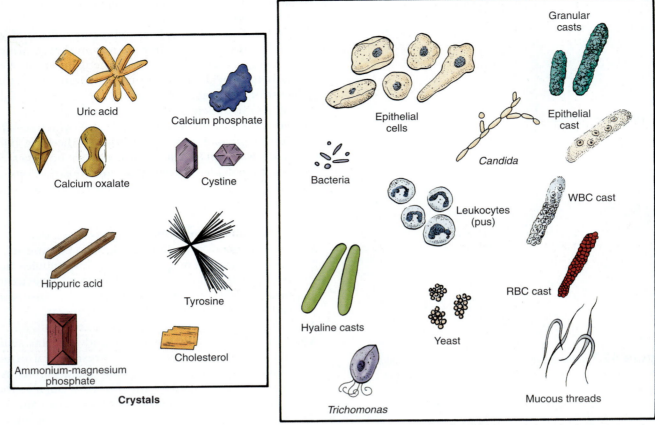

Figure 45.2   Types of urine sediment. Healthy individuals lack many of these sediments and possess only occasional to trace amounts of others. (*Note:* Shades of white to purple sediments are most characteristic when using Sedi-stain.)

The ⬅ corresponds to the indicated outcome(s) found at the beginning of the laboratory exercise.

Urinalysis

Laboratory Report

45

Part A Assessments

1. Enter your observations, test results, and evaluations in the following table. ⬅1 ⬅2 ⬅3

Urine Characteristics	Observations and Test Results	Normal Values	Evaluations
Color		Light yellow to amber	
Transparency		Clear	
Specific gravity (corrected for temperature)		1.003–1.035	
pH		4.6–8.0	
Glucose		0 (negative)	
Protein		0 to trace	
Ketones		0	
Bilirubin		0	
Hemoglobin/occult blood		0	
(Other)			
(Other)			

2. Summarize the results of the urinalysis. ⬅5 _____

_____

_____

_____

_____

_____

_____

Critical Thinking Application

Why do you think it is important to refrigerate a urine sample if an analysis cannot be performed immediately after collecting it?

_____

_____

_____

_____

Part B Assessments

1. Make a sketch for each type of sediment you observed. Label any from those shown in figure 45.2. ◄**4**

2. Summarize the results of the urine sediment study. ◄**5** _____

_____

_____

_____

_____

_____

_____

Male Reproductive System

The organs of the male reproductive system are specialized to produce and maintain the male sex cells, to transport these cells together with supporting fluids to the female reproductive tract, and to produce and secrete male sex hormones.

These organs include the testes, in which sperm cells and male sex hormones are produced, and sets of internal and external accessory organs. The internal organs include various tubes and glands, whereas the external structures are the scrotum and the penis.

Purpose of the Exercise

To review the structure and functions of the male reproductive organs and to microscopically examine some of these organs.

Learning Outcomes

After completing this exercise, you should be able to

1. Locate and identify the organs of the male reproductive system.
2. Describe the functions of these organs.
3. Sketch and label the major features from microscopic sections of the testis, epididymis, and penis.

Procedure A—Male Reproductive Organs

1. Review the sections entitled "Testes," "Male Internal Accessory Organs," and "Male External Reproductive Organs" in chapter 19 of the textbook.
2. As a review activity, label figures 46.1 and 46.2.
3. Observe the human torso model, the model of the male reproductive system, and the anatomical chart of the male reproductive system. Locate the following features:

   testes
       seminiferous tubules
       interstitial cells (Leydig cells)
   epididymis
   ductus deferens (vas deferens)
   ejaculatory duct
   seminal vesicles
   prostate gland
   bulbourethral glands
   scrotum
   penis
       corpora cavernosa
       corpus spongiosum
       glans penis
       prepuce (foreskin)
       external urethral orifice

4. Complete Part A of Laboratory Report 46.

Procedure B—Microscopic Anatomy

1. Obtain a microscope slide of a human testis section and examine it, using low-power magnification. Locate the thick *fibrous capsule* on the surface and the numerous sections of *seminiferous tubules* inside (fig. 46.3).
2. Focus on some of the seminiferous tubules, using high-power magnification. Locate the *basement*

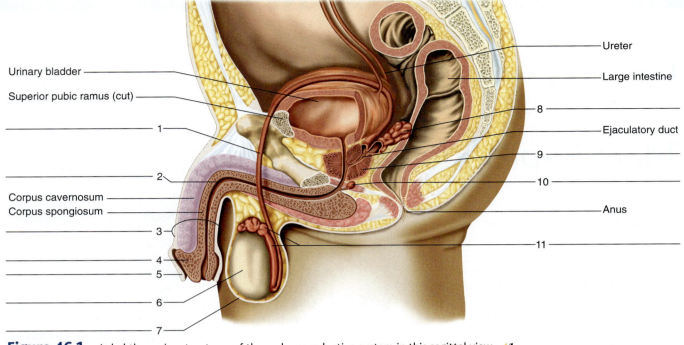

Urinary bladder

Superior pubic ramus (cut)

Corpus cavernosum

Corpus spongiosum

1

2

3

4

5

6

7

Ureter

Large intestine

8

Ejaculatory duct

9

10

Anus

11

Figure 46.1 Label the major structures of the male reproductive system in this sagittal view. ◀**1**

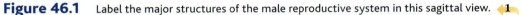

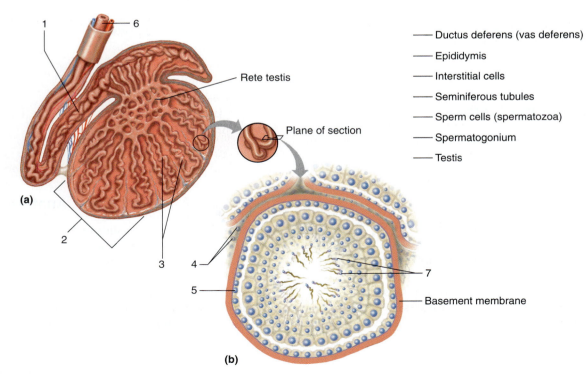

Rete testis

Plane of section

Ductus deferens (vas deferens)

Epididymis

Interstitial cells

Seminiferous tubules

Sperm cells (spermatozoa)

Spermatogonium

Testis

Basement membrane

(a)

(b)

Figure 46.2 Label the diagram of (a) the sagittal section of a testis and (b) a cross section of a seminiferous tubule by placing the correct numbers in the spaces provided. ◀**1**

membrane and the layer of *spermatogonia* just beneath the basement membrane. Identify some *supporting cells* (sustentacular cells; Sertoli cells), which have pale, oval-shaped nuclei, and some *spermatogenic cells,* which have smaller, round nuclei. Spermatogonia give rise to spermatogenic cells that are in various stages of spermatogenesis as they are forced toward the lumen. Near the lumen of the tube, find some darkly stained, elongated heads of developing sperm cells. In the spaces between adjacent seminiferous tubules, locate some isolated *interstitial cells* of the endocrine system (fig. 46.4). Interstitial cells (Leydig cells) produce the hormone testosterone, transported by the blood.

3. Prepare a labeled sketch of a representative section of the testis in Part B of the laboratory report.
4. Obtain a microscope slide of a cross section of an *epididymis.* Examine its wall, using high-power magnification. Note the elongated, *pseudostratified columnar epithelial cells,* which comprise most of the inner lining. These cells have nonmotile cilia on their free surfaces (fig. 46.5). Also note the thin layer of smooth muscle and connective tissue surrounding the tube.
5. Prepare a labeled sketch of the epididymis wall in Part B of the laboratory report.
6. Obtain a microscope slide of a *penis* cross section and examine it with low-power magnification (fig. 46.6). Identify the following features:

corpora cavernosa

corpus spongiosum

urethra

skin

7. Prepare a labeled sketch of a penis cross section in Part B of the laboratory report.
8. Complete Part B of the laboratory report.

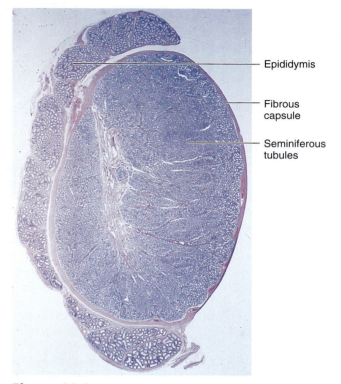

Figure 46.3   Micrograph of a human testis (1.7×).

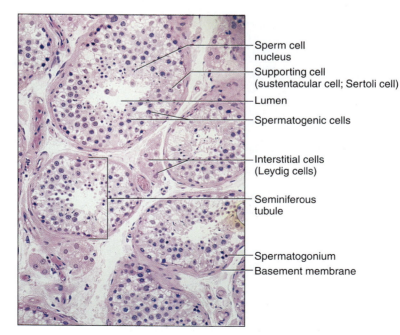

Figure 46.4   Micrograph of seminiferous tubules (50× micrograph enlarged to 135×).

347

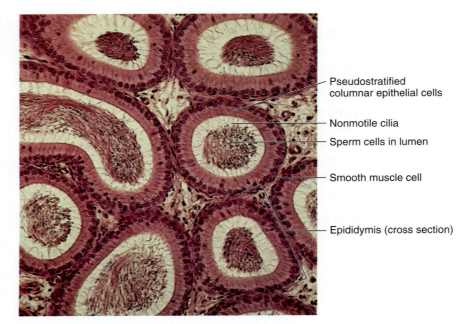

Figure 46.5 Micrograph of a cross section of a human epididymis (50× micrograph enlarged to 145×).

Pseudostratified columnar epithelial cells
Nonmotile cilia
Sperm cells in lumen
Smooth muscle cell
Epididymis (cross section)

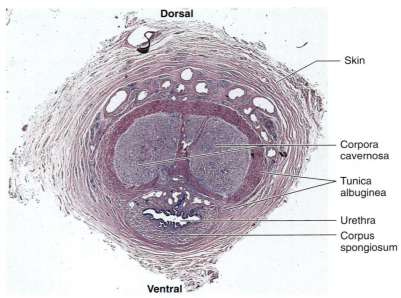

Dorsal

Skin

Corpora cavernosa

Tunica albuginea

Urethra

Corpus spongiosum

Ventral

Figure 46.6 Micrograph of a cross section of the body of the penis (5×).

Name _____

Date _____

Section _____

The ⬅ corresponds to the indicated outcome(s) found at the beginning of the laboratory exercise.

Male Reproductive System

Part A Assessments

Complete the following statements:

1. Connective tissue subdivides a testis into many lobules, which contain _____ . ⬅**1**

2. The _____ is a highly coiled tube on the outer surface of the testis. ⬅**1**

3. The _____ cells of the epithelium that lines the seminiferous tubules give rise to sperm cells. ⬅**2**

4. Undifferentiated sperm cells in the male embryo are called _____ . ⬅**1**

5. _____ is the process by which sperm cells are formed. ⬅**2**

6. The number of chromosomes normally present in a human sperm cell is _____ . ⬅**1**

7. The anterior end of a sperm head, called the _____ , contains enzymes that aid in the penetration of an egg cell at the time of fertilization. ⬅**2**

8. Sperm cells undergo maturation while they are stored in the _____ . ⬅**2**

9. The secretion of the seminal vesicles is rich in the monosaccharide called _____ . ⬅**2**

10. Prostate gland secretion helps neutralize seminal fluid because the prostate secretion is _____ . ⬅**2**

11. The secretion of the _____ glands lubricates the end of the penis in preparation for sexual intercourse. ⬅**2**

12. The sensitive, cone-shaped end of the penis is called the _____ . ⬅**1**

Part B Assessments

1. Sketch and label a representative section of the testis. ⬅**3**

2. Sketch and label a region of the epididymis wall. ◄3

3. Sketch and label a penis cross section. ◄3

4. Briefly describe the function of each of the following: ◄2
   a. supporting cells (sustentacular cells; Sertoli cells) in seminiferous tubules

   _____

   _____

   b. spermatogenic cells

   _____

   _____

   c. interstitial cells (Leydig cells)

   _____

   _____

   d. epididymis

   _____

   _____

   e. corpora cavernosa and corpus spongiosum

   _____

   _____

Laboratory Exercise 47

Female Reproductive System

The organs of the female reproductive system are specialized to produce and maintain the female sex cells, to transport these cells to the site of fertilization, to provide a favorable environment for a developing offspring, to move the offspring to the outside, and to produce female sex hormones.

These organs include the ovaries, which produce the egg cells and female sex hormones, and sets of internal and external accessory structures. The internal accessory structures include the uterine tubes, uterus, and vagina. The external structures are the labia majora, labia minora, clitoris, and vestibular glands.

Purpose of the Exercise

To review the structure and functions of the female reproductive organs and to examine some of their features microscopically.

Learning Outcomes

After completing this exercise, you should be able to

1. Locate and identify the organs of the female reproductive system.
2. Describe the functions of these organs.
3. Sketch and label the major features from microscopic sections of the ovary, uterine tube, and uterine wall.

Procedure A—Female Reproductive Organs

1. Review the sections entitled "Ovaries," "Female Internal Accessory Organs," "Female External Reproductive Organs," and "Mammary Glands" in chapter 19 of the textbook.
2. As a review activity, label figures 47.1, 47.2, and 47.3.
3. Observe the human torso model, the model of the female reproductive system, and the anatomical chart of the female reproductive system. Locate the following features:

ovaries
    medulla
    cortex
uterine tubes (oviducts; fallopian tubes)
    infundibulum
    fimbriae
uterus
    body
    cervix
    endometrium
    myometrium
    perimetrium (serous membrane)
vagina
    vaginal orifice
    hymen
    mucosal layer
    muscular layer
    fibrous layer
vulva (external accessory reproductive organs)
    mons pubis
    labia majora
    labia minora
    vestibule
    vestibular glands
    clitoris

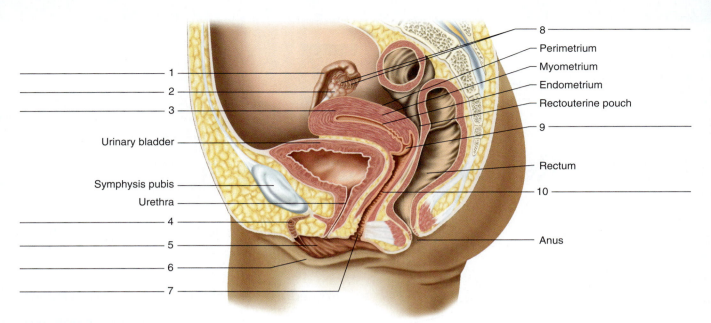

Figure 47.1 Label the structures of the female reproductive system in this sagittal view. ◀1

Labels on figure 47.1:
- 8
- Perimetrium
- Myometrium
- Endometrium
- Rectouterine pouch
- 9
- Rectum
- 10
- Anus
- 1
- 2
- 3
- Urinary bladder
- Symphysis pubis
- Urethra
- 4
- 5
- 6
- 7

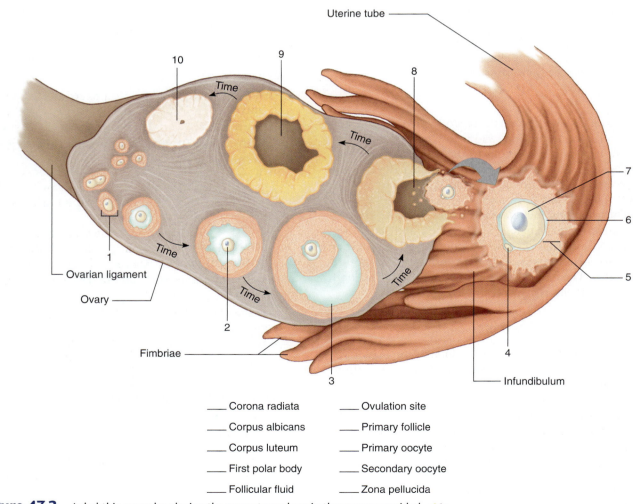

Labels on figure 47.2:
- Uterine tube
- 10
- 9
- 8
- 7
- 6
- 5
- 4
- Time
- 1
- Ovarian ligament
- Ovary
- 2
- Fimbriae
- 3
- Infundibulum

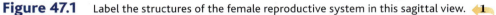

_____ Corona radiata   _____ Ovulation site

_____ Corpus albicans   _____ Primary follicle

_____ Corpus luteum   _____ Primary oocyte

_____ First polar body   _____ Secondary oocyte

_____ Follicular fluid   _____ Zona pellucida

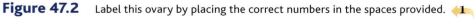

Figure 47.2 Label this ovary by placing the correct numbers in the spaces provided. ◀1

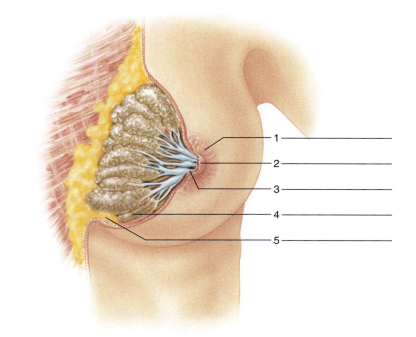

Terms:
Adipose tissue
Alveolar glands
Areola
Lactiferous duct
Nipple

1
2
3
4
5

Figure 47.3   Using the terms provided, label the structures of the breast (anterior view). ◀1

breasts

  nipple

  areola

  alveolar glands (compose milk-producing parts of
    mammary glands)

  lactiferous ducts

  adipose tissue

4. Complete Part A of Laboratory Report 47.

Procedure B—Microscopic Anatomy

1. Obtain a microscope slide of an ovary section with maturing follicles and examine it with low-power magnification (fig. 47.4). Locate the outer layer, or *cortex,* of the ovary, composed of densely packed cells, and the inner layer, or *medulla,* which largely consists of loose connective tissue.

2. Focus on the cortex of the ovary, using high-power magnification (fig. 47.5). Note the thin layer of small cuboidal cells on the free surface. These cells comprise the *germinal epithelium.* Also locate some *primordial follicles* just beneath the germinal epithelium. Each follicle consists of a single, relatively large *primary oocyte* with a prominent nucleus and a covering of *follicular cells.*

3. Prepare a labeled sketch of the ovarian cortex in Part B of the laboratory report.

4. Use low-power magnification to search the cortex for maturing follicles in various stages of development. Prepare three labeled sketches in Part B of the laboratory report to illustrate the changes that occur in a follicle as it matures.

5. Obtain a microscope slide of a cross section of a uterine tube. Examine it, using low-power magnification (fig. 47.6). The shape of the lumen is very irregular.

6. Focus on the inner lining of the uterine tube, using high-power magnification. The lining is composed of *simple columnar epithelium,* and some of the epithelial cells are ciliated on their free surfaces.

7. Prepare a labeled sketch of a representative region of the wall of the uterine tube in Part B of the laboratory report.

8. Obtain a microscope slide of the uterine wall section (fig. 47.7). Examine it, using low-power magnification, and locate the following:

   endometrium (inner mucosal layer)

   myometrium (middle, thick muscular layer)

   perimetrium (outer serosal layer)

9. Prepare a labeled sketch of a representative section of the uterine wall in Part B of the laboratory report.

10. Complete Part B of the laboratory report.

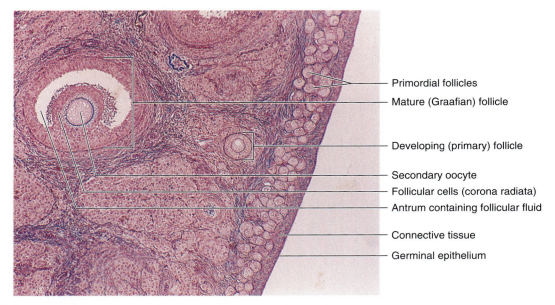

Figure 47.4   Micrograph of the ovary (30× micrograph enlarged to 80×).

Primordial follicles

Mature (Graafian) follicle

Developing (primary) follicle

Secondary oocyte

Follicular cells (corona radiata)

Antrum containing follicular fluid

Connective tissue

Germinal epithelium

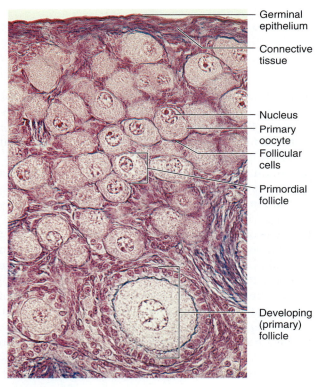

Germinal epithelium

Connective tissue

Nucleus

Primary oocyte

Follicular cells

Primordial follicle

Developing (primary) follicle

Figure 47.5   Micrograph of the ovarian cortex (100× micrograph enlarged to 200×).

Demonstration

Observe the slides in the demonstration microscopes. Each slide contains a section of uterine mucosa taken during a different phase in the reproductive cycle. In the *early proliferative phase,* note the simple columnar epithelium on the free surface of the mucosa and the many sections of tubular uterine glands in the tissues beneath the epithelium. In the *secretory phase,* the endometrium is thicker and the uterine glands appear more extensive and they are coiled. In the *early menstrual phase,* the endometrium is thinner because its surface layer has been lost. Also, the uterine glands are less apparent, and the spaces between the glands contain many leukocytes. What is the significance of these changes?

_____

_____

_____

_____

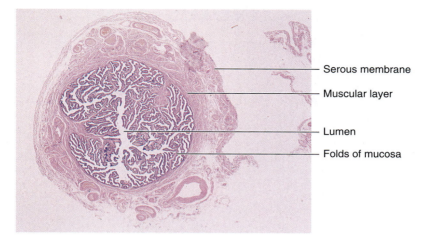

Figure 47.6   Micrograph of a cross section of the uterine tube (8×).

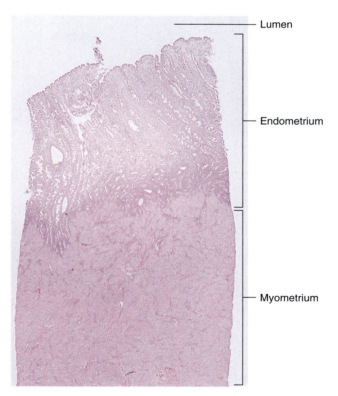

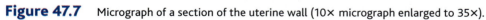

Figure 47.7   Micrograph of a section of the uterine wall (10× micrograph enlarged to 35×).

Notes

Name _____

Date _____

Section _____

The ⬅ corresponds to the indicated outcome(s) found at the beginning of the laboratory exercise.

Female Reproductive System

Part A Assessments

Complete the following statements:

1. The ovaries are located in the lateral wall of the _____ cavity. ⬅**1**

2. The ovarian cortex appears granular because of the presence of _____ . ⬅**1**

3. A primary oocyte is closely surrounded by flattened epithelial cells called _____ cells. ⬅**1**

4. When a primary oocyte divides, a secondary oocyte and a(n) _____ are produced. ⬅**2**

5. Primordial follicles are stimulated to develop into primary follicles by the hormone called _____ . ⬅**2**

6. _____ is the process by which a secondary oocyte is released from the ovary.

7. Uterine tubes are also called _____ . ⬅**1**

8. The _____ is the funnel-shaped expansion at the end of a uterine tube. ⬅**1**

9. A portion of the uterus called the _____ extends downward into the upper portion of the vagina. ⬅**1**

10. The inner mucosal lining of the uterus is called the _____ . ⬅**1**

11. The myometrium is largely composed of _____ tissue. ⬅**1**

12. The vaginal orifice is partially closed by a thin membrane called the _____ . ⬅**1**

13. The group of external accessory organs that surround the openings of the urethra and vagina comprise the _____ . ⬅**1**

14. The rounded mass of fatty tissue overlying the symphysis pubis of the female is called the _____ . ⬅**1**

15. The female organ that corresponds to the male penis is the _____ . ⬅**1**

16. The _____ of the female correspond to the bulbourethral glands of the male. ⬅**1**

Part B Assessments

1. Sketch and label a representative region of the ovarian cortex. ⬅**3**

2. Sketch and label a series of three changes to illustrate follicular maturation. ◀**3**

3. Sketch and label a representative section of the wall of a uterine tube. ◀**3**

4. Sketch and label a representative section of the uterine wall. ◀**3**

5. Complete each of the following:

   a. Describe the fate of a mature follicle. ◀**2**

_____

_____

_____

   b. Describe the function of the cilia in the lining of the uterine tube. ◀**2**

_____

_____

_____

   c. Briefly describe the changes that occur in the uterine lining during the reproductive cycle. ◀**2** _____

_____

_____

_____

Laboratory Exercise  48

Genetics

Many examples of human genetics in this laboratory exercise are basic, external features to observe. Some traits are based upon simple Mendelian genetics. As our knowledge of genetics continues to develop, traits such as tongue roller and free earlobe may no longer be considered examples of simple Mendelian genetics. We may need to continue to abandon some simple Mendelian models. New evidence may involve polygenic inheritance, effects of other genes, or environmental factors as more appropriate explanations. Therefore, the analysis of family genetics is not appropriate nor is it the purpose of this laboratory exercise.

G enetics is the study of the inheritance of characteristics. The genes that transmit this information are coded in segments of DNA in chromosomes. Homologous chromosomes possess the same gene at the same *locus*. These genes may exist in variant forms, called *alleles*. If a person possesses two identical alleles, the condition is *homozygous*. If a person possesses two different alleles, the condition is *heterozygous*. The particular combination of these gene variants (alleles) represents the person's *genotype;* the appearance of the individual that develops as a result of the way the genes are expressed represents the person's *phenotype*.

If one allele determines the phenotype by masking the expression of the other allele in a heterozygous individual, the allele is termed *dominant*. The allele whose expression is masked is termed *recessive*. If the heterozygous condition determines an intermediate phenotype, the inheritance represents *incomplete dominance*. However, different alleles are *codominant* if both are expressed in the heterozygous condition. Some characteristics inherited on the sex chromosomes result in phenotype frequencies that might be more prevalent in males or females. Such characteristics are called sexlinked (X-linked or Y-linked) characteristics.

As a result of meiosis during the formation of eggs and sperm, a mother and father each transmit an equal number of chromosomes (the haploid number 23) to form the zygote (diploid number 46). An offspring will receive one allele from each parent. These gametes combine randomly in the formation of each offspring. Hence, the *laws of probability* can be used to predict possible genotypes and phenotypes of offspring. A genetic tool called a *Punnett square* simulates all possible combinations (probabilities) that can occur in offspring genotypes and resulting phenotypes.

Purpose of the Exercise

To observe some selected human traits, to use pennies and dice to demonstrate laws of probability, and to solve some genetic problems using a Punnett square.

Learning Outcomes

After completing this exercise, you should be able to

1. Examine and record twelve genotypes and phenotypes of selected human traits.
2. Demonstrate the laws of probability using tossed pennies and dice and interpret the results.
3. Predict genotypes and phenotypes of complete dominance, codominance, and sex-linked problems using Punnett squares.

Procedure A—Human Genotypes and Phenotypes

A complete set of genetic instructions in one human cell constitutes one's *genome*. The human genome contains about 2.9 billion base pairs representing approximately 20,500 protein-encoding genes. These instructions represent our genotypes and are expressed as phenotypes sometimes clearly observable on our bodies. Some of these traits are listed in table 48.1 and are discernible in figure 48.1.

Table 48.1 Examples of Some Common Human Phenotypes

Dominant Traits and Genotypes	Recessive Traits and Genotypes
Tongue roller (R__)	Nonroller (rr)
Freckles (F__)	No freckles (ff)
Widow's peak (W__)	Straight hairline (ww)
Dimples (D__)	No dimples (dd)
Free earlobe (E__)	Attached earlobe (ee)
Normal skin coloration (M__)	Albinism (mm)
Astigmatism (A__)	Normal vision (aa)
Natural curly hair (C__)	Natural straight hair (cc)
PTC taster (T__)	Nontaster (tt)
Blood type A (I^A __), B (I^B __), or AB ($I^A I^B$)	Blood type O (ii)
Normal color vision ($X^C X^C$), ($X^C X^c$), or ($X^C Y$)	Red-green colorblindness ($X^c X^c$) or ($X^c Y$)

A dominant trait might be homozygous or heterozygous, so only one capital letter is used along with a blank for the possible second dominant or recessive allele. For a recessive trait, two lowercase letters represent the homozygous recessive genotype for that characteristic. Dominant does not always correlate with the predominance of the allele in the gene pool; dominant means one allele will determine the appearance of the phenotype.

1. **Tongue roller/nonroller:** The dominant allele (R) determines the person's ability to roll the tongue into a U-shaped trough. The homozygous recessive condition (rr) prevents this tongue rolling (fig. 48.1). Record your results in the table in Part A of Laboratory Report 48.

2. **Freckles/no freckles:** The dominant allele (F) determines the appearance of freckles. The homozygous recessive condition (ff) does not produce freckles (fig. 48.1). Record your results in the table in Part A of the laboratory report.

3. **Widow's peak/straight hairline:** The dominant allele (W) determines the appearance of a hairline above the forehead that has a distinct downward point in the center, called a widow's peak. The homozygous recessive condition (ww) produces a straight hairline (fig. 48.1). A receding hairline would prevent this phenotype determination. Record your results in the table in Part A of the laboratory report.

4. **Dimples/no dimples:** The dominant allele (D) determines the appearance of a distinct dimple in one or both cheeks upon smiling. The homozygous recessive condition (dd) results in the absence of dimples (fig. 48.1). Record your results in the table in Part A of the laboratory report.

5. **Free earlobe/attached earlobe:** The dominant allele (E) codes for the appearance of an inferior earlobe that hangs freely below the attachment to the head. The homozygous recessive condition (ee) determines the earlobe attaching directly to the head at its inferior border (fig. 48.1). Record your results in the table in Part A of the laboratory report.

6. **Normal skin coloration/albinism:** The dominant allele (M) determines the production of some melanin, producing normal skin coloration. The homozygous recessive condition (mm) determines albinism due to the inability to produce or use the enzyme tyrosinase in pigment cells. An albino does not produce melanin in the skin, hair, or the middle tunic (choroid coat, ciliary body, and iris) of the eye. The absence of melanin in the middle tunic allows the pupil to appear slightly red to nearly black. Remember that the pupil is an opening in the iris filled with transparent aqueous humor. An albino human has pale white skin, flax-white hair, and a pale blue iris. Record your results in the table in Part A of the laboratory report.

7. **Astigmatism/normal vision:** The dominant allele (A) results in an abnormal curvature to the cornea or the lens. As a consequence, some portions of the image projected on the retina are sharply focused, and other portions are blurred. The homozygous recessive condition (aa) generates normal cornea and lens shapes and normal vision. Use the astigmatism chart and directions to assess this possible defect described in Laboratory Exercise 32. Other eye defects, such as nearsightedness (myopia) and farsightedness (hyperopia), are different genetic traits due to genes at other locations. Record your results in the table in Part A of the laboratory report.

Dominant Traits

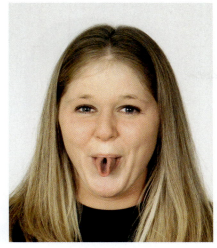

(a) Tongue roller

Recessive Traits

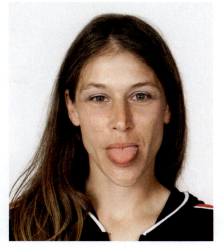

(b) Nonroller

(c) Freckles

(d) No freckles

(e) Widow's peak

(f) Straight hairline

Figure 48.1 Representative genetic traits comparing dominant and recessive phenotypes: (*a*) tongue roller; (*b*) nonroller; (*c*) freckles; (*d*) no freckles; (*e*) widow's peak; (*f*) straight hairline; (*g*) dimples; (*h*) no dimples; (*i*) free earlobe; (*j*) attached earlobe.

(g) Dimples

(h) No dimples

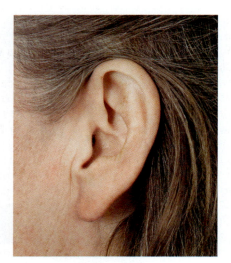

(i) Free earlobe

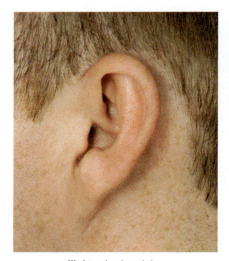

(j) Attached earlobe

Figure 48.1   (*Continued*).

8. **Curly hair/straight hair:** The dominant allele (*C*) determines the appearance of curly hair. Curly hair is somewhat flattened in cross section, as the hair follicle of a similar shape served as a mold for the root of the hair during its formation. The homozygous recessive condition (*cc*) produces straight hair. Straight hair is nearly round in cross section from being molded into this shape in the hair follicle. In some populations (Caucasians) the heterozygous condition (*Cc*) expresses the intermediate wavy hair phenotype (incomplete dominance). This trait determination assumes no permanents or hair straightening procedures have been performed. Such hair alterations do not change the hair follicle shape, and future hair growth results in original genetic hair conditions. Record your unaltered hair appearance in the table in Part A of the laboratory report.

9. **PTC taster/nontaster:** The dominant allele (*T*) determines the ability to experience a bitter sensation when PTC paper is placed on the tongue. About 70% of people possess this dominant gene. The homozygous recessive condition (*tt*) makes a person unable to notice the substance. Place a piece of PTC (phenylthiocarbamide) paper on the upper tongue surface and chew it slightly to see if you notice a bitter sensation from this harmless chemical. The nontaster of the PTC paper does not detect any taste at all from this substance. Record your results in the table in Part A of the laboratory report.

10. **Blood type A, B, or AB/blood type O:** There are three alleles (I^A, I^B, and i) in the human population affecting RBC membrane structure. These alleles are located on a single pair of homologous chromosomes, so a person could possess either two of the three alleles or two of the same allele. All of the possible combinations of these alleles of genotypes and the resulting phenotypes are depicted in table 48.2. The expression of the blood type AB is a result of both codominant alleles located in the same individual. Possibly you have already determined your blood type in Laboratory Exercise 34 or have it recorded on a blood donor card. (If simulated blood-typing kits were used for Laboratory Exercise 34, those results would not be valid for your genetic factors.) Record your results in the table in Part A of the laboratory report.

11. **Sex determination:** A person with sex chromosomes XX displays a female phenotype. A person with sex chromosomes XY displays a male phenotype. Record your results in the table in Part A of the laboratory report.

12. **Normal color vision/red-green color-blindness:** This condition is a sex-linked (X-linked) characteristic. The alleles for color vision are also on the X chromosome, but absent on the Y chromosome. As a result, a female might possess both alleles (C and c), one on each of the X chromosomes. The dominant allele (C) determines normal color vision; the homozygous recessive condition (cc) results in red-green colorblindness. However, a male would possess only one of the two alleles for color vision because there is only a single X chromosome in a male. Hence a male with even a single recessive gene for colorblindness possesses the defect. Note all the possible genotypes and phenotypes for this condition (table 48.1). Review the color vision test in Laboratory Exercise 32 using the color plates in figure 32.4 and Ichikawa's or Ishihara's book. Record your results in the table in Part A of the laboratory report.

13. Complete Part A of Laboratory Report 48.

Procedure B—Laws of Probability

The laws of probability provide a mathematical way to determine the likelihood of events occurring by chance. This prediction is often expressed as a ratio of the number of results from experimental events to the number of results considered possible. For example, when tossing a coin there is an equal chance of the results displaying heads or tails. Hence the probability is one-half of obtaining either a heads or a tails (there are two possibilities for each toss). When all of the probabilities of all possible outcomes are considered for the result, they will always add up to a 1. To predict the probability of two or more events occurring in succession, multiply the

Table 48.2 Genotypes and Phenotypes (Blood Types)

Genotypes	Phenotypes (Blood Types)
$I^A I^A$ or $I^A i$	A
$I^B I^B$ or $I^B i$	B
$I^A I^B$ (codominant)	AB
ii	O

probabilities of each individual event. For example, the probability of tossing a die and displaying a 4 two times in a row is $1/6 \times 1/6 = 1/36$ (there are six possibilities for each toss). Each toss in a sequence is an *independent event* (chance has no memory). The same laws apply when parents have multiple children (each fertilization is an independent event). Perform the following experiments to demonstrate the laws of probability:

1. Use a single penny (or other coin) and toss it 20 times. Predict the number of heads and tails that would occur from the 20 tosses. Record your prediction and the actual results observed in Part B of the laboratory report.

2. Use a single die (*pl.* dice) and toss it 24 times. Predict the number of times a number below 3 (numbers 1 and 2) would occur from the 24 tosses. Record your prediction and the actual results in Part B of the laboratory report.

3. Use two pennies and toss them simultaneously 32 times. Predict the number of times two heads, a heads and a tails, and two tails occur. Record your prediction and the actual results in Part B of the laboratory report.

4. Use a pair of dice and toss them simultaneously 32 times. Predict the number of times for both dice coming up with odd numbers, one die an odd and the other an even number, and both dice coming up with even numbers. Record your prediction and the actual results in Part B of the laboratory report.

5. Obtain class totals for all of the coins and dice tossed by adding your individual results to a class tally location as on the blackboard.

6. A Punnett square can be used for a visual representation to demonstrate the probable results for two pennies tossed simultaneously. For the purpose of a genetic comparison, an *h* (heads) will represent one "allele" on the coin; a *t* (tails) will represent a different "allele" on the coin.

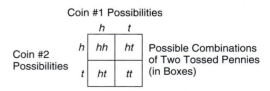

7. Complete Part B of the laboratory report.

Procedure C—Genetic Problems

1. A Punnett square can be constructed to demonstrate a visual display of the predicted offspring from parents with known genotypes. Recall that in complete dominance, a dominant allele is expressed in the phenotype, as it can mask the other recessive allele on the homologous chromosome pair. Recall also that during meiosis, the homologous chromosomes with their alleles separate (Mendel's Law of Segregation) into different gametes. An example of such a cross might be a homozygous dominant mother for dimples (*DD*) has offspring with a father homozygous recessive (*dd*) for the same trait. The results of such a cross, according to the laws of probability, would be represented by the following Punnett square:

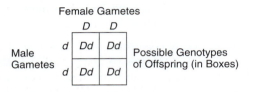

Results: Genotypes: 100% *Dd* (all heterozygous)

Phenotypes: 100% dimples

In another example, assume that both parents are heterozygous (*Dd*) for dimples. The results of such a cross, according to the laws of probability, would be represented by the following Punnett square:

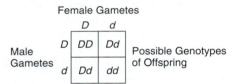

Results: Genotypes: 25% *DD* (homozygous dominant); 50% *Dd* (heterozygous); 25% *dd* (homozygous recessive) (1:2:1 genotypic ratio)

Phenotypes: 75% dimples; 25% no dimples (3:1 phenotypic ratio)

2. Work the genetic problems 1 and 2 in Part C of the laboratory report.
3. The ABO blood type inheritance represents an example of codominance. Review table 48.2 for the genotypes and phenotypes for the expression of this trait. A similar technique to that used previously can be used to predict the offspring of parents of known genotypes. In this example, assume the genotype of the mother is $I^A I^B$, and the father is *ii*.

The results of such a cross would be represented by the following Punnett square:

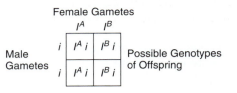

Results: Genotypes: 50% $I^A i$ (heterozygous for A); 50% $I^B i$ (heterozygous for B) (1:1 genotypic ratio)

Phenotypes: 50% blood type A; 50% blood type B (1:1 phenotypic ratio)

Note: In this particular cross, all of the children would have blood types unlike either parent.

4. Work the genetic problems 3 and 4 in Part C of the laboratory report.
5. Review the inheritance of red-green colorblindness, an X-linked characteristic, in table 48.1. A similar technique to that used to identify complete dominance can be used to predict the offspring of parents of known genotypes. In this example, assume the genotype of the mother is heterozygous $X^C X^c$ (normal color vision, but a carrier for the colorblindness defect), and the father is $X^C Y$ (normal color vision; no allele on the Y chromosome). The results of such a cross would be represented by the following Punnett square:

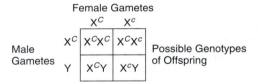

Results: Genotypes: 25% $X^C X^C$; 25% $X^C X^c$; 25% $X^C Y$; 25% $X^c Y$ (1:1:1:1 genotypic ratio)

Phenotypes for sex determination: 50% females; 50% males (1:1 phenotypic ratio)

Phenotypes for color vision: Females 100% normal color vision (however, 50% are heterozygous carriers for colorblindness)

Males 50% normal; 50% with red-green colorblindness (1:1 phenotypic ratio)

Note: In X-linked inheritance the colorblind males received the recessive gene from their mothers.

6. Complete Part C of the laboratory report.

Genetics

48

The ⬅ corresponds to the indicated outcome(s) found at the beginning of the laboratory exercise.

Name _____

Date _____

Section _____

Part A Assessments

1. Enter your test results for genotypes and phenotypes in the table. Circle your particular phenotype and genotype for each of the twelve traits. ⬅**1**

Trait	Dominant Phenotype	Genotype	Recessive Phenotype	Genotype
Tongue movement	Roller	R__	Nonroller	rr
Freckles	Freckles	F__	No freckles	ff
Hairline	Widow's peak	W__	Straight	ww
Dimples	Dimples	D__	No dimples	dd
Earlobe	Free	E__	Attached	ee
Skin coloration	Normal (some melanin)	M__	Albinism	mm
Vision	Astigmatism	A__	Normal	aa
Hair shape*	Curly	C__	Straight	cc
Taste	PTC taster	T__	Nontaster for PTC	tt
Blood type	A, B, or AB	I^A__; I^B__; or $I^A I^B$	O	ii
Sex		XX or XY		
Color vision	Normal	$X^C X$__ or $X^C Y$	Red-green colorblindness	$X^c X^c$ or $X^c Y$

*In some populations (Caucasians) the heterozygous condition (Cc) results in the appearance of wavy hair, which actually represents an example of incomplete dominance for this trait.

2. Analyze all the genotypes that you circled for the dominant phenotypes. If it is feasible to observe your biological parents and siblings for any of these traits, are you able to determine if any of your dominant genotypes are homozygous dominant or heterozygous? _____ If so, which ones? Explain the rationale for your response.

_____

_____

_____

_____

_____

Part B Assessments

1. Single penny tossed 20 times and counting heads and tails: ⬅**2**

   Probability (prediction): ____/20 heads ____/20 tails

   (Note: Traditionally, probabilities are converted to the lowest fractional representation.)

   Actual results: ____ heads ____ tails

   Class totals: ____ heads ____ tails

2. Single die tossed 24 times and counting the number of times a number below 3 occurs: ◂**2**

    Probability: ____/24 number below 3 (numbers 1 and 2)

    Actual results: ____ number below 3

    Class totals: ____ number below 3 ____ total tosses by class members

3. Two pennies tossed simultaneously 32 times and counting the number of two heads, a heads and a tails, and two tails: ◂**2**

    Probability: ____/32 of two heads ____/32 of a heads and a tails ____/32 of two tails

    Actual results: ____two heads ____heads and tails ____two tails

    Class totals: ____two heads ____heads and tails ____two tails

4. Two dice tossed simultaneously 32 times and counting the number of two odd numbers, an odd and an even number, and two even numbers: ◂**2**

    Probability: ____/32 of two odd numbers ____/32 of an odd and an even number ____/32 of two even numbers

    Actual results: ____ two odd numbers ____ an odd and an even number ____ two even numbers

    Class totals: ____ two odd numbers ____ an odd and an even number ____ two even numbers

5. Use the example of the two dice tossed 32 times and construct a Punnett square to represent the possible combinations that could be used to determine the probability (prediction) of odd and even numbers for the resulting tosses. Your construction should be similar to the Punnett square for the two coins tossed that is depicted in Procedure B of the laboratory exercise. ◂**2**

6. Complete the following:

    a. Are the class totals closer to the predicted probabilities than your results? ____ Explain your response.

    _____

    _____

    _____

    _____

    b. Does the first toss of the penny or the first toss of the die have any influence on the next toss? _____
    Explain your response.

    _____

    _____

    _____

    _____

    c. Assume a family has two boys or two girls. They wish to have one more child, but hope for the child to be of the opposite sex from the two they already have. What is the probability that the third child will be of the opposite sex? ____ Explain your response.

    _____

    _____

    _____

    _____

d. What is the probability (prediction) that a couple without children will eventually have four children, all girls? _____ Explain your response.

_____

_____

_____

_____

Part C Assessments

For each of the genetic problems, (a) determine the parents' genotypes, (b) determine the possible gametes for each parent, (c) construct a Punnett square, and (d) record the resulting genotypes and phenotypes as ratios from the cross. Problems 1 and 2 involve examples of complete dominance; problems 3 and 4 are examples of codominance; problem 5 is an example of sex-linked (X-linked) inheritance.

1. Determine the results from a cross of a mother who is heterozygous (*Rr*) for tongue rolling with a father who is homozygous recessive (*rr*). ◂3

2. Determine the results from a cross of a mother and a father who are both heterozygous for freckles. ◂3

3. Determine the results from a mother who is heterozygous for blood type B and a father who is homozygous dominant for blood type A. ◂3

4. Determine the results from a mother who is heterozygous for blood type A and a father who is heterozygous for blood type B. ◄**3**

5. Colorblindness is an example of X-linked inheritance. Hemophilia is another example of X-linked inheritance, also from a recessive allele (*h*). The dominant allele (*H*) determines whether the person possesses normal blood clotting. A person with hemophilia has a permanent tendency for hemorrhaging due to a deficiency of one of the clotting factors (VIII—antihemophilic factor). Determine the offspring from a cross of a mother who is a carrier (heterozygous) for the disease and a father with normal blood coagulation. ◄**3**

 ## Critical Thinking Application

Assume that the genes for hairline and earlobes are on different pairs of homologous chromosomes. Determine the genotypes and phenotypes of the offspring from a cross if both parents are heterozygous for both traits. (1) First determine the genotypes for each parent. (2) Determine the gametes, but remember each gamete has one allele for each trait (gametes are haploid). (3) Construct a Punnett square with 16 boxes that has four different gametes from each parent along the top and the left edges. (This is an application to demonstrate Mendel's Law of Independent Assortment.) (4) List the results of genotypes and phenotypes as ratios. ◄**3**

Appendix 1

Preparation of Solutions

Amylase solution, 0.5%

Place 0.5 g of bacterial amylase in a graduated cylinder or volumetric flask. Add distilled water to the 100 mL level. Stir until dissolved. See the Instructor's Manual for a supplier of amylase that is free of sugar. (Store amylase powder in a freezer until mixing this solution.)

Benedict's solution

Prepared solution is available from various suppliers.

Glucose solutions

1. *1.0% solution.* Place 1 g of glucose in a graduated cylinder or volumetric flask. Add distilled water to the 100 mL level. Stir until dissolved.
2. *10% solution.* Place 10 g of glucose in a graduated cylinder or volumetric flask. Add distilled water to the 100 mL level. Stir until dissolved.

Iodine-potassium-iodide (IKI solution)

Add 20 g of potassium iodide to 1 L of distilled water, and stir until dissolved. Then add 4.0 g of iodine, and stir again until dissolved. Solution should be stored in a dark stoppered bottle.

Methylene blue

Dissolve 0.3 g of methylene blue powder in 30 mL of 95% ethyl alcohol. In a separate container, dissolve 0.01 g of potassium hydroxide in 100 mL of distilled water. Mix the two solutions. (Prepared solution is available from various suppliers.)

Physiological saline solution

Place 0.9 g of sodium chloride in a graduated cylinder or volumetric flask. Add distilled water to the 100 mL level. Stir until dissolved.

Ringer's solution (frog)

Dissolve the following salts in 1 L of distilled water:
6.50 g sodium chloride
0.20 g sodium bicarbonate
0.14 g potassium chloride
0.12 g calcium chloride

Sodium chloride solutions

1. *0.9% solution.* Place 0.9 g of sodium chloride in a graduated cylinder or volumetric flask. Add distilled water to the 100 mL level. Stir until dissolved.
2. *3.0% solution.* Place 3.0 g of sodium chloride in a graduated cylinder or volumetric flask. Add distilled water to the 100 mL level. Stir until dissolved.

Starch solutions

1. *0.5% solution.* Add 5 g of cornstarch to 1 L of distilled water. Heat until the mixture boils. Cool the liquid, and pour it through a filter. Store the filtrate in a refrigerator.
2. *1.0% solution.* Add 10 g of cornstarch to 1 L of distilled water. Heat until the mixture boils. Cool the liquid, and pour it through a filter. Store the filtrate in a refrigerator.
3. *10% solution.* Add 100 g of cornstarch to 1 L of distilled water. Heat until the mixture boils. Cool the liquid, and pour it through a filter. Store the filtrate in a refrigerator.

Appendix 2

Assessments of Laboratory Reports

Many assessment models can be used for laboratory reports. A rubric, which can be used for performance assessments, contains a description of the elements (requirements or criteria) of success to various degrees. The term *rubric* originated from *rubrica terra,* which is Latin for the application of red earth to indicate anything of importance. A rubric used for assessment contains elements for judging student performance, with points awarded for varying degrees of success in meeting the objectives. The content and the quality level necessary to attain certain points are indicated in the rubric. It is effective if the assessment tool is shared with the students before the laboratory exercise is performed.

Following are two sample rubrics that could easily be modified to meet the needs of a specific course. Some of the elements for these sample rubrics may not be necessary for every laboratory exercise. The generalized rubric needs to contain the possible assessment points that correspond to outcomes for a specific course. The point value for each element may vary. The specific rubric example contains performance levels for laboratory reports. The elements and the point values could easily be altered to meet the value placed on laboratory reports for a specific course.

Assessment: Generalized Laboratory Report Rubric

Element	Assessment Points Possible	Assessment Points Earned
1. Figures are completely and accurately labeled.		
2. Sketches are accurate, contain proper labels, and are of sufficient detail.		
3. Colored pencils were used extensively to differentiate structures on illustrations.		
4. Matching and fill-in-the-blank answers are completed and accurate.		
5. Short-answer/discussion questions contain complete, thorough, and accurate answers. Some elaboration is evident for some answers.		
6. Data collected are complete, accurately displayed, and contain a valid explanation.		

Total Points _____

Assessment: Specific Laboratory Report Rubric

Element	Excellent Performance (4 points)	Proficient Performance (3 points)	Marginal Performance (2 points)	Novice Performance (1 point)	Points Earned
Figure labels	Labels completed with ≥ 90% accuracy.	Labels completed with 80%–89% accuracy.	Labels completed with 70%–79% accuracy.	Labels <70% accurate.	
Sketches	Accurate use of scale, details illustrated, and all structures labeled accurately.	Minor errors in sketches. Missing or inaccurate labels on one or more structures.	Sketch is not realistic. Missing or inaccurate labels on two or more structures.	Several missing or inaccurate labels.	
Matching and fill-in-the-blanks	All completed and accurate.	One to two errors or omissions.	Three to four errors or omissions.	Five or more errors or omissions.	
Short-answer and discussion questions	Answers are complete, valid, and contain some elaboration. No misinterpretations are noted.	Answers are generally complete and valid. Only minor inaccuracies were noted.	Marginal answers to the questions and contains inaccurate information.	Many answers are incorrect or fail to address the topic. There may be misinterpretations.	
Data collection and analysis	Data are complete and displayed with a valid interpretation.	Only minor data missing or a slight misinterpretation exists.	Some omissions. Not displayed or interpreted accurately.	Data are incomplete or show serious misinterpretations.	

TOTAL POINTS EARNED _____

Credits

Index

abdominal wall
  arteries of, 278, 280
  muscles of, 153, 154
  surface anatomy of, 172
  veins of, 283, 285
abdominoplevic cavity, 12, 14
ABO blood typing, 253–54, 363
accommodation pupillary
        reflex, 233
accommodation test, 230–31
acid, bone treatment with, 80, 82
acinar cells, 243
actin, 136
adipose cell, 76
adipose tissue, 66
adrenal gland, 240, 241
agonist, 137
albinism, 360
alpha cells, 242, 243
alveoli, 320
Amoeba, 42
amylase, 313–16
amylase solutions, 369
anatomical position, 13
ankle-jerk reflex, 193–94, 195
antagonist, 137
aorta, 278, 279
appendix, 309
areolar tissue, 66
arm. See upper limb
arteries, 278–82, 290
articulations, 129–34
astigmatism, 360
astigmatism test, 230
astrocyte, 184
atlas, 101, 103
atom, 22
auditory acuity test, 214
auditory system, 213–20
axis, 101, 103
axon, 183, 184

Babinski reflex, 194
basophils, 251, 252
Benedict's solution, 369
Benedict's test, 24
beta cells, 241, 242, 243
biceps-jerk reflex, 194
biceps reflex, 194, 195
bilirubin, urinary, 341
Biuret test, 23–24
blind spot, 224, 233
blood, 66, 249–57
blood cells, 249–52
blood pressure, 292–94
blood slide, 250–52
blood type, 363
blood typing, 253
blood vessels
  arteries, 278–82, 290
  renal, 332, 333, 334
  structure of, 275–77
  temperature effect on, 277
  veins, 283–88, 290
body
  anatomical positions of, 13
  cavities of, 9, 10, 11, 12
  membranes of, 9, 11
  organ systems of, 10, 12

planes of, 13
regions of, 15–16
sections of, 16
bone. See also skeleton
  acid treatment of, 80, 82
  heat treatment of, 80, 82
  joints of, 129–34
  structure of, 66, 79–84
brachial pulse, 292
brain, 199–200, 201, 202, 204
  sheep, 205–12
breast, 353
breathing, 323–24
bronchi, 318
bronchioles, 318, 319, 320

cabbage water test, 22
calcaneal reflex, 193–94, 195
canaliculus, 81
cardiac cycle, 267, 273
cardiac muscle, 70. See also heart
cardiovascular system, 10
carpal bones, 114, 117, 118
cell(s), 39–44
  blood, 249–52
  nerve, 181–85
  structure of, 39–41
cell cycle, 53–60
cell membrane, 39, 41
  movement through, 48–52
cervical vertebrae, 101, 103,
        104, 108
cheek cell, 41
chemistry, 21–38
  organic, 23–24
  unknown compound
        identification with, 24
Chemstrip, 341
chest
  bones of, 105–6
  muscles of, 145, 146, 151, 152
chromosomes
  karyotype of, 56
  slide of, 55
cilia, 62, 319, 320
circulation, 277, 279, 280
clavicle, 109, 110, 116
coccyx, 105
cochlea, 216, 217, 220
color blindness, 231, 232
color vision, 363
color vision test, 231
connective tissue, 65–68, 76
convergence reflex, 233
cornea, 226
cranial bones, 91, 92, 93, 94, 98,
        99, 100
cranial nerves, 200
curly hair, 362
cytokinesis, 54
cytoplasm, 39

dense connective tissue, 66
dermis, 73, 75, 76
diabetes mellitus, 237–38, 243
differential white blood cell count,
        250–52
diffusion, 46, 48
digestive enzymes, 313–16

digestive system, 10
  organs of, 301–12
dimples, 360, 362
directional terminology, 13
dissecting microscope, 35, 36
dorsal root ganglion, 184

ear, 213, 214
  inner, 214, 215, 216
  middle, 217
earlobe, 360, 362
eggshell membrane, 46
elastic cartilage, 66
electrocardiogram, 267, 268–70
emphysema, 320
endocrine system, 10, 237–48
endomysium, 136
enzymes, 313–16
eosinophils, 251, 252
epidermis, 73, 75, 76
epididymis, 345, 347, 348
epithelial tissue, 61–64
esophagus, 303–5
excretory system. See urinary
        system
expiratory reserve volume, 325
extracellular matrix, 81
eye, 221–28
  dissection of, 225–26
  examination of, 223, 224
  structure of, 221–23

face
  bones of, 94, 98, 99, 100
  muscles of, 140, 142–44
facial expression, muscles of,
        142–44
farsightedness, 229
female reproductive system, 351–58
  microscopic anatomy of,
        353–55
femur, 121, 122, 126
fetal skeleton, 95
fibrocartilage, 66
fibula, 121, 123, 126
filtration, 47–48
fontanel, 95
foot
  bones of, 121, 124, 127
  surface anatomy of, 175
freckles, 360, 361
functional residual capacity (FRC),
        325, 326

genetics, 359–68
  problems in, 364
genitourinary system. See
        reproductive system;
        urinary system
genome, 359
genotype, 359–63
glucose, urinary, 341
glucose solution, 369

hair follicle, 75, 76
hairline, 360, 361
hand
  bones of, 114, 117, 118
  muscles of, 146

head
  arteries of, 278, 281
  bones of, 91, 92, 93, 94, 98, 99, 100
  muscles of, 141–44
  surface anatomy of, 170
  veins of, 283
hearing tests, 214–15
heart, 259–66
  cardiac cycle of, 267, 273
  electrocardiogram of, 267, 268–70
  sheep, 261–64
  structure of, 259–61, 262
heart sounds, 267–68
heat
  amylase reaction to, 314
  bone treatment with, 80, 82
hemoglobin, urinary, 341
hepatic portal system, 283, 285
hip, 119–20, 126
humerus, 111, 112, 116
hyaline cartilage, 66
hydrometer, 340, 341
hyperopia, 229
hypertonic solution, 47
hypotonic solution, 47

inspiratory capacity (IC), 325, 326
inspiratory reserve volume (IRV),
        325, 326
insulin resistance, 238
insulin shock, 241
integumentary system, 10, 73–78
interphase, 54
intestine, 307–10
iodine-potassium-iodide solution
        (IKI solution), 369
iodine test, 24
ionic bond, 22
Ishihara's color plates, 231, 232
islets of Langerhans, 241, 242, 243
isotonic solution, 47

joints, 129–34
  movements of, 130, 134

karyotype, 56
ketones, urinary, 341
kidneys
  nephron of, 334, 335
  structure of, 331–33, 337–38
knee, 126, 129
knee-jerk reflex, 193, 195

laboratory report assessment, 370–71
lacuna, 81
large intestine, 307–10
larynx, 318, 319
laws of probability, 363
leg. See lower limb
lens, 225, 226
lipid test, 24
liver, 307
lower limb
  arteries of, 278, 282, 290
  bones of, 121–27
  measurement of, 2–3
  muscles of, 159–68
  surface anatomy of, 175, 176
  veins of, 283, 286, 290

lumbar vertebrae, 104
lung, 318
    blood vessels of, 279, 290
    micrograph of, 319–20
lymphatic pathways, 295, 296
lymphatic system, 10, 295–300
lymph nodes, 295–97
lymphocytes, 251, 252

male reproductive system,
    345–50
    microscopic anatomy of, 345,
        347–48
mastication, muscles of, 141, 142
measurement, 1, 2–3
melanin, 74
membranes
    body, 9, 11
    cell, 39, 41, 48–52
    eggshell, 46
meninges, 190
metacarpal bones, 114, 117, 118
metatarsal bones, 121, 123, 127
methylene blue, 369
metric system, 1, 2–3
micrometer scale, 33, 34
microscope, 29–38
    adjustment of, 32–33
    basics of, 29–33
    field of view of, 31–32
    lens of, 29, 30, 31
    lens-cleaning for, 29
    micrometer scale of, 33, 34
    parts of, 30–31
    slide preparation for, 33, 35
minute respiratory volume, 326
mitosis, 54, 60
molecule, 22, 23
monocytes, 252
mouth, 301–3
Multistix, 341
muscle(s)
    abdominal wall, 153, 154
    arm, 145, 146, 151, 152
    chest, 145, 151, 152
    eye, 221, 222
    facial, 140, 141, 144
    foot, 159
    forearm, 145, 148, 152
    hand, 146
    hip, 159, 160, 161, 162
    leg, 159, 163, 164, 165, 168
    neck, 140, 141, 151
    pectoral girdle, 145
    pelvic outlet, 153, 155
    shoulder, 146, 147, 152
    thigh, 159, 160, 161, 162
muscle fiber, 135, 136, 137
muscle spindle, 193
muscle tissue, 69–72, 135–47
muscular system, 10
myelin sheath, 185
myopia, 229
myosin, 136

nasal cavity, 317, 318
nearsightedness, 229
neck
    arteries of, 278
    bones of, 101, 103, 104, 108
    muscles of, 140, 141, 151
    veins of, 283
nephron, 334, 335
nerve(s), 181–88
    cranial, 199–204
    reflex and, 193–98
    sheep brain, 205–12
    spinal, 189–92
    structure of, 181–85
nervous system, 10
nervous tissue, 69–72
neuron, 70, 181–85
neutrophils, 251, 252
nucleus, cell, 39

ophthalmoscopy, 223, 224
optic disc, 225
organ of Corti, 216, 217, 220
organic molecules, 23–24
organ systems, 10, 12. See also
        specific systems
osmosis, 46, 47, 48
osteon, 81
ovary, 351, 352, 353–55

pain, referred, 20
palpation, 169
pancreas, 240–43, 307
Paramecium, 42
paranasal sinuses, 317, 318
parathyroid gland, 240
patella, 121
patellar reflex, 193, 195
pectoral girdle, 109–11
pelvic girdle, 119–20, 126
pelvic outlet, muscles of, 154, 155
penis, 345, 346, 348
pH, 22–23, 313
    urine, 341
phalanges
    foot, 121, 123, 127
    hand, 114, 117, 118
pharynx, 303–5, 318
phenotype, 359–63
photopupillary reflex, 233
pituitary gland, 238, 239
planes, 12–14
    body, 13
plantar reflex, 194, 196
plant tissue, 42
platelets, 66, 250, 252
pleura, 319
positions, body, 13
probability, laws of, 363
protein
    Biuret test for, 23–24
    urine, 341
pseudostratified columnar
        epithelium, 62
PTC taster, 362
pulse rate, 291–92

radial pulse, 292
radius, 111, 113, 116
Ranvier, node of, 185
reagent test strips, 341
red blood cell, 66, 250, 252
red-green color-blindness, 363
red pulp, 298
referred pain, 20
reflex arc, 193, 194
reflexes, 193–98
regions, body, 15–16
renal blood vessels, 332, 333, 334
renal cortex, 333–35
renal medulla, 333–35
reproductive system
    female, 12, 351–58
    male, 12, 345–50
residual volume, 325, 326
respiratory system, 12
    air volumes and capacities of,
        324–30
    organs of, 317–22
resting tidal volume, 324
retina
    blind spot of, 233
    examination of, 223, 224
Rh blood typing, 253–54
rib, 105, 106
Ringer's solution, 369
Rinne test, 214–15, 218

sacrum, 105
safety, 39, 48
saline solution, 369
salivary amylase, 313–16
salivary glands, 302, 303, 304
sarcolemma, 136, 137

sarcomere, 140
scapula, 109, 111, 116
Schwann cell, 184, 185
scientific law/principle, 1
scientific method, 1–3
sebaceous gland, 76
sections, 16
seminiferous tubules, 345, 347
sex chromosomes, 363
sheep brain, 205–12
    dorsal surface in, 206
    lateral surface in, 207
    midsagittal section of, 209
    ventral surface in, 208
sheep heart, 261–64
    dorsal surface of, 264
    ventral surface of, 263
shoulder
    arteries of, 278
    bones of, 110, 111, 112, 116
    muscles of, 145–47
    surface anatomy of, 170, 171,
        172, 173
    veins of, 283, 284
simple columnar epithelium, 62
simple cuboidal epithelium, 62
simple squamous epithelium, 62
skeletal system, 10
skeleton
    appendicular, 85
    axial, 85
    fetal, 95
    lower extremity, 121–27
    organization of, 85–90
    pectoral girdle, 109–11, 116
    pelvic girdle, 110, 120, 126
    skull, 91–95
    thoracic cage, 105–6
    upper limb, 110, 111–14, 116
    vertebral column, 101–5, 108
skin, 61–64, 73–78
skin color, 360
skull, 91–95
    fetal, 95
slide preparation, 33, 35
small intestine, 307–10
smoking, 320
smooth muscle, 70
Snellen eye chart, 230
sodium chloride solutions, 369
solutions, 47, 369
sound localization test, 214
specific gravity, urine, 340
sphygmomanometer, 292
spinal cord, 189, 190, 192
spirometer, 323, 324, 325
spleen, 297–98
starch
    digestion of, 314
    iodine test for, 24
starch solutions, 369
stereomicroscope, 35, 36
sternum, 105, 106
stomach, 305–7
straight hair, 362
stratified squamous epithelium, 62
stratum basale, 75
stratum corneum, 75
stratum granulosum, 75
stratum spinosum, 75
stretch reflex, 193
strip tests, 340–41
Sudan IV, 24
sugar, Benedict's test of, 24
surface anatomy, 169–80
    anterior, 178, 180
    arm, 173, 174
    chest, 172
    foot, 175
    forearm, 174
    hand, 174
    head and neck, 170
    leg, 175, 176
    posterior, 179, 180

shoulder, 171, 173
torso, 171, 172
sweat gland, 76

tapetum fibrosum, 226
tarsal bones, 121, 123, 127
teeth, 303
testis, 345–47
theory, 1
thoracic cage, 105–6
thoracic vertebrae, 104
thymus, 297–98
thyroid gland, 238, 240
tibia, 121, 123, 126
tidal volume, 324, 325
tissue
    connective, 65–68
    epithelial, 61–64
    muscle, 69–72, 135–47
    nervous, 69–72, 181–88
    respiratory, 319–20
tongue roller, 360, 361
total lung capacity, 325
trachea, 318, 319
transitional epithelium, 62
transparency, urine, 340
triceps reflex, 194, 196
tunica externa, 276, 277
tunica interna, 276, 277
tunica media, 276, 277
tuning fork tests, 214–15, 218

ulna, 111, 113, 116
unknown compounds, 24
upper limb
    arteries of, 278, 281, 290
    bones of, 111–14
    measurement of, 2–3
    muscles of, 145–52
    surface anatomy of, 173, 174
    veins of, 283, 284, 290
urinalysis, 339–44
urinary system, 12
    kidney of, 331–38
    tests for, 339–44
urine
    bilirubin in, 341
    cells in, 341, 342
    color of, 340
    hemoglobin in, 341
    ketones in, 341
    pH of, 341
    protein in, 341
    sediments of, 341, 342
    specific gravity of, 340
    strip tests of, 340–41
    transparency of, 340
uterine tube, 353, 355
uterus, 353, 355

veins, 283–88, 290
vertebral column, 101–5
visual acuity test, 229–30
visual tests, 229–36
vital capacity, 325, 327–28

Weber test, 215, 218
wet mount, 33–35
white blood cells, 66, 250–52
whitefish blastula, 60
white pulp, 298
widow's peak, 360, 361
withdrawal reflex, 193

X ray, 88
    elbow, 116
    foot, 126
    hand, 117
    knee, 126
    lower extremity, 126, 127
    neck, 108
    pelvis, 126
    shoulder, 116
    upper limb, 116